BIOMATERIALS

From Molecules to Engineered Tissues

ADVANCES IN EXPERIMENTAL MEDICINE AND BIOLOGY

BIOMATERIALS

From Molecules to Engineered Tissues

Edited by

Nesrin Hasırcı and Vasıf Hasırcı

Middle East Technical University
Biomaterials Research Labs
Ankara, Turkey

Kluwer Academic/Plenum Publishers
New York, Boston, Dordrecht, London, Moscow

Library of Congress Cataloging-in-Publication Data

Biomaterials: from molecules to engineered tissues/edited by Nesrin Hasırcı and Vasıf Hasırcı.
    p.   cm. — (Advances in experimental medicine and biology; v. 553)
    Includes bibliographical references and index.
    ISBN 0-306-48583-4
    1. Biomedical materials.   I. Hasırcı, Nesrin.   II. Hasırcı, Vasıf.   III. Series.

    R857.M3B5725 2004
    610'.28'4—dc22

                                                                    2004042180

Proceedings of BIOMED 2003, the 10th International Symposium on Biomedical Science and Technology, held October 10–12, 2003, in Northern Cyprus

Sponsored by the Middle East Technical University and the Scientific and Technical Research Council of Turkey

ISSN 0065 2598
ISBN 0-306-48583-4
ISBN E-book 0-306-48584-2

©2004 Kluwer Academic/Plenum Publishers, New York
233 Spring Street, New York, New York 10013

http://www.wkap.nl/

10 9 8 7 6 5 4 3 2 1

A C.I.P. record for this book is available from the Library of Congress

Permissions for books published in Europe: *permissions@wkap.nl*
Permissions for books published in the United States of America: *permissions@wkap.com*

Printed in the United States of America

CONTRIBUTORS

KIMIO ABE
Department of Functional Bioscience
Fukuoka Dental College, Fukuoka
JAPAN

NUREDDIN ASHAMMAKHI
Institute of Biomaterials
Tampere University of Technology, Tampere
FINLAND

ALPTEKIN AYDIN
Hipokrat A.Ş
407/6 Sok, No:10, Pınarbaşı, Izmir
TURKEY

NAZIFE BAYKAL
Middle East Technical University
Informatics Institute, Ankara 06531
TURKEY

MURAT BENGISU
Department of Industrial Engineering
Eastern Mediterranean University, Famagusta
TRNC, Mersin 10
TURKEY

IWONA MASZCZYNSKA BONNEY
Department of Anesthesia
Tufts University-New England Medical Center
Boston, MA
USA

ELIZABETH BURGER
Vrije Universiteit Medical Center
Amsterdam
THE NETHERLANDS

AYER BURKE
European University of Lefke
Faculty of Architecture and Engineering
Department of Electrical and Electronic Engineering
TRNC, Mersin 10
TURKEY

SARAH BRODY
National Centre for Biomedical Engineering Science
Department of Mechanical and Biomedical Engineering
National University of Ireland, Galway
IRELAND

J.CARLOS RODRIGUEZ-CABELLO
Department of Condensed Matter Physics (BIOFORGE Group)
E.T.S.I.I., University of Valladolid
Paseo del Cauce , s/n 47011 Valladolid
SPAIN

MARIO CARENZA
Istituto per la Sintesi Organica e la Fotoreattività
CNR, v.le dell'Università
2- 35020 Legnaro-Padova
ITALY

DANIEL B. CARR
Department of Anesthesia
Tufts University-New England Medical Center
Boston, MA
USA

EMRAH ÇELIK
Materials Science and Engineering Program
Izmir Institute of Technology
Gulbahce Koyu, Urla, Izmir
TURKEY

SINAN ÇETİNER
Hipokrat A.Ş
407/6 Sok, No:10, Pınarbaşı, Izmir
TURKEY

PHILLIP CHOI
Department of Chemical and Materials Engineering
Faculty of Engineering, University of Alberta
Edmonton, Alberta
CANADA

MARIETA CONSTANTIN
'Petru Poni' Institute of Macromolecular Chemistry
700487 Iasi
ROMANIA

FILIPA J. COSTA
3B's Research Group – Biomaterials, Biodegradables, Biomimetics
University of Minho, Campus de Gualtar, 4710-057 Braga
PORTUGAL

GETA DAVID
'Petru Poni' Institute of Macromolecular Chemistry
700487 Iasi
ROMANIA

EMİR B. DENKBAŞ
Hacettepe University, Chemistry Department,
Biochemistry Division, Beytepe, Ankara
TURKEY

MARTIJN VAN DIJK
Vrije Universiteit Medical Center
Amsterdam
THE NETHERLANDS

Y. MURAT ELÇİN
Ankara University,
Faculty of Science and Biotechnology Institute
Tissue Engineering and Biomaterials Laboratory, Ankara 06100
TURKEY

ALİ ERKAN ENGİN
Department of Mechanical Engineering
University of South Alabama, Mobile, AL
USA

MUZAFFER EROĞLU
SSK İhtisas Hospital
Ankara
TURKEY

LUCA FAMBRI
Dipartimento di Ingegneria dei Materiali e Tecnologie Industriali
Università di Trento via Mesiano, Trento
ITALY

GHEORGHE FUNDUEANU
'Petru Poni' Institute of Macromolecular Chemistry
700487 Iasi
ROMANIA

BORA GARİPCAN
Hacettepe University-Center of Bioengineering and Bioengineering
Division, and TÜBİTAK: Center of Excellence-BİYOMÜH
Beytepe, Ankara
TURKEY

İSMAİL HALUK GÖKÇORA
Department of Pediatric Surgery, Ankara University
School of Medicine, Cebeci Hastanesi, Dikimevi, 06100 Ankara
TURKEY

MUSTAFA GUDEN
Department of Mechanical Engineering and Center for Materials Research
Izmir Institute of Technology, Gulbahce Koyu, Urla, Izmir
TURKEY

THIERRY HAMAIDE
Laboratoire de Chimie et Procédés de Polymérisation
CNRS, ESCPE Lyon, 43, Bd. du 11 Novembre
69616 Villeurbanne Cedex
FRANCE

VALERIA HARABAGIU
'Petru Poni' Institute of Macromolecular Chemistry
700487 Iasi
ROMANIA

NESRIN HASIRCI
Middle East Technical University
Faculty of Arts and Sciences
Departments of Chemistry and Polymer Science and Technology
Ankara 06531
TURKEY

VASIF HASIRCI
Middle East Technical University
Departments of Biological Sciences and Biotechnology
Biotechnology Research Unit
Ankara 06531
TURKEY

PETER HERRLICH
Institute of Molecular Biotechnology
D-7708 Jena
GERMANY

PAUL HIGHAM
Stryker Orthopaedics
300 Commerce Court, Mahwah, NJ 07430
USA

ASIF JALIL
Department of Biomedical Engineering
Faculty of Medicine and Dentistry
University of Alberta
Edmonton, Alberta T6G 2G6
CANADA

SÜHEYLA KAŞ
Department of Pharmaceutical Technology
University of Hacettepe
Sıhhiye, Ankara
TURKEY

DILEK SENDIL KESKIN
Department of Engineering Sciences
Middle East Technical University
Ankara 06531
TURKEY

GAMZE TORUN KÖSE
Department of Genetics and Bioengineering
Yeditepe University, Istanbul 34755
TURKEY

JOSEPH KOST
Department of Chemical Engineering
Ben-Gurion University of the Negev
P.O.B. 653, Beer-Sheva 84105
ISRAEL

JOHN R. LAKEY
Department of Surgery, Surgical-Medical Research Institute
Faculty of Medicine and Dentistry
University of Alberta
Edmonton, Alberta T6G 2G6
CANADA

ANKA LETIC-GAVRILOVIC
International Clinic for Neo-Organs, ICNO
Via Cardinal P. Parente 26
San Cesareo, 00030 Rome
ITALY

ANDRZEJ W. LIPKOWSKI
Neuropeptide Laboratory
Medical Research Centre
Polish Academy of Sciences, Warsaw
POLAND

SILVANO LORA
Istituto per la Sintesi Organica e la Fotoreattività
CNR, v.le dell'Università, 2- 35020 Legnaro-Padova
ITALY

NUNO M. NEVES
3B's Research Group – Biomaterials, Biodegradables, Biomimetics
and Department of Polymer Engineering
University of Minho, Campus de Gualtar
4710-057 Braga,
PORTUGAL

BRYAN NORRIE
Department of Biomedical Engineering
Faculty of Medicine and Dentistry
University of Alberta, Edmonton
Alberta T6G 2G6
CANADA

NALAN ÖZDEMİR
Hacettepe University,
Chemistry Dept., Biochemistry Division
Beytepe, Ankara
TURKEY

EYLEM ÖZTÜRK
Hacettepe University
Chemistry Dept., Biochemistry Division
Beytepe, Ankara,
TURKEY

ABHAY PANDIT
National Centre for Biomedical Engineering Science
Department of Mechanical and Biomedical Engineering
National University of Ireland, Galway
IRELAND

MARIANA PINTEALA
'Petru Poni' Institute of Macromolecular Chemistry
700487 Iasi
ROMANIA

ERHAN PİŞKİN
Hacettepe University-Center of Bioengineering and Bioengineering
Division, and TÜBİTAK: Center of Excellence-BİYOMÜH
Beytepe, Ankara
TURKEY

RAY V. RAJOTTE
Department of Surgery, Surgical-Medical Research Institute
Faculty of Medicine and Dentistry
University of Alberta
Edmonton, Alberta T6G 2G6
CANADA

RUI L. REIS
3B's Research Group – Biomaterials, Biodegradables, Biomimetics
University of Minho, Campus de Gualtar
4710-057 Braga
PORTUGAL

HUBERT SCHORLE
Bonn University
Department of Developmental Pathology
D-53105 Bonn
GERMANY

GABRIELA A. SILVA
3B's Research Group – Biomaterials, Biodegradables, Biomimetics,
University of Minho, Campus de Gualtar
4710-057 Braga
PORTUGAL

BOGDAN C. SIMIONESCU
Department of Macromolecules,
'Gh. Asachi' Technical University
700050 Iasi,
ROMANIA

THEO SMIT
Vrije Universiteit Medical Center
Amsterdam,
THE NETHERLANDS

ESA SUOKAS
Linvatec Biomaterials Ltd
Tampere
FINLAND

TAMAR TRAITEL
Department of Chemical Engineering
Ben-Gurion University of the Negev
P.O.B. 653, Beer-Sheva 84105
ISRAEL

LJUBOMIR TODOROVIC
Department of Oral Surgery
Faculty of Stomatology
University of Belgrade
YUGOSLAVIA

PERTTI TÖRMÄLÄ
Institute of Biomaterials
Tampere University of Technology
Tampere
FINLAND

DEGER C. TUNC
Stryker Howmedica OSTEONICS
300 Commerce Court
Mahwah, NJ 07430
USA

KEZBAN ULUBAYRAM
Department of Basic Pharmaceutical Sciences
Faculty of Pharmacy, Hacettepe University
Sıhhiye, Ankara 06100
TURKEY

HASAN ULUDAG
Department of Chemical and Materials Engineering
Faculty of Engineering, University of Alberta
Edmonton, Alberta T6G 2G6
CANADA

RESAT UNAL
Research Center Karlsruhe
Institute of Toxicology and Genetics
D-76021 Karlsruhe
GERMANY

MINNA VEIRANTO
Institute of Biomaterials
Tampere University of Technology
Tampere,
FINLAND

FALK WEIH
Research Center Karlsruhe
Institute of Toxicology and Genetics
D-76021 Karlsruhe
GERMANY

PAUL WUISMAN
Vrije Universiteit Medical Center,
Amsterdam
THE NETHERLANDS

JENNIFER E.I. WRIGHT
Department of Chemical and Materials Engineering
Faculty of Engineering, University of Alberta
Edmonton, Alberta
CANADA

ELVAN YILMAZ
Department of Chemistry, Faculty of Arts and Sciences
Eastern Mediterranean University, Famagusta
TRNC, Mersin 10
TURKEY

LIYAN ZHAO
Department of Chemical and Materials Engineering
Faculty of Engineering, University of Alberta
Edmonton, Alberta
CANADA

Preface

The interdisciplinary fields of Biomaterials and Biomedicine are founded on basic sciences, engineering, medicine, pharmacy, dentistry, and other health-related fields. The aim of these areas is to solve health related problems using the scientific and technological tools available. Although wooden legs or metallic hands were known since the ancient times, the field started to take shape in the last quarter of the 20th Century. A special branch of biomaterials, Tissue Engineering, started in the late 1980's and then stem cell research became the hottest topic of the first decade of the 21st Century. Today with these developments, we are almost ready for generic implants that could be converted into the implants needed when the appropriate biological, chemical and physical signals are provided. Responsive materials used in drug delivery systems, sensors, and other biomedical devices constitute the intelligent biomaterials. Biodegradable materials that are gradually replaced by the natural tissue during the healing process, on the other hand, are the third generation biomaterials. Miniaturization of electronic systems and developments in the field of computers helped to create hybrid biomedical systems employing high-tech biomaterials.

This book is based on papers presented in Biomed-2003, the 10th International Symposium on Biomedical Science and Technology, held in Northern Cyprus from October 10-12, 2003, and covers a broad spectrum of biomedical research, from molecules to engineered artificial organs. In the initial chapters basic and intelligent biomaterials are discussed, followed by specific applications in analysis, separation and tissue design. The development of drug delivery systems is the next topic covered. Tissue repair and engineering using differentiated cells and stem cells are studied in several exciting chapters. The subjects of medical informatics and ethics constitute the two final topics that complement and guide the scientific research making the results of the research worthwhile.

As the editors we believe that the chapters presented here give a condensed knowledge about a wide spectrum of biomedical applications and will be very beneficial for the researchers and graduate students involved in research in the fields of biomaterials and biomedicine.

We wish to express our gratitude to the authors for their cooperation and very valuable contributions to the book. It would not be possible to complete this book without them. We also would like to thank the President of the Middle East Technical University and the Scientific and Technical Research Council of Turkey for their support during the organization of 10th International Symposium on Biomedical Science and Technology.

Editors
Dr. Nesrin Hasırcı and Dr. Vasıf Hasırcı
Middle East Technical University
Departments of Chemistry and Biological Sciences
Biomaterials Research Labs,
Ankara, Turkey

Contents

VIII.   HARD-TISSUE SYSTEMS

IX.   TISSUE ENGINEERING

X.   BIOMEDICAL INFORMATICS AND ETHICAL ISSUES

Functional Micro- and Nanoparticles Based on Poly[(N-acylimino)ethylene]

BOGDAN C. SIMIONESCU[*,#] and GETA DAVID[*]
[*] *Department of Macromolecules, "Gh. Asachi" Technical University, 700050 Iasi, Romania*
[#] *"Petru Poni" Institute of Macromolecular Chemistry, 700487 Iasi, Romania*

1.   INTRODUCTION

The investigation of structures with nanometer dimensions and their use in the fabrication of nanoscale devices is of topical, intense scientific and technological interest. In this context, colloidal dispersions of polymer micro- or nanoparticles, with controlled size and surface functionality, have received much attention. Different applications are possible in environmental management (i.e. sorbents), technical areas (microelectronics, biotechnology) and biomedical domains. Especially the last ones, implying commercial advantages, have determined a rapid development of this research subject. Such polymer materials are used for cell label/targeting, as support for controlled drug delivery, for latex diagnostics or heterogeneous immunoassay, for physico-chemical or affinity bioseparation, as magnetic carriers[1]. More, new techniques are now developing making use of self-organizing block copolymers, block copolymer micelles or polymer micro-/nanoparticles as templates for the preparation of nanostructured patterns with sizes smaller than those produced by electron beam or photolithography. These techniques are the subject of many investigations due to their use in the preparation of high selectivity membranes, of molecule separation

Biomaterials: *From Molecules to Engineered Tissues,* edited by
N. Hasırcı and V. Hasırcı, Kluwer Academic/Plenum Publishers, 2004

devices, of nanowires with special electro-optical and magnetic properties or of biosensors. As an example, nanosphere lithography is an efficient tool for the preparation of gold nanostructured films for surface Raman-spectroscopy[2], or of surface–confined triangular Ag nanoparticles for nanoscale affinity optical biosensors of high sensitivity and selectivity[3].

Polymer particles can be prepared through two approaches[1,4]. The first one supposes the fabrication of solid particles from existing polymers by solvent-in-emulsion evaporation, phase separation, spray drying, etc. This alternative offers the advantage of using natural polymers, on which most drug carriers are still relied. Generally, the resulted microspheres have a broad size distribution. However, many applications in medical or biomedical domains require the use of microparticles of controlled size and narrow size distribution, or even monodisperse, in order to achieve a reproducible behavior. These ones make use of the polymeric functional particles resulted by heterogeneous polymerization techniques (emulsion polymerization, dispersion polymerization, soap-free emulsion polymerization etc.). The features of the prepared microspheres (size, size distribution, surface functionality) are mainly decided by the selected polymerization technique and recipe formulation.

The polymers derived from 2-substituted-2-oxazolines, characterized by precisely controlled dimension, functionality and hydrophile/lipophile ratio, were extensively used in the synthesis of polymer microspheres usually presenting a core-shell structure, with biocompatible poly[(N-acetylimino) ethylene] (PMOZO) or poly[(N-propionylimino)ethylene] (PEOZO) hydrophile chains disposed at the surface[5,6].

Considering our interest in poly[(N-acylimino)ethylene]–based polymeric materials[7] water soluble, reactive PMOZO derivatives (macroazoinitiators and macromonomers) were synthesized[8] and tested as stabilizers of heterogeneous polymerization systems. The factors of influence were studied in order to find the most appropriate conditions for the preparation of micro– and nanoparticles with prerequisite characteristics.

2.   SYNTHESIS OF POLY[(N-ACETYLIMINO) ETHYLENE] DERIVATIVES

One of the topical requirements in synthetic **macromolecular chemistry is** the control of the polymer structure. Appropriate approaches for the synthesis of such materials were developed, most of them based on the peculiarities of the living polymerization. The living nature of the cationic ring–opening polymerization of cyclic imino ethers in appropriate reaction conditions allowed the preparation of a large number of poly(2-alkyl-2-oxazoline) (PROZO) derivatives as intermediates for functional polymers[5,6]. Macromonomers with styryl, butadiene or vinyl ester groups were obtained through the initiation of the living cationic polymerization of oxazolines with suitable functional compounds. End–capping of living growing chains with electrophiles or nucleophiles is another alternative tool for the preparation of functional polymers with controlled architecture. Thus, by making use of the terminator method, i. e. by quenching the oxazolinium species with selected nucleophiles, (meth)acryl– or methacrylamide–type polymerisable groups were quantitatively introduced at the end of the PROZO living chains[6]. Functional polymers acting as nonionic emulsifiers were prepared by a similar approach using fatty acids as quenching compounds. The production of PROZO telechelics was achieved in good yields by using a bifunctional initiator and water, ammonia or alkylamine as end–capping compounds[6]. This convenient method was also applied using cinnamic or maleic acid as nucleophiles, in order to synthesize macromonomers with cinnamoyl or maleic moieties in their structure[7,8]. Different halogenated initiating systems (methyl iodide, benzyl bromide, 1,4-dibromo-2-butene) were used to verify the effect of counterion nature and growing chain functionality on functionalization efficiency. The end-capping reaction was carried out in the presence of a low molecular weight proton scavenger (triethylamine) or of a macromolecular one (beads of poly(vinylpyridine-co-divinylbenzene)). A second approach – the coupling of hydroxylated PMOZO prepolymers with an appropriate acyl chloride – was also developed. A comparative study performed in order to determine the optimum reaction conditions led to the following conclusions[8b-8d]:

1. the functionalization proceeds easier in the systems initiated with methyl iodide as compared to those initiated with benzyl bromide due to an increased contribution of ionic against covalent species to the chain growth in the presence of an iodine counterion (almost exclusively ionized centers) as compared to the chain growth promoted by a bromine one;

2. depending on the nature of the initiating system and on nucleophile/initiator ratio mono–, bi– or even multifunctional macromonomers may be obtained, i.e.,

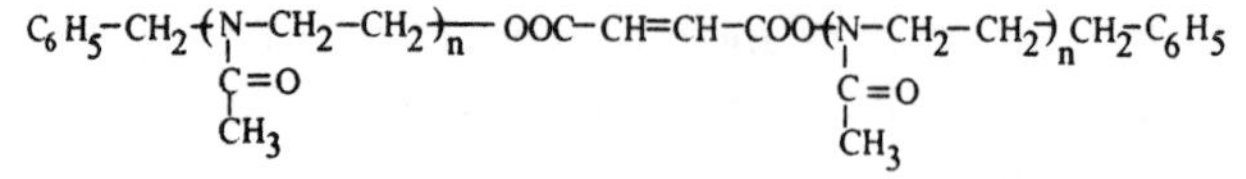

$$C_6H_5-CH_2-(N-CH_2-CH_2)_n-OOC-CH=CH-COO-(N-CH_2-CH_2)_n-CH_2-C_6H_5$$

with pendant groups $C=O$ / CH_3 on both nitrogen atoms

BCin

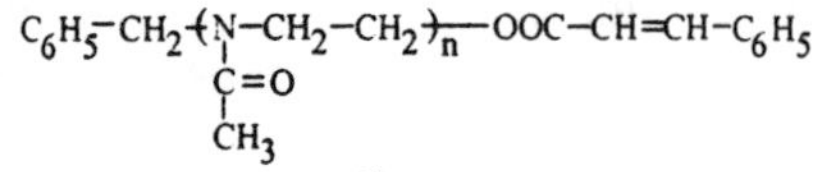

$$C_6H_5-CH_2-(N-CH_2-CH_2)_n-OOC-CH=CH-C_6H_5$$

Cin

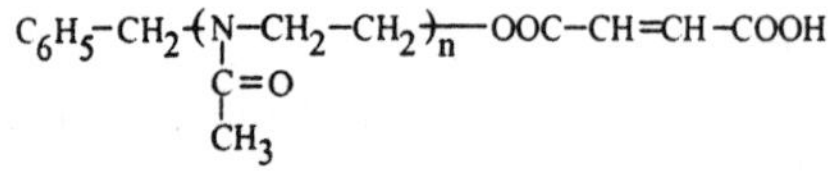

$$C_6H_5-CH_2-(N-CH_2-CH_2)_n-OOC-CH=CH-COOH$$

MA

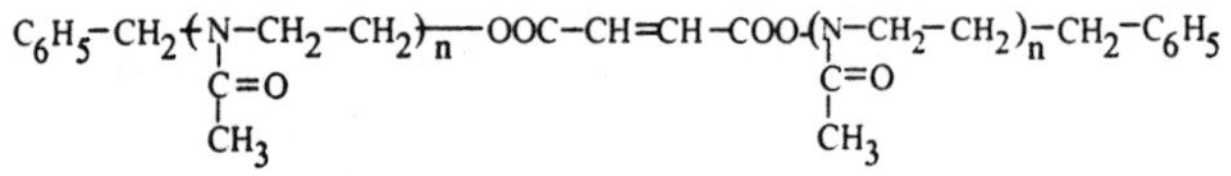

$$C_6H_5-CH_2-(N-CH_2-CH_2)_n-OOC-CH=CH-COO-(N-CH_2-CH_2)_n-CH_2-C_6H_5$$

BMA

$$[-OC-CH=CH-COO-(N-CH_2-CH_2)_n-CH_2-CH=CH-CH_2-(N-CH_2-CH_2)_n-O-]_x$$

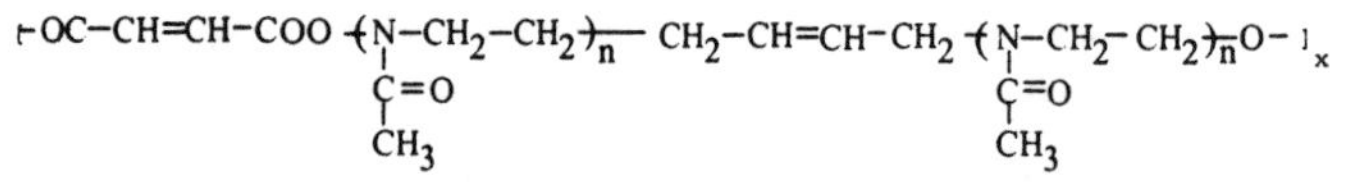

PEN

3. no differences have been observed between the two end–active centres in bifunctional growing chains in the coupling reaction;

4. the most convenient approach, i.e. yielding the highest functionalization efficiency with fewer preparation steps and lowest quantitative looses ($\eta >$ 85%) supposes the application of the end–capping method and the use of a macromolecular scavenger.

Generally, water soluble macromers or macroinitiators based on biocompatible, functional polymers are used in the formulation of heterogeneous polymerization systems in order to attain the colloidal stability and functionality of the system, with minimum contamination when application in the biomedical domain is envisaged[1,4]. As a consequence, a water soluble poly[(N-acetylimino)ethylene] macroazoinitiator with PMOZO sequences of controlled dimension inserted between the cleavable –N=N– groups, I, was also prepared[8a].

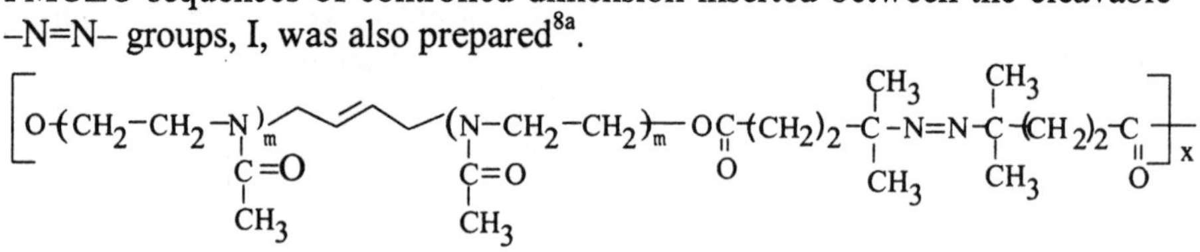

$$\left[O-(CH_2-CH_2-N)_m \diagup\diagup (N-CH_2-CH_2)_m-OC-(CH_2)_2-\underset{CH_3}{\overset{CH_3}{C}}-N=N-\underset{CH_3}{\overset{CH_3}{C}}-(CH_2)_2-C \right]_x$$

with pendant $C=O$ / CH_3 groups on the nitrogen atoms

(I)

The synthesis pathway consisted in the coupling of PMOZO glycols with controlled average molecular weights with 4,4'-azobis(4-cyanopentanoyl chloride).

3.   PREPARATION OF MICRO- AND NANOPARTICLES

3.1   Soapless Emulsion Polymerization

Among the synthetic routes for the preparation of polymer microparticles to be considered, the soapless emulsion polymerization is one of the most studied techniques. Both theoretical and practical reasons justify this choice[9]. Particle cleanness and their facile functionalization through the selection of recipe components recommend this technique especially for the preparation of polymer microparticles for biomedical uses. Polymer latexes have been successfully prepared by the soapless emulsion polymerization method using the following components: (a) ionizable initiator; (b) amphiphilic macromer (surfmer) or macroinitiator (inisurfmer); (c) hydrophilic comonomers of ionic or nonionic type; (d) hydrophilic macromers.

The inclusion of the synthesized PMOZO macroinitiator in the soapless polymerization recipe of vinyl monomers (styrene, methyl methacrylate, butyl methacrylate) gave spherical microparticles with diameters in the nanometer range (100-200 nm), with a core-shell structure, having PMOZO hydrophilic chains disposed at the surface[10]. The performed studies evidenced that the colloidal stability of the polymerization system, particle characteristics and the final solid yield strongly depend on the nature and concentration of both monomer and macroinitiator (Table 1).

Table 1. Soapless emulsion polymerization of vinyl monomers in the presence of PMOZO macroinitiators

	Polymerization				
	without silica			in the presence of silica	
Monomer	η (%)	Stability	PI	η (%)	Stability
St[a]	17	+	1.04	9	-
BMA[a]	52	+	1.5	18	-
MMA[a]	20	-	> 2	64	+
MMA[b]	60	+	1.02		

a - 0.0075 g I_7/g-w, 0.045 g monomer/g-w, 5 h – 80 °C, 1 h – 90 °C, Ar
b - 0.01 g I_7/g-w, 0.09 g monomer/g-w, 4 h, 80 °C, Ar; silica – 0.002 g/g-w

Particle size diminishes and size distribution becomes narrower with the increase of macroazoinitiator concentration and for longer PMOZO sequences inserted between the –N=N– groups. To reach maximum solid yield, the optimum values of 0.0125g I/g-w and $DP_{PMOZO} = 7$ were found for macroinitiator concentration and PMOZO dimension, respectively. This feature was explained by:

(a) the competition between the increase of the nucleation rate with initiator concentration and the concomitant increase of the content in allylic groups – with inhibitory effect – inserted in the macroinitiator backbone,

(b) the usual lowering of emulsion polymerization rate with increasing concentration of the nonionic emulsifier,

(c) the lower decomposition efficiency of the macromolecular initiators as compared to the low – molecular weight ones,

(d) the generation of an increased amount of water soluble block copolymers at higher PMOZO concentration or higher PMOZO chain length.

The nucleation was found to proceed through *in situ* micellization, the early formed oligomeric amphiphilic block copolymers acting as emulsifiers.

Considering these experimental results as well as literature data on soapless emulsion polymerization in the presence of inorganic powders[11] and the peculiarities of the PROZO based amphiphilic materials, emerging in their application as dispersants or adhesives[12], silica particles were introduced in the recipe of the soapless polymerization in the presence of PMOZO macroinitiator[13] (Table 1). Different papers point out the advantages offered by this technique as compared to the conventional procedure of mechanical blending or to the laborious technique of chemical binding (grafting) at the interface between the polymer and the filler, in the obtaining of composites with a homogeneous dispersion of the inorganic filler in the polymer matrix. Studies demonstrated that the choice of an appropriate monomer (more hydrophilic) along with the use of a preliminary treatment with suitable compounds, including nonionic surfactants, are required to achieve an uniform covering of the fine powder with polymer[14,15].

As a consequence, methyl methacrylate (MMA) was selected as the vinyl monomer. As expected, the early formed water soluble amphiphilic oligomers acted as efficient dispersants (Fig. 1) giving rise to the preparation of a homogeneous composite.

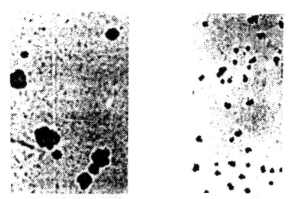

Figure 1. Modification of silica dispersibility in the earlier stage of the reaction
a) before and b) after 5 min from reaction beginning
Recipe: 0.094 g MMA/g-w, 0.01 g I/g-w, 0.01 g S/g-w, T = 80 °C

The increase of the polymerization rate with silica amount in the polymerization system yielded the increase of the final solid content (Fig. 2). The known higher adhesiveness of PMOZO to polar supports, together with its ability to develop hydrogen bonds with the carboxylic groups on the silica surface favored this behavior.

Thus, the soapless emulsion polymerization of MMA in the presence of PMOZO macroinitiator and silica powder yielded an efficient bonding of the resulted polymer on the surface of inorganic particles favoring their uniform encapsulation.

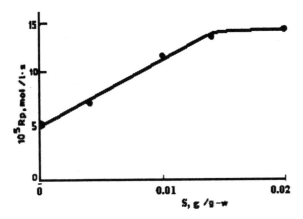

Figure 2. Dependence of the initial polymerization rate on silica powder concentration.
Recipe : 0.094 g MMA/g-w, 0.01 g I/g-w, T = 80 °C, t = 4 h

The main disadvantage of the low solid yield attained with hydrophilic macroinitiators may be also avoided by use of hydrophilic macromonomers and potassium persulfate as initiator[16]. A thorough comparative study performed on the polymerization systems based on the before mentioned synthesized macromonomers evidenced differences in the kinetic behavior, colloidal stabilization performances, and implicitly in products characteristics.

Generally, to provide higher latex stability, commercial polymer latexes are produced using mixed emulsifiers of ionic and nonionic types. In the studied systems, the presence of both ionic groups ($-SO_4^-$), originated from the decomposition of the initiator, and nonionic hydrophilic PMOZO grafts on particle surface provides an electrostatic repulsion and also a steric effect. Thus, the introduction of the PMOZO macromonomers in the polymerization system improved latex stability, as evidenced by a decrease of the particle size, an increased particle number density and a narrowing of particle size distribution (Table 2). The stabilization efficiency was found to increase in the order:

PEN < BMA < Cin ~ BCin < MA

The experimental data evidenced the possibility to increase the monomer content in the polymerization system up to 15% for the macromers of MA type. A slight increase in overall monomer conversion with macromer concentration in the system was registered, peculiarity explained by an increased facility of the initiation of the copolymerization in the aqueous medium in the presence of the water soluble macromer and by the possible emulsifying function of the oligomers formed in the early stage of the polymerization process.

Table 2. The effect of the macromonomer presence on the characteristics of the polymerization system

Macromer type	-	MA_{25}	Cin_{25}	PEN_{25}	
Dn (nm)	390	282	286	443	↘
Dw (nm)	554	339	371	620	↘
PI	1.4	1.2	1.3	1.4	↘
10^{13}Np (cm^{-3})	0.19	0.39	0.34	-	↗
η (%)	81	93	88	90	↗
System stability	-+	++	++	-+	↗

Recipe: 5 g St, 0.25 g PMOZO macromer, 0.1 g $K_2S_2O_8$, 95 g H_2O, 75 °C, 6 h, Ar
Dn = ΣNiDi/ΣNi, Dw = ΣNiDi^4/ΣNiDi^3, PI = Dw/Dn
Np = $6M_0Xm/(\pi Dn^3\rho_p)$, where Xm is monomer conversion per cm^3 water, ρ_p is polymer density,
Mo is the weight of styrene monomer

The increase of the macromer concentration in the polymerization system and of the PMOZO grafts dimension resulted in a significant reduction in the average size of polymer particles and their size polydispersity. Monodisperse latex particles (Fig. 3) can be obtained for a content of PMOZO macromer as high as 10 wt % relative to total monomer or a polymerization degree of 40 units for the PMOZO grafts (PI = 1.002).

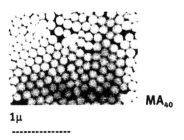

Figure 3. Typical TEM micrograph of the final micron-size polymer particles

However, the values of these parameters are limited by the concomitant formation of larger amounts of water soluble polymers that decrease the final solid content.

The increase of macromer concentration gives rise to an increase in the polymerization rate, due to an increase of the number of primary particles and thus of the polymerization loci. On the contrary, high macromonomer dimensions yield a lowering of the polymerization rate. The decrease proceeds monotonously when macromers of MA type are used, while for Cin macromonomers the variation plot presents a maximum (Fig. 4). This behavior was ascertained to the peculiarities of the stabilization mechanism. In the steric stabilization, the thickness of the polymer shell is a major parameter[17] that assures the colloidal stability of the system. Thus, a low polymerization degree of the nonionic PMOZO grafts results in a lower stabilization efficiency and implicitly in a low polymerization rate, the effect being compensated for the MA macromers by the increased content in carboxylic groups (electrosteric stabilization). A higher macromer dimension facilitates the formation of increased amounts of water soluble oligomers in the early stage of the polymerization, implying the increase of the number of primary particles, but also the increase of nucleation duration that determines a lowering of the polymerization rate. The competition of these two effects gives rise to the registration of a maximum in the graphic representation of this dependence for Cin type macromonomers.

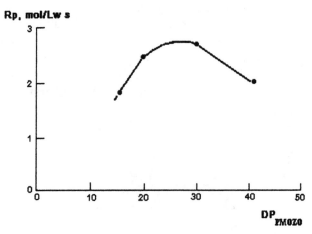

Figure 4. Dependence of the polymerization rate on the dimension of the involved water soluble macromer

According to these results, a macromer of MA type with DP_{PMOZO} in the range of 25 – 35 seems to be the best alternative in order to obtain the highest system stability and the narrowest particle size distribution (i.e. for 10 wt% MA_{25} a value of 1.01 was attained for the size polydispersity index), the macromer concentration being a variable value related to the required particle size for the envisaged application.

3.2   Dispersion Copolymerization

In addition to soapless emulsion polymerization, the one step dispersion polymerization also offers a facile, convenient route to prepare functional microparticles with a large size variety in nanometer range and narrow size distribution. The method involves the use of a polymeric stabilizer. Particle size is controlled by changing polymerization parameters, i.e., solvent polarity, concentration and chemical nature of the stabilizer, monomer or initiator.

The experimental results obtained for the polymerization systems containing the mentioned macromonomers and styrene as a main monomer (dispersion medium: alcohol – water mixtures) clearly evidenced the ability of the PMOZO derivatives to assure the colloidal stability of such disperse systems[8b,18]. Microparticles with diameters in the 400 – 1100 nm range and size distributions of 1.02 – 1.05 were obtained. The macromonomer acted as both co-monomer and stabilizer.

The experimental results underline some conclusions. The stabilization efficiency is essentially influenced by the length of the PMOZO sequence included in the macromonomer structure, the system becoming unstable under a limiting value (i.e. $DP_{PMOZO} > 15$). An increased concentration of the polymeric stabilizer implies an improved stabilization of the polymerization system, higher solid yields and lower dimensions of the polymer particles. This is explained by the increase of the number of generated nuclei with increasing stabilizer concentration, yielding in an increased number of polymer particles of smaller size. The size distribution is narrower for a concentration domain dependent on macromer structure, the stabilization ability decreasing in the order

Cin >> MA > BMA >> PEN

in accordance with the relative reactivity of the macromonomers in the copolymerization. The latter one affects the grafting efficiency, an important feature of a stabilizer in dispersion polymerization[17]. The multiple functionality of PEN type macromers was found to favor the formation of a coagulum by interparticle bridging. The use of a mixture of mono- and bifunctional PMOZO macromonomers gave rise to an improved particle size distribution. Colloidal stable systems with near monodisperse polymer particles (Dn = 462 nm, PI = 1.02) were obtained for a Cin_{40} macromonomer concentration of only 3 wt% relative to total monomer.

The main factor of influence is the polarity of the dispersion medium, the domain of narrowest size distribution being in the same range of solvent composition, i.e. of water content in the ethanol – water mixture (10 – 20 v/v %), independent on PMOZO macromer structure (i.e. nature of the inserted polymerizable group and its position in the PMOZO chain). The diameter of the particles is diminishing with the increase of the water amount in the dispersion medium, water being a poorer solvent for the resulting graft copolymer as compared to ethanol and thus permitting the decrease of the critical chain length of oligomer precipitation/nuclei generation.

The analysis of the peculiarities of this technique pointed on the difficulty to choose the appropriate synthesis conditions. To achieve the desired particle characteristics, numerous factors of influence must be taken into account. The specific variation of the size polydispersity index with medium composition makes this analysis particularly difficult. To solve this aspect, the *Table Curve 3D programme*[19] was applied and proved to be an efficient tool for the selection of the appropriate macromer type and concentration (Fig. 5).

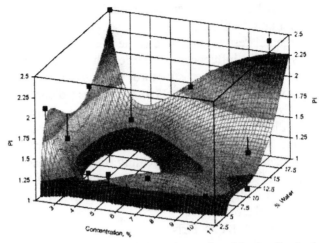

Figure 5. Graphical representation of the dependence of particle size distribution on PMOZO macromer concentration and on the polarity of the dispersion medium (Table Curve 3D program, Cin_{25} samples)

Considering the self assembling ability of monodisperse particle colloids and the ability of polyethylenimine to immobilize heavy metals and bivalent metal ions[5,6], the synthesized microparticles provide an interesting material for microsphere lithography. Studies on this subject are under way.

3.3   Microemulsion Photopolymerization

Microemulsions are isotropic, optically transparent or translucent and thermodynamically stable systems, formed by nanoparticles with diameters lower than 60 nm[20]. Their size range, similar to those of viruses, allows the application in the biomedical area. The principle of microemulsions formation consists in the use of an important amount of surfactant or of a mixture of surface active agents (surfactant and cosurfactant) due to the imperative of stabilizing a large overall interfacial area. This peculiarity restricts the potential applications, since in most cases a high solid content and a minimum contamination with surfactant are required. Thus, a serious interest exists for new effective surfactant systems able to facilitate the formation of smaller and non-toxic lattices. Systems containing nonionic surfactants were tested in the stabilization of fine emulsions as a result of the delicate balance between synergistic stabilization and poisoning of a microemulsion by the most ionic surfactants (i. e. standard cationic emulsifiers deeply influence proteins structure and even denature them), making difficult applications in medical area.

The before mentioned water soluble PMOZO derivatives were used as co-surfactants in the microemulsion polymerization of acrylic monomers[21] in order to improve the quality of the resulted nanoparticles (surface functionality and biocompatibility, ability to form continuous polymer films due to the known compatibility of PMOZO with common polymers[5]).

Table 3. Microemulsion copolymerization data

Exp No	η (%)	PMOZO costabilizer (g/ml)	SDS (g/ml)	SDS/Monomer weight ratio	Dn (nm)	Dw (nm)	PI
1	85	0.002	0.010	0.40	17	20.5	1.2
2	75	0.002	0.006	0.24	27	30	1.1
3	72	0.002	0.002	0.08	50	56	1.1
4	70	0.002	0.000	0.00	200	225	1.1
5	70	0.004	0.002	0.08	40	45	1.1
6	90	0.000	0.002	0.08	54	65	1.2
7	90	0.002	0.010	0.40	9	12.5	1.4

Recipe: monomers - BMA, MMA (50/50 %)
*samples 1 - 5: monomers, macromer Cin_{40}, SDS, $K_2S_2O_8$, hv, 25°C, 4h; sample 7: monomers, macroazoinitiator (Pn = 6.3, Mn = 10000), SDS, hv, 25 °C, 4 h; sample 6: monomers, SDS, $K_2S_2O_8$, 70 °C, 6 h; SDS – sodium dodecyl sulfate

Considering system transparency, the photochemical initiation – more appropriate for polymer quality – was chosen as a preparation procedure. The average diameters of latex particles were in the range of 12.5 – 60 nm (Table 3). A value of about 200 nm was obtained in the absence of the main surfactant, SDS. The experimental data evidenced that particle size is mainly dependent on SDS/PMOZO ratio. As compared to the usual systems using ionic surfactants/monomer ratios higher than 1, in the presence of PMOZO macromer the SDS/monomer ratio can be drastically lowered. The polymerization yields were around 75-85%. The synthesized copolymers had viscosity average molecular weights in the range of $2.1 – 2.4 \times 10^6$ and glass transition temperatures of 38.0 – 43.5 °C. The investigation by FTIR and ^{1}H-NMR techniques revealed that PMOZO was incorporated into the nanoparticles. As already known, PMOZO may be easily modified to polyethylenimine, a polymer able to immobilize biocompounds, metal nanoparticles or metal ions. Such materials with controlled functionality are thus recommended for various biomedical applications like receptive binding, biosensors, a.s.o.

4.   IMMOBILIZATION ABILITY

Polymer particles prepared by soapless emulsion polymerization of styrene in the presence of PMOZO macromonomers were subjected to acid hydrolysis, emerging in the partial modification of the PMOZO hydrophilic shell to polyethylenimine ($\sim$ 3 wt%). They were submitted to test the immobilization of inorganic or organic compounds. The experimental data proved that the specific behaviour of the polymer particles depends on the presence and nature of the hydrophilic grafts at their surface, medium pH, contact duration.

Thus, by comparison with common polystyrene microspheres, the polymer particles with partially hydrolyzed PMOZO shell issued through the use of MA macromonomers (i.e., containing also carboxylic groups), after a preliminary treatment with Zn^{2+} aqueous solution presented a 12 times higher ability to immobilize lysozyme.

The synthesized particles also present an improved ability to retain Fe^{3+} ions at their surface (increased Fe^{3+} retained amount, higher rate of response). The immobilization selectivity was proved to be drastically dependent on medium pH (Fig. 6).

In the recent years, the growing technological interest in the utilization of nano-sized materials has also drawn attention on nanocatalysts immobilized on latex supports. The latex dispersions based on PMOZO macromonomers were investigated with respect to their ability to absorb platinum nanoparticles obtained by refluxing an alcoholic solution of H_2PtCl_6. Such metal – latex –protective polymer systems present advantages that can be summarized as follows:

(a) agglomeration of nanoparticles is prevented

(b) both polymer–supported and polymer–protected colloidal catalysts characteristics are combined, giving rise to increased catalytic selectivity and improved stability

(c) latex support itself may influence the surface properties of the nanocatalysts.

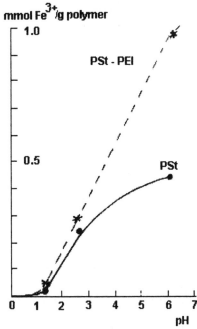

Figure 6. Fe^{3+} immobilization on polymer particle surface: (∗) polymer particles with hydrophilic PMOZO shell partially hydrolysed to PEI; (•) polystyrene particles

The immobilization of the metal particles on the latex supports was demonstrated by transmission electron microscopy (Fig. 7). PMOZO hydrophilic grafts acted as a protective polymer layer.

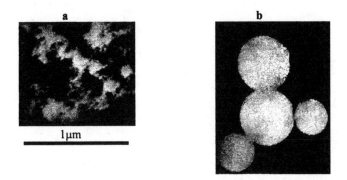

Figure 7. Typical TEM micrographs of agglomerated Pt nanoparticles (a) and of latex supported Pt catalyst (b)

The partial hydrolysis of the grafts to polyethylenimine resulted in an increased attachment efficiency and higher colloidal stability of the system (Fig. 8). The catalytic properties of such Pt – polystyrene latex – polyethylenimine protected systems are under study.

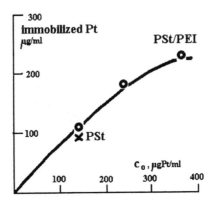

Figure 8. Pt immobilization ability of the latex particles: (○) polymer particles with hydrophilic PMOZO shell partially hydrolysed to PEI; (x) polystyrene latex

Graft copolymers obtained by the copolymerization of styrene and MA type macromonomers were found to be appropriate for the preparation of magnetic carriers by a modified solvent evaporation method[22]. The presence of both PMOZO grafts and COOH groups facilitates the complete covering of the magnetic powder (ilmenit, ~ 1μm) and determines an improved colloidal stability of the resulted composite particles (Fig. 9).

Figure 9. Magnetic carrier (cross section)
Recipe: PSt-g-PMOZO 7 % wt/v in $CHCl_3$/water mixture; 12.5 wt% ilmenit and 1.2 wt% poly(viny alcohol) relative to polymer

These examples clearly show that PMOZO functional derivatives may be successfully used to develop new macromolecular compounds, with properties in accordance with the topical interests in nanostructured materials for technological and biomedical areas.

REFERENCES

1. Kawaguchi H., 2000, Functional Polymer Microspheres. *Prog Polym Sci.* **25**: 1171-1210.
2. Tessier P. M., Velev O. D., Kalambur A. T., Rabolt J. F.,. Lenhof A. M., and Kaler E. W., 2000, Assembly of Gold Nanostructured Films Templated by Colloidal Crystals and Use in Surface – Enhanced Raman Spectroscopy. *J Am Chem Soc.* **122**: 9554-9555.
3. Haes A. J., and Van Duyne R. P., 2002, A Nanoscale Optical Biosensor: Sensitivity and Selectivity of an Approach Based on the Localized Surface Plasmon Resonance Spectroscopy of Triangular Silver Nanoparticles. *J Am Chem Soc.* **124**: 10596-10604.
4. Pichot C., 1995, Position Paper: Functional Polymer Latexes. *Polym Adv Techn.* **6**: 427-434.
5. Kobayashi S., and Uyama H., 2002, Polymerization of Cyclic Imino Ethers: From Its Discovery to the Present State of the Art. *J Polym Sci: Part A: Polym Chem.* **40**: 192-209.
6. Kobayashi S., and Uyama H., 1994, New functional polymers based on poly(2-oxazoline)s. *Trends in Macromol Res.* **1**: 121-131.
7. Simionescu C. I., and David G., 1996, Poly(N-acetylethylenimine)s, In *The Polymeric Materials Encyclopedia* vol. 7 (J. C. Salamone, ed.), CRC Press, Boca Raton, Fl., pp 5334- 5343.
8. a) Simionescu C. I., David G., Ioanid A., Paraschiv V., Riess G., and Simionescu B. C., 1994, Solution Polymerization of Vinyl Monomers in the Presence of Poly(N-acetyliminoethylene) Macroazoinitiators. *J Polym Sci: Part A: Polym Chem.* **32**: 3123-3132.
   b) David G., and Ioanid A., 2001, Synthesis and Dispersion Copolymerization of Poly(2-methyl-2-oxazoline) Macromers with Cinnamoyl End Groups. *J Appl Polym Sci.* **80**: 2191-2199.
   c) David G., Alupei V., and Simionescu B. C., 2001, End-capping of Living Poly(2-methyl-2-oxazoline) with Maleic Acid. *Eur Polym J.* **37**: 1353-1358.
   d) Alupei V., David G., and Simionescu B. C., 2001, Factors of Influence in Coupling Reaction of Poly(2-oxazolinium) Cations with Maleic Acid. *Bull. Techn. Univ. Jassy*, **XLVII**, fasc. **3-4**: 251-256.
9. Wang Q., Fu S., and Yu T., 1994, Emulsion Polymerization. *Prog Polym Sci.* **19**: 703-753.
10. Simionescu C. I., Paraschiv V., David G., and Simionescu B. C., 1995, Soap-free Emulsion Polymerization of Vinyl Monomers in the Presence of Poly(N-acetyliminoethylene) Macroazoinitiators. *Macromol Reports.* **A 32**: 1095-1101.
11. Arai M., Arai K., and Saito S., 1982, Soapless Emulsion Polymerization of MMA in Water in the Presence of Calcium Sulfite. *J Polym Sci, Polym Chem Ed.* **20**: 1021-1029.

12. Akasaki I., 1996, Polyethylenimine, In *The Polymeric Materials Encyclopedia* vol. 7 (J. C. Salamone, ed.), CRC Press, Boca Raton, Fl., pp. 6124-6132.
13. Simionescu C. I., David G., Alupei V., Ioan C., and Simionescu B. C., 1998, Soapless Emulsion Polymerization of MMA at the Surface of Silica Powder Particles. *Synth Polym J.* **5**: 260-269.
14. a) Espiard Ph., Revillon A., Guyot A., and Mark J. E., 1992, Nucleation of Emulsion Polymerization in the Presence of Small Silica Particles. *ACS Symposium Series* **492**: 387-404;
    b) Bourgeat – Lami E., Espiard Ph., and Guyot A., 1995, Poly(ethylacrylate) Latexes Encapsulating Nanoparticles of Silica: 1. Functionalization and Dispersion of Silica. *Polymer* **36**: 4385-4389.
15. Alexeev V. I., Ilekti P., Persello J., Lambard J., Gulik T., and Cabane B., 1996, Dispersions of Silica Particles in Surfactant Phases. *Langmuir* **12**: 2392-2401.
16. David G., Simionescu B. C., and Ioanid A., 2004, *Mol Cryst Liq Cryst.* (in press)
17. Piirma I. J., 1992, *Polymeric Surfactants*, Surfactant Sci. Series, Marcel Dekker, Inc., New York.
18. Alupei V., David G., Abadie M. J. M., and Simionescu B. C., 2003, Poly[(N-acetylimino)ethylene Macromonomers with Maleic Moieties in the Dispersion Copolymerization with Styrene. *J Macromol Sci Part A: Pure Appl Chem.* **40**: 547-561.
19. David G., and Ibanescu B., 2003, Controlled Design of Polymer Materials Based on Poly[(N-acetylimino)ethylene] Derivatives. In *BRAMAT 2003 Proceedings -IV*, "Transilvania" University Publishing House, Brasov, pp 203-208.
20. Antonietti M., 1995, Polymerization in Microemulsions – a New Approach to Ultrafine, Highly Functionalized Polymer Dispersions. *Macromol Chem Phys.* **196**: 441-466.
21. David G., Ozer F., Simionescu B. C., Zareie H., and Piskin E., 2002, Microemulsion Photopolymerization of Methacrylates Stabilized with Sodium Dodecyl Sulfate and Macromonomers. *Eur Polym J.* **38**: 73-78.
22. David G., Alupei V., Simionescu B. C., Pricop L. et Badescu V., 2002, Proprietes de quelques macromonomeres a base de poly[(N-acetylimino)ethylene] et leurs atouts pour des applications biomedicales. In *Actes du COFrRoCA*, Alma Mater Bacau & Tehnica - Info Chisinau, Bacau, pp 69-72.

In-Situ Crosslinking Thermoreversible Polymers for Cell Entrapment

BRYAN NORRIE[*], ASIF JALIL[*], JOHN R. LAKEY[#], RAY V. RAJOTTE[#], and HASAN ULUDAG[*,†]

[*]*Department of Biomedical Engineering, Faculty of Medicine and Dentistry,* [#]*Department of Surgery, Surgical-Medical Research Institute, Faculty of Medicine and Dentistry,* [†]*Department of Chemical and Materials Engineering, Faculty of Engineering, University of Alberta, Edmonton, Alberta T6G 2G6, CANADA*

1.   INTRODUCTION

Immunoisolation technologies are being explored to enable cell transplantation without the risks of chronic systemic immuno-suppression[1]. Polymeric hydrogels have been utilized in immunoisolation techniques as synthetic extracellular matrices essential for maintenance of cell viability. The synthetic matrix helps to prevent excessive cell aggregation and provide a three-dimensional support for the cells. The stability of the polymeric hydrogel is critical for long-term cell viability. Hydrogels that are formed by physical association, such as alginate gels held together by the Ca^{+2} ions or agarose gels held together by polymer chain entanglements, are not suitable for long-term applications, due to instability of this type of associations. A

Biomaterials: *From Molecules to Engineered Tissues,* edited by
N. Hasırcı and V. Hasırcı, Kluwer Academic/Plenum Publishers, 2004

Bryan Norrie et al.

hydrogel with chemically-crosslinked linkages is more desirable since it offers better stability for long-term cell viability. However, the need to maintain the normal physiology of mammalian cells, which are highly sensitive to environmental conditions, severely restricts the choice of available methods for hydrogel crosslinking. The essential elements of conventional processes, such as the free radicals from the initiators or the chemical crosslinkers with reactive moieties, are not compatible with the mammalian cells.

To develop cell-compatible, *in-situ* crosslinking hydrogels, thermo-reversible polymers that contain crosslinkable moieties were designed in our lab. Thermoreversible polymers in aqueous buffers undergo a solubility change as a function of temperature[2]. Polymers of N-isopropylacrylamide (NiPAM) were chosen for our purposes due to well-known cell-compatibility of this class of polymers[3]. NiPAM-based polymers exhibit a solubility transition at ~30 °C (so called, Lower Critical Solution Temperature, LCST), so that they can be suspended in aqueous solutions along with mammalian cells at a low temperature (typically 4 °C). The polymers can then entrap the cells in a hydrogel upon increasing the temperature to the physiological temperature (37 °C). Mammalian cells readily tolerate such a temperature change without adverse effects. We envisioned incorporating amine-reactive N-acryloxysuccinimide (NASI) into the polymers so as to utilize diamine reagents to crosslink polymer chains after hydrogel formation (Figure 1). Because the reactive moieties are on polymeric materials that are typically impermeable to cells, the cell toxicity of the reactive polymers should be lower than the smaller reactive crosslinkers.

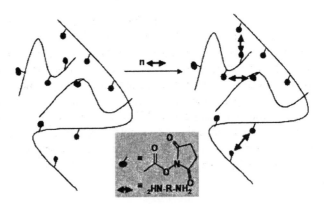

Figure 1. Schematic representation of NASI-containing thermoreversible polymers being crosslinked by a diamine reagent. The diamine reagent can crosslink either the NASI groups within a polymer chain or NASI groups between two polymer chains.

This study explored the feasibility of using NASI-containing polymers to construct crosslinked hydrogels that contained mammalian cells. Highly sensitive pancreatic islets, which are actively explored as transplants for treatment of Type I Diabetes[4], were utilized as model cells. The islets were entrapped in NiPAM/NASI hydrogels and their viability was determined after crosslinking the hydrogels with diamines. Our studies demonstrated the feasibility of maintaining islet viability after crosslinking the hydrogels containing pancreatic islets.

2.   MATERIALS AND METHODS

2.1.   Polymer Synthesis and Characterization

The polymers were synthesized by a free-radical process[5], using anhydrous 1,4-dioxane as the solvent and benzoylperoxide as the initiator (70 °C for 24 h; yield: 30-50%). The polymers were based on NiPAM, NASI and the hydrophobic monomers, methylmethacrylate (MMA) and ethylmethacrylate (EMA). After precipitation of the polymer in diethyl ether, polymers were dried under vacuum at 50 °C for 1 week. The composition of the polymers was determined by ^{1}H-NMR, elemental analysis, and a spectroscopic assay for NASI, as described before[5]. The LCSTs and weight-average molecular weights (M_w) were determined by spectroscopy and gel permeation chromatography with a static light scattering detector, respectively[5].

2.2.   Water Uptake of Hydrogels

Hydrogels were formed by incubating polymer solutions (prepared at 4 °C as 10 mg/mL in 0.1 M phosphate buffer, pH 7.4) in glass vials at 37 °C. Polymers were incubated in either phosphate buffer or in phosphate buffer containing the diamines, the amino acid lysine (Lys) and ethylenediamine (EA). The concentrations of the diamine reagents are given in figure legends. The hydrogel formation was assessed after 24 hours. The gel was dabbed dried with tissue paper and the gel wet weight (ww) was determined. Gels that did not remain as a single piece (i.e., disintegrated) during this procedure were designated as 'unstable'. After weighing, gels were returned to the vials at 37 °C with fresh buffer. At the end of the study, the gels were baked at 65 °C for 3 days to completely dehydrate them so as to obtain the dry weight (dw). The water uptake was calculated by: $\{(ww-dw)/dw\}$[4].

2.3.   Islet Culture and Entrapment in Gels

Islets were obtained from 200-250 g Wistar-Lewis male rats, by collagenase digestion of the pancreas[7]. The islets were purified by a Ficoll™ gradient and transferred into 100 mm dishes in CMRL medium (GIBCO). The desired number of islets was manually counted under a microscope and placed in 24-well plates using micropipettes. The islets were maintained at 37 °C in a humidified atmosphere of 95/5% air/CO_2. To determine the compatibility of the diamines with pancreatic islets, 100 islets were placed in CMRL with varying concentrations of Lys and EA, and allowed to incubate overnight. To entrap islets in gels, 100 islets were placed into the wells of a 24-well plate with 0.3 mL CMRL, and 0.5 mL CMRL containing 20 mg/mL polymer solution (at 4 °C) was added to the islets. After 1 h incubation at 37 °C, 0.2 mL of CMRL medium (with and without the diamines) was added to the wells, and the plates were then incubated at 37 °C.

2.4.   MTT Assay for Islet Viability

Islet viability was assessed by the 3-(4,5-dimethylthiazol-2-yl)-2,5-diphenyltetrazolium bromide (MTT) assay, which measured the activity of mitochondrial dehydrogenases, and was used as a quantitative measure of metabolic activity[8,9]. The islets were incubated in CMRL medium containing 1 mg/mL MTT (SIGMA) at 37 °C. After 4 h, islets were centrifuged at low speed for 3 min. The supernatant was removed, the islets were washed with 1 mL of 100 mM phosphate buffer and centrifuged (x3). The islets were dissolved in 300 mL DMSO and the absorbance was determined at 570 nm (ref: 630 nm). A similar procedure was used for islets entrapped in un-crosslinked gels, except that the gels were first incubated at 4 °C to dissolve the polymers, after which MTT absorbance was determined.

A different procedure was required for the islets entrapped in crosslinked gels, where the gels did not dissolve at 4 °C. Islets were incubated with MTT as above and the gels were incubated in phosphate buffer for 30 min to remove the medium. The gels were then centrifuged at low speed in tubes that had been pierced at the bottom to allow fluid drainage, but gel retention (x3). 300 mL DMSO was then added to the gels for overnight dissolution and the absorbance was determined as above.

A control procedure, in which hydrogels without islets were utilized, indicated this processing technique yielded an insignificant increase in the measured absorbance values.

3. RESULTS AND DISCUSSION

The properties of the polymers used in this study are summarized in Table 1. Incorporating MMA and EMA reduced the LCST of the NiPAM homopolymer from 26.7 °C to as low as 19.9 °C. The NASI also affected the LCST, lowering it by ~3 °C at the highest concentration tested in this study (8.5 mol%). The NASI groups provided reactive groups that were amenable to crosslinking by diamines. The choice of a neutral buffer and small diamines with no bulky side-groups and relatively high -NH$_2$ pK$_a$s were intended to minimize the hydrolysis of succinimide ester[10].

Table 1. The composition, LCST and M$_w$ of thermoreversible polymers used in this study.

Polymer	%NiPAM	%NASI	%MMA	%EMA	MW (kD)	LCST (°C)
A	100				ND*	26.7
D	97.3	2.7			171	26.5
F	94.9	5.1			183	23.5
H	91.5	8.5			383	23.0
J	90.4		9.6		578	23.4
K	93.1	1.3	5.6		622	25.2
N	89.1	1.2	9.7		551	24.2
S	78.0	1.1	20.9		465	19.9
U	84.6			15.4	404	20.3
W	75.7	1.0		23.3	596	23.2
X	71.8	0.9		27.2	422	21.5

*ND: not determined.

Assuming the relative reactivity (i.e., hydrolysis:aminolysis ratio) of succinimide esters was not altered in a gelled state, aminolysis rate in our experimental system was expected to exceed hydrolysis rate by more than 10-fold[5]. Whereas a semi-stable gel was obtained with NiPAM homopolymer A, NiPAM/NASI polymers D, F and H were not capable of forming a gel in the absence of diamines. This was presumably due to hydrolysis of NASI groups to yield ionized carboxyl groups (at the experimental pH of 7.4), which was likely (i) to hinder polymer-polymer associations needed for the gel formation, or (ii) to elevate the polymer LCST (>37 °C). The NiPAM/NASI polymer D gave a stable gel in the presence of Lys and EA, but F and H, which contained a higher fraction of NASI, did not. NiPAM/NASI polymers containing MMA and EMA (K, N and W) were able to form gels without diamines.

The incorporation of hydrophobic MMA and EMA was presumably sufficient to enhance the necessary polymer-polymer association for gel formation. Treatment with both diamines reduced the water uptake of the gels (Figure 2A) of polymers D and K, but not for polymers N and W. The water uptake of the latter polymers was lower in accordance with hydrophobic character (i.e., lower LCST) of the polymers. The water uptake of the polymers without NASI (A, J and U) was not affected by diamine treatment (not shown). To determine hydrogel stability, the water uptake of the gels was determined over a 15-day period. There were no significant changes in water uptake with time for uncrosslinked gels (Figure 2B), as well as gels crosslinked with Lys (Figure 2B) or EA (not shown). In a parallel set of gels prepared, lowering the temperature to 4 °C led to the dissolution of uncrosslinked gels (K, N and W) after 15 days, but the same polymers treated with Lys and EA did not dissolve and remained intact upon lowering the temperature.

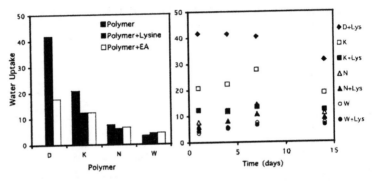

Figure 2. (A) Water uptake of hydrogels of polymers D, K, N and W in the absence and presence of Lys and EA on day 1. For clarity, only the average water uptake values are shown. D by itself was not able to form a stable hydrogel. Polymers containing MMA/EMA were able to form stable hydrogels, and in the case of K, the water uptake was reduced after crosslinking. (B) Changes in water uptake of uncrosslinked and crosslinked hydrogels of D, K, N and W. There were no significant changes in water uptake during the study period.

This was indicative of the amide linkages (i.e., product of amine-NASI reaction) to be stable under experimental conditions.

Not all NASI groups are expected to participate in crosslinking. Some are expected to react with only one of the $-NH_2$ groups of diamines, where the other $-NH_2$ remains free, and some will hydrolyze to -COOH, resulting in charged moieties in the polymer chain. Charged moieties may elevate the LCST, possibly leading to network destabilization[11]. The above studies indicated a stable network formation, even at temperatures below the LCST, indicating that the side-reactions were relatively insignificant.

The cell compatibility of selected polymers and the crosslinking process were determined using pancreatic islets. Primary pancreatic islets are fragile cells that are sensitive to environmental conditions more than the cells adopted for proliferation under culture conditions. A linear relationship between the islet number and the MTT signal was obtained (Figure 3A), indicating the positive correlation of the MTT signal with the total metabolic activity. To determine the cell compatibility of the diamine reagents, islets were treated with a non-physiological and a physiological diamine, EA and lysine, respectively. EA did not influence the islet viability at 10 mM but a significant loss of viability was observed at 30 mM (Figure 3B). Lys, on the other hand, did not influence islet viability even at 30 mM (Figure 3C). This was in accordance with a report that indicated no effect of Lys on insulin secretion of islets[12]. Based on the compatibility results, Lys was chosen as the preferred reagent for polymer crosslinking studies.

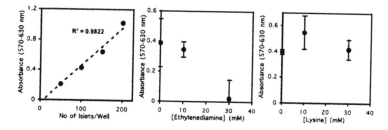

Figure 3. (A) MTT assay for pancreatic islets. There was a linear relationship between the MTT signal and the number of islets per well. Viability of islets treated with EA (B) and Lys (C). The results are summarized as mean±SD formazan absorbance of triplicate measurements. Whereas all Lys concentrations were compatible with the islets in gels, high EA concentration (30 mM) was incompatible with islet viability.

Pancreatic islets were entrapped in 3 polymers, N, S and X, and the viability was compared with and without crosslinking. No significant difference in islet viability was noted whether the gels were crosslinked with Lys or not (Figure 4). The obtained metabolic activities were similar to the activity of islets alone (Figure 3C), indicating lack of polymer toxicity on the cells. Similar to the results obtained in studies without cells, Lys-treated gels containing islets did not dissolve upon incubation at 4 °C during the MTT assay, further confirming the formation of stable polymeric network in culture medium.

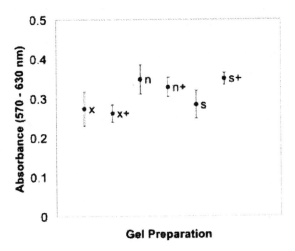

Figure 4. Viability of islets in hydrogels of polymers N, S and X, either uncrosslinked or crosslinked with Lys (indicated by +). The results are summarized as mean±SD formazan absorbance of triplicate measurements. There were no differences in islet viability whether the hydrogels were Lys crosslinked or not.

Since our initial focus was to determine the islet viability after crosslinking process, islet viability was determined in short-term (~ 1 day) studies. Long term studies on the entrapped islet viability remains to be performed. Others have shown that NiPAM-based hydrogels were able to sustain islet viability in culture in long terms studies (>2 weeks) and we anticipate that crosslinked NiPAM gels will maintain the cell viability on the long term as well[13].

4.   CONCLUSIONS

These results indicate that NASI-containing thermoreversible polymers can be utilized to form crosslinked hydrogels that were compatible with sensitive mammalian cells. The crosslinked hydrogels can be envisioned as an 'artificial' extracellular matrix[13,14], which will serve to prolong the viability of entrapped islets, or as an immunoisolation barrier[3], which will serve to exclude humoral and cellular components of an immune attack against the entrapped islets. Further research will be needed to optimize the hydrogel properties to achieve desired performance criteria in particular applications. Nevertheless, the hydrogels chemically crosslinked by the proposed approach should provide superior properties as compared to the hydrogels held together by physical means.

LIST OF ABBREVIATIONS

EA: Ethylenediamine; EMA: Ethylmethacrylate; LCST: Lower Critical Solution Temperature; Lys: Lysine; MMA: Methylmethacrylate; MTT: 3-(4,5-dimethylthiazol-2-yl)-2,5-diphenyltetrazolium bromide; NASI: N-acryloxysuccinimide; NiPAM: N-isopropylacrylamide.

ACKNOWLEDGEMENTS

This project was supported by operating funds from Natural Sciences and Engineering Research Council (NSERC) of Canada and infrastructure grants from Alberta Heritage Foundation for Medical Research (AHFMR) and Canadian Foundation for Innovation (CFI). The authors thank Dr. Jiang Bai for critical review of this chapter.

REFERENCES

1. Uludag, H., De Vos, P., and Tresco, P.A., 2000, Technology of Mammalian Cell Encapsulation. *Adv Drug Del Rev* 42: 29-64.
2. Bromberg, L.E., and Ron, E.S., 1998, Temperature-responsive gels and thermogelling polymer matrices for protein and peptide delivery. *Adv Drug Del Rev* 31: 197-221.
3. Hisano, N., Morikawa, N., Iwata, N., and Ikada, Y., 1998, Entrapment of islets into reversible disulfide hydrogels. *J Biomed Mater Res* 40: 115-123.
4. Shapiro A.M., Lakey J.R., Ryan E.A., Korbutt G.S., Toth E., Warnock G.L., Kneteman, N.M., and Rajotte, R.V., 2000, Islet transplantation in seven patients with type 1 diabetes mellitus using a glucocorticoid-free immunosuppressive regimen. *New Eng J Med* 343: 230-238.
5. Uludag, H., and Fan, X.D., 2000, Synthesis and characterization of thermoreversible, protein-conjugating copolymers based on N-isopropylacrylamide. *In Drug Delivery in the 21st Century*, Vol. 752 (K. Park, and R. Msryn Eds.:), ACS Publishers, Washington, pp. 253-262.
6. Uludag, H., Norrie, B., Kouisinioris, N., and Gao, T.J., 2001, Engineering temperature-sensitive poly(N-isopropylacrylamide) polymers as carriers of therapeutic proteins. *Biotech Bioeng* 73: 510-521.
7. Liu C., McGann L.E., Gao D., Haag B.W., and Critser J.K., 1996, Osmotic separation of pancreatic exocrine cells from crude islet cell preparations. *Cell Transplant* 5: 31-39.
8. Kumar, P., Delfino, V., McShane, P., Gray, D.W., and Morris, P.J., 1994, Rapid assessment of islet cell viability by MTT assay after cold storage in different solutions. *Transplant Proc* 26: 814.
9. Janjic, D., and Wollheim, C.B., 1992, Islet cell metabolism is reflected by the MTT (tetrazolium) colorimetric assay. *Diabetologia* 35: 482-485.
10. Cline, G.W., and Hanna, S.B., 1998, Kinetics and mechanisms of the aminolysis of N-hydroxysuccinimide esters in aqueous buffers. *J Org Chem* 53: 3583-3586.

11. Feil, H., Bae, Y.H., Feijen, J., and Kim, S.W., 1993, Effect of comonomer hydrophilicity and ionization on the lower critical solution temperature of N-isopropylacrylamide copolymers. *Macromol* 26: 2496-2500.
12. Sener, A., Blachier, F., Rasschaert, J., Mourtada, A., Malaisse-Lagae, F., and Malaisse, W.J., 1989, Stimulus-secretion coupling of arginine-induced insulin release: comparison with lysine-induced insulin secretion. *Endocrinology* 124: 2558-2567
13. Bae, Y.H., Vernon, B., Han, C.K., and Kim, S.W., 1998, Extracellular matrix for a rechargeable cell delivery system. *J Cont Rel* 53: 249-258.
14. Rowley, J.A., Madlambayan, G., and Mooney, D.J., 1999, Alginate hydrogels as synthetic extracellular matrix materials. *Biomat.* 20: 45-53.

pH-Responsive Hydrogels: Swelling Model

TAMAR TRAITEL and JOSEPH KOST [*,#]
Department of Chemical Engineering, Ben-Gurion University of the Negev, P.O.B. 653,
Beer-Sheva 84105, ISRAEL

1. INTRODUCTION

Polymers that alter their characteristics in response to changes in their environment have been of great recent interest. Several research groups have been developing drug delivery systems based on these responsive polymers that intend mimic the normal physiological process. In these devices drug delivery is regulated by means of an interaction with the surrounding environment (feedback information) without any external intervention. The most commonly studied polymers having environmental sensitivity are either pH or temperature sensitive; there are also inflammation-sensitive systems. In this chapter we will concentrate on pH-sensitive systems.

1.1 pH-Sensitive Systems

Generally, pH-sensitive systems are hydrogels that contain ionic co-monomers in their polymeric backbone. The main properties of these hydrogels that make them applicable in the medical field are biocompatibility and their ability to be manipulated to be responsive to specific physiological stimuli[1-13]. Still the clinical applications for these responsive hydrogels are limited because there is little fundamental understanding of the systems' physical non-steady state behavior[14].

In this chapter we will exhibit a mathematical model of pH-responsive hydrogels swelling. This system (Figure 1) is based on the hydrogel poly(2-hydroxyethyl methacrylate-co-N,N-dimethylaminoethyl methacrylate), also

Biomaterials: *From Molecules to Engineered Tissues,* edited by
N. Hasırcı and V. Hasırcı, Kluwer Academic/Plenum Publishers, 2004   29

called poly (HEMA-co-DMAEMA), crosslinked with tetraethylene glycol dimethacrylate (TEGDMA). By entrapping glucose oxidase, catalase and insulin into the matrix it becomes glucose-responsive insulin release system. This system has the capability of adapting the rate of insulin release in response to changes in glucose concentration. We have shown this polymer to be biocompatible in-vivo[12]. When exposed to physiological fluids, glucose diffuses into the hydrogel; glucose oxidase catalyzes the glucose conversion to gluconic acid, causing swelling of the pH-sensitive hydrogel and subsequently increased insulin release. The higher/lower the glucose concentration in the medium, the higher/lower and faster/slower the swelling and release rates. However, the glucose response time is slow, at approximately three hours, to be used for diabetes treatment. A number of parameters affect the glucose sensitivity including: crosslinking density in the hydrogel, concentration of ionizable groups in the hydrogel, oxygen availability, concentration of enzymes in the hydrogel, and enzymes' activity. Optimization of all the parameters is a labor and resource intensive project. To deepen understanding the glucose-sensitive system's swelling/deswelling mechanism, we decided to concentrate on the swelling mechanism of pH-sensitive hydrogel (i.e. the system without incorporated enzymes). In the following paragraphs we will describe the parameters that affect the swelling mechanism of pH-sensitive hydrogels.

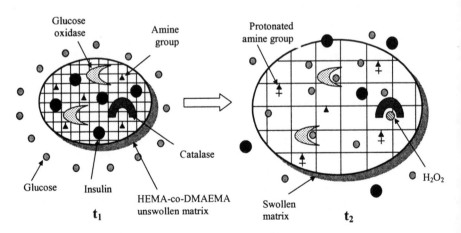

Figure 1. Schematic presentation of a matrix system based on poly (HEMA-co-DMAEMA): (t_1) unswollen matrix at time t=0, (t_2) swollen matrix at time t, due to exposure to a medium containing glucose[15].

1.2   Swelling of Ionic Hydrogels

The ionic hydrogels' water absorption, the dominant property[16-18], is impacted by parameters[16-20] such as: (a) monomer composition (hydrophilicity/hydrophobicity), (b) crosslinking density (diffusion coefficient), and (c) ionic charge (Donnan effect).

Several studies[19,21-33] have been performed on polymers containing acidic or basic groups in the polymeric backbone. The charge density of the polymers depends on pH and ionic composition of the outer solution[34,35] (concentration and pK$_a$). Altering the pH of the solution will cause swelling or deswelling of the polymer. Thus drug release from devices made from these polymers will display release rates that are pH dependent. For example, for polybasic polymers, in acidic solution, the basic groups (such as amine groups), which are bound to the polymer chains, combining with protons, which result in positively charged groups. Charge neutrality is maintained by anions that enter the hydrogel, from the outer solution, together with the protons. The increased anions concentration gives rise to an osmotic pressure that causes the hydrogel to swell. The flow of ions in and out of the hydrogel is not instantaneous. The diffusion is a main process, but the electrical field created due to the movement of the charged ions inside the hydrogel is found to play a significant role [33]. As the concentration of different ions inside the gel rises, the osmotic pressure rises and thus the deformation of the hydrogel, which generates the elastic restoring force of the network, also rises. Finally, an equilibrium is established when the elastic restoring force of the network balances the osmotic forces. In other words, whereby the free ions distribute themselves between the inside and outside of the hydrogels, the equilibrium is set up (i.e. Donnan equilibrium).

Optimization of the parameters is a major focus of hydrogel research [1,6,36], and can be carried out efficiently by computational methods[37,38]. Most of the mathematical models reported, focus on the equilibrium swelling state[21,39-47]. However, for many stimuli-responsive hydrogel applications, especially for controlled drug release devices, it is the dynamic non-steady state that is significant[1,4,6,8,10-13,19,45]. In order to elucidate the mechanisms of these systems, it is vital to dissect the influence of the various parameters. Kinetic analysis of experimental data, using compartmental modeling[48-52], enables dissection of each parameter's influence on hydrogel swelling. Furthermore, models allow testing of various non-steady state experimental conditions, and simulations and design of future experiments[37,38,48-50].

2.   MATHEMATICAL COMPARTMENTAL MODEL

Mathematical model is a very important tool in developing controlled drug delivery systems. Many controlled delivery systems release the incorporated drug over a long period of time (weeks or months). A mathematical model that enables simulation of release rates under different conditions will decrease the time, the cost and the number of *in-vitro* and *in-vivo* experiments needed for developing a release system with defined specifications. Moreover, a mathematical model can aid in the understanding of the release mechanisms from different systems such as swelling/deswelling and biodegradable[37,38,48-50].

Taking into account the numerous parameters affecting the function of hydrogel devices is an ominous task. Therefore, several mathematical models (such as SAAM II software) have been developed that enable the investigator to begin the bench research with an "educated guess".

SAAM II software is based on compartmental models, in that the events in the system are described by a finite number of state variables (compartments) with specified interactions among them. A state variable (a chemical species in a physical place) is an amount of material that acts as though it is well mixed and kinetically homogeneous. The interactions represent a flux of material, which physically represents transport from one location to another or a chemical transformation, or both. Equation 1 represents the change in the amount of compartment i over time (sum of fluxes related to compartment i):

$$flux = \frac{dq_i}{dt} = -\sum_{j=1, j \neq i}^{n} K_{ji} q_i + \sum_{j=1, j \neq i}^{n} K_{ij} q_j \qquad \dots\dots\dots\dots\dots\dots 1$$

where q_i is the amount of material in compartment i; q_j is the amount of material in compartment j; K_{ij} is the rate coefficient or constant governing the process that leaves compartment i to j; K_{ji} is the rate coefficient or constant governing the transport process entering compartment i from j. With the help of the SAAM II software, the modeller creates a system of ordinary differential equations among all the state variables, and characterizes the rate coefficients according to the theoretical hypothesis upon which the model is based.

In this chapter we concentrate on development of mathematical compartmental model of the swelling dynamics of the pH-responsive copolymer poly(HEMA-co-DMAEMA) that crosslinked with TEGDMA[12,14].

2.1   Experimental Methods

Preparation of the hydrogels: The methods for preparation of poly (HEMA-co-DMAEMA), a pH-responsive hydrogel, and for swelling experiments, were carried out as described previously[12]. Briefly, hydrogel matrices were prepared by mixing ethylene glycol, water, the monomers HEMA and DMAEMA, and the crosslinking agent TEGDMA in a glass vial. When a homogeneous solution was obtained, aqueous solutions of initiators were added: 0.2 g/ml ammonium persulphate and 0.75 g/ml sodium metabisulfite. The solution was then poured between two glass plates (15x15 cm) separated by isolating film, resulting in matrix thickness of 0.05 ± 0.005 cm. The plates with the polymeric solution were placed at room temperature for several hours until the polymerisation process was completed.

Swelling experiments: Disks (1.7 cm diameter and 0.05 cm thick) were cut from the polymerised hydrogel slabs and placed in a glass beaker containing 100 ml of the swelling medium (phosphate buffer saline, PBS). The beakers were placed into a shaking incubator at 37°C and 150 rpm. The disks were weighed periodically throughout the experiment. Water uptake was calculated using the following relationship:

$$\text{Water Uptake} = (W_{wet} - W_{initial})/W_{initial} \dots\dots\dots\dots\dots\dots\dots(2)$$

where, W_{wet} is the weight of swollen matrix at time t and $W_{initial}$ is the weight of unswollen matrix at time t=0.

Kinetic Analysis: The data collected was analyzed by compartmental analysis using the SAAM II software package version 1.1 (The SAAM Institute, University of Washington, Seattle, Website: www.saam.com).

2.2   Development of Mathematical Modeling of Ionic Hydrogel Swelling

The development of the mathematical model of the hydrogel swelling was based upon the swelling theory of ionic hydrogels, as mention before[16,19,21,43,44]. Briefly, this theory states that swelling of ionic hydrogels is due to three contributions: (1) Thermodynamic mixing between the net polymer and the solvent, (2) The elastic contribution of the polymer, and (3) The interaction between immobilized and free ions.

The first contribution was tested by comparing the swelling of a cationic hydrogel with immobilized amine groups (Figure 2a, unfilled squares points) to a non-cationic hydrogel without amine groups (Figure 2a, filled diamonds points). The second contribution is determined by testing the water uptake by polymers with different amount of crosslinking agent (Figure 2a, unfilled points). The third contribution was tested by immersing the matrix in media of different pH's (Figure 2b).

In the model developed here, the matrix was described as a well-mixed volume and this resulted in consistency between the model solution and the experimental data. The composition ranges used were: 0–20.4 vol% of DMAEMA, 0 – 0.95 vol% of crosslinking agent (TEGDMA), and the range of pH buffer solution (PBS) was from 5.8 to 7.4. Where, the limitations of this model, based on the experimental ranges are: 0–40.5 vol% of DMAEMA, 0–2.0 vol% of crosslinking agent, and the pH solution range is 5.8–7.4 (phosphate buffer with pK_b=7.2).

Based upon this series of experiments, we developed a mathematical model made up of six compartments that describe matrix swelling (Figure 3). The model is also based on three assumptions that made it simpler. (1) We assumed that the swelling rate is governed by the ionization rate at the swelling front because the non-cationic hydrogel did not swell in PBS at pH 5.8 (Figure 2a). This assumption is also based on the proposed mechanism that acidic protons (protons attached to acidic buffers) are transferred to uncharged tertiary amine groups because of the higher proton affinity of the amine groups, whose pK_a is approximately 7.8, relative to the buffer's proton affinity (pK_b=7.2)[45,53]. Therefore, the buffer was treated as a source of protons.

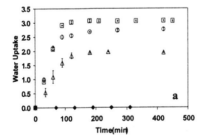 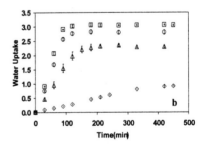

Figure 2. Swelling kinetics of pH-responsive hydrogel under different experimental conditions. **Panel a:** Different concentrations of crosslinker: 0 vol%, 0.27 vol% and 0.95 vol%, with 91.54 µmol of amine groups, and external media at pH 5.8. The experimental results represented by squares, circles, and triangles, respectively. **Panel b:** Different pH of external media: 5.8, 6.2, 6.6, and 7.4 with 0 vol% crosslinker and 91.54 µmol amine groups. The experimental results represented by squares, circles, triangles, and diamonds, respectively[14].

The following two assumptions were based on earlier experiments that tested the influence of dividing the matrix and the medium into internal and external layers, respectively, in terms of accuracy of model solutions (results not shown). (2) The matrix was treated as a series of concentric shells yielding results indistinguishable from the simpler model. In other words, diffusion of water and protons in this system occurs rapidly (6×10^{-5} cm^2/min for water and 1×10^{-6} cm^2/min for protons), whereas this model is looking at events that occur over many minutes and hours, therefore, we assume that diffusion is rapid on the time-scale of modelled events and this constant does not play a major role in our model. (3) In addition, we used an effective volume of 20 ml as the external media initial condition, instead of dividing the external media (100 ml, Experimental Methods) into five layers since the concentration gradient in the external media is not of high interest in this case. In conclusion, it may be more physically realistic to account for layers; however, when we did this, the model had too many degrees of freedom, making the variables not uniquely identifiable or statistically viable. In addition, following the laws of parsimony, it is more elegant and more correct to develop a model as simple as possible, which will describe the results in acceptable accuracy similar to the more complex model.

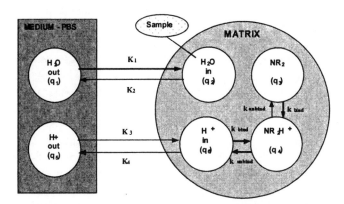

Figure 3. Schematic representation of the matrix, detailing the relationships described by the mathematical model. The arrows interconnecting the compartments represent material flux, which physically can be the transport from one physical location to another, or the chemical transformation from one state to another, or both. Also represented in the schematic drawing is the location of experimental measurement (labelled sample)[14].

Using the SAAM II software, all of the differential equations of the model are solved for all of the experimental conditions simultaneously, and then fit to the data using a least squares routine, in order to obtain the model solution[14].

2.3   Model Solution and Comparison to the Experimental Data

As described in the section above, the developed compartmental model focuses on the three main contributions to ionic hydrogel swelling: the concentration of amine groups incorporated in the polymer, the crosslinking density, and the pH of the external media. Figure 4 shows the results of experimental data and the model solution for these parameters. It is important to note that all three panels of Figure 4 were solved simultaneously. In other words, all of the parameters that describe the transport relationships among the state variables remain constant. The only differences are those directly reflective of changes in the experimental conditions.

In addition to model solutions that are consistent with the experimental data (seen as the close fit between the experimental data points and the model solution line in each case), it is important to note that the solutions are well-defined, with fractional standard deviations (FSD) of less than 10% for all adjustable parameters, and FSD's set to 32% for Bayesian parameters[14] (Bayesian parameter statistics is set by the modeller based on prior information from the literature or experimental results).

Figure 4a shows the results of experimental data and the model solution for water uptake in matrices with 0 μmol amine groups (non-cationic hydrogel) and 91.54 μmol amine groups (cationic hydrogel), respectively. The experimental conditions were: approximately 50% water in the matrices after polymerization process, and swelling medium composed of PBS at pH 5.8 with ionic strength of 0.15 M. Here it is demonstrated that, under the above experimental conditions, the non-cationic hydrogel did not swell while the cationic hydrogel did swell due to ionization of unprotonated amine groups. Therefore, in this case, the first contribution is likely negligible, since a polymer with no immobilized amine groups did not swell

Figure 4b depicts the experimental results and the model solution for water uptake in matrices with different volume percentages of crosslinking agent (0%, 0.27% and 0.95%, respectively), reflecting the influence of the second contribution. Previously[12], we showed that the less crosslinker in the matrix, the more responsive the matrix, yet even at 0% crosslinker, the hydrogel is stable (i.e. does not degrade) when exposed to water, even for extended periods of a few months.

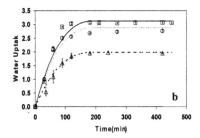

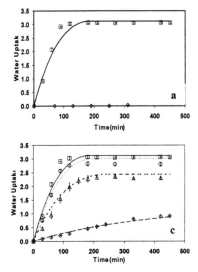

Figure 4. Swelling kinetics of pH-responsive hydrogel under different experimental conditions. Comparison of experimental data (points) and model solutions (lines). **Panel a:** Different amounts of amine groups 0 and 91.54 μmol in the matrices with 0 vol% crosslinker, and external media at pH 5.8. The experimental results represented by diamonds and squares, respectively, and the model solutions are dashed and solid lines, respectively. **Panel b:** Different concentrations of crosslinker: 0 vol%, 0.27 vol% and 0.95 vol% with 91.54 μmol of amine groups, and external media at pH 5.8. The experimental results represented by squares, circles, and triangles, respectively, and the model solutions are solid line, light dashed line, and dark dashed line, respectively. **Panel c:** Different pH of external media: 5.8, 6.2, 6.6, and 7.4, matrix with 0 vol% crosslinker and 91.54 μmol amine groups. The experimental results represented by squares, circles, triangles, and diamonds, respectively, and the model solutions are solid line, light dash line, dark dashed line, and long-dashed line, respectively[14].

Figure 4c shows the results, both experimental and the model solutions for water uptake in the matrices without crosslinker and with 91.54 μmol amine groups, when they are immersed in media of 4 different pH's (5.8, 6.2, 6.6 and 7.4). This result demonstrates the contribution of the interaction between immobilized and free ions. The behavior of the matrices immersed in acidic pH's are all similar, and also the matrix in pH 7.4 shows mild swelling; this result is expected since the pH is below the pKa of amines (7.8) in all cases. The fundamental influence of the pH, i.e. concentration gradient driven diffusion of protons into the matrix, is lowered by the ionic strength of the solution, but increasing as the pH is lowered. Ionic strength impacts the osmotic pressure of the system [17,18,45]. Thus, the relationship between the water uptake and the external proton concentration is not linear.

The development of the mathematical model for the pH-sensitive hydrogel increased our understanding of the mechanisms functioning when the matrix swells. For example, here we learned that the rate of proton (μmol H^+/min) entry into the matrix is not equal to the rate of water (μmol H_2O/min) entry into the matrix. When the rate laws for entrance of water and protons were correlated to each other directly by a factor indicating the ratio of protons to water (μmol H^+/μmol H_2O) in each experimental situation, there was no possibility for the model solution to be consistent with the experimental data (results not shown). However, this finding is consistent, as mentioned earlier, with osmotic pressure being the driving force for swelling. In addition, this is consistent with the finding (results not shown) that when the matrix is immersed in aqueous buffer solutions of the same pH, but of reduced ionic strength, due to the type of buffer used, the swelling equilibrium and kinetics increases[45].

2.4   Discussion of the Predictive Capability of the Developed Model

One significant advantage of mathematical modelling is the ability to deepen our understanding of the mechanisms involved by calculation of unmeasurable parameters. Figure 5 shows the simulated results of the protonation of the amine groups versus time in the pH-sensitive hydrogel, when immersed in media of varying pH levels. This value cannot be measured directly, experimentally. However, seeing the simulated result aids in our understanding of the swelling process. In this case, the amount of amine groups used experimentally (91.54 μmol/matrix disk) was based upon earlier reports[2,12,54,55], and considered in excess. It was thought that swelling equilibrium was attained when a certain percentage of the amine groups were protonated. However, the model solution for protonated amine groups in different pH's, shown in Figure 5, indicates that swelling equilibrium is attained when all of the amine groups are protonated. At the pH values studied, in the range of 5.8 – 6.6, it is reasonable to assume that all amine groups in the hydrated region are charged, since pH < pK_a of the amines. In order for swelling to progress, protons must be transported from the outer solution to uncharged amine groups at the swelling front. The protons may be either bound to water as hydronium ions, or bound to the buffer in the acid form. Hydronium ions will be Donnan excluded by the positively charged hydrogel, but the protonated neutral buffer will be able to enter the matrix, diffuse to the swelling front, and deliver the proton to the unionized amine[45]. Furthermore, different amounts of water are absorbed by the matrix possibly because of the different starting points of pH, and not solely because different quantities of amine groups were protonated. These results relate to matrix without crosslinker; we therefore think that the swelling acceleration, as the pH decreases, may open physical crosslink junctions and therefore increase the overall equilibrium swelling.

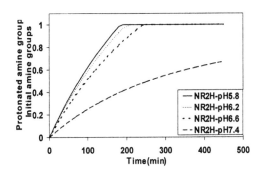

Figure 5. Time course for accumulation of protonated amine groups during the swelling experiments of matrix with 0 vol% crosslinker and 91.54 μmol unprotonated amine groups) at different pH's in the external media[14].

The second significant advantage of modelling is the predictive capability, as is demonstrated in the next set of experiments. Figure 6a shows the model simulation for a matrix with a chemical make-up of 0.47 vol.% crosslinker, and the comparison of the model prediction with the subsequent experimental results. In this case, the consistency between the experimental data and the model solution (0.47 vol.% crosslinker) is demonstrated by the closeness between the model solution (line) and the experimental data (points).

In order to evaluate the predictive capacity of the model for conditions that were beyond those in the model developed (i.e. extrapolation), we tested the influence of amine group concentration (inside the hydrogel) on the swelling kinetics of matrix with 137.3 μmol amine groups per one disk and 0.27 vol.% crosslinker (Figure 5b). Figure 6b demonstrates the predictive capability of the model, which is a significant advantage for design of future experiments. It can be seen that the prediction for chemically crosslinked hydrogel fits well to the experimental data.

It is of importance to note that the simulations of the model were carried out prior to the collection of the experimental data and are shown here with no modifications whatsoever. Based on predicted results, it should be possible to make "educated guesses" about how to improve specifications required for a drug delivery system, such as response time to step changes, and then test the new chemical make up *in-vitro*. Ultimately, the mathematical model developed could be used to simulate various combinations of the parameters, thus enabling better design of bench experiments in the optimization process of developing new drug delivery systems.

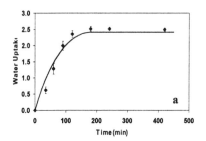 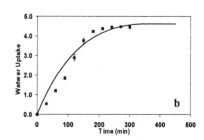

Figure 6. Model prediction of water uptake for various experimental conditions that were not part of the model development. Note that the same model, with no changes in parameters was used for all solutions. **Panel a:** Comparison of simulation result (solid line) and experimental data (filled diamonds) of swelling kinetics of matrix with 0.47 vol% crosslinker and 91.5 μmol amine groups at external media of pH 5.8. **Panel b:** Comparison of simulation result (solid line) and experimental data (filled square) of swelling kinetics of matrix with 137.3 μmol amine groups and 0.27 vol% crosslinker at external media of pH 5.8[14].

3.   CONCLUSIONS

The combination of kinetic analysis by compartmental modeling with experimental data proved to be very valuable. First, we obtain enhanced understanding of the swelling mechanisms, such as the differing rate of entry of water and protons into the matrix, and equilibrium swelling of the hydrogel being attained when protonation of the amine groups is complete. Second, questions that help determine the direction of future experimentation arise. And third, the model enables simulation of results of future experiments thus hastening the development process of new controlled delivery devices.

ACKNOWLEDGEMENTS

This work was supported by JDFI grant # 1-2000-564.

REFERENCES

1. Chen, G., and Hoffman, A. S., 1995, Graft copolymers that exhibit temperature-induced phase transitions over a wide range of pH, *Nature* 373(6509):49-52.
2. Goldraich, M., and Kost, J., 1993, Glucose-sensitive polymeric matrices for controlled drug delivery, *Clin. Mater.* 13(1-4):135-42.
3. Holtz, J. H., and Asher, S. A., 1997, Polymerized colloidal crystal hydrogel films as intelligent chemical sensing materials, *Nature* 389(6653):829-32.
4. Kiser, P. F., Wilson, G., and Needham, D., 1998, A synthetic mimic of the secretory granule for drug delivery, *Nature* 394(6692):459-62.
5. Kost, J., and Langer, R., 2001, Responsive polymeric delivery systems, *Adv. Drug. Deliv. Rev.* 46(1-3):125-48.
6. Miyata, T., Asami, N., and Uragami, T., 1999, A reversibly antigen-responsive hydrogel, *Nature* 399:766-769.
7. Osada, Y., Okuzaki, H., and Hori, H., 1992, A polymer gel with electrically driven motility, *Nature* 355:242-244.
8. Podual, K., Doyleiii, F. J., and Peppas, N. A., 2000, Glucose-sensitivity of glucose oxidase-containing cationic copolymer hydrogels having poly(ethylene glycol) grafts, *J. Controlled Release* 67(1):9-17.
9. Stayton, P. S., Shimoboji, T., Long, C., Chilkoti, A., Chen, G. H., Harris, J. M., and Hoffman, A. S., 1995, Control of protein-ligand recognition using a stimuli-responsive polymer, *Nature* 378:472-474.
10. Taillefer, J., Jone, M. C., Brasseur, N., Van Leir, J. E., and Leroux, J. C., 2000, Preparation and characterization of pH-responsive polymeric micelles for the delivery of photo-sensitizing anti-cancer drugs, *J. Pharm. Sci.* 89:52-62.
11. Tanihara, M., Suzuki, Y., Nishimura, Y., Suzuki, K., Kakimaru, Y., and Fukunishi, Y., 1999, A novel microbial infection-responsive drug release system., *J. Pharm. Sci.* 88:510-514.
12. Traitel, T., Cohen, Y., and Kost, J., 2000, Characterization of glucose-sensitive insulin release systems in simulated in vivo conditions, *Biomaterials* 21(16):1679-1687.
13. Yoshida, R., Uchida, K., Kaneko, Y., Sakai, K., Kikuchi, A., Sakurai, Y., and Okano, T., 1995, Comb-type grafted hydrogels with rapid de-swelling response to temperature changes, *Nature* 374:240-242.
14. Traitel, T., Kost, J., and Lapidot, S. A., 2003, Modeling ionic hydrogels swelling: characterization of the non-steady state, *Biotechnol. Bioeng.* 84(1):20-8.
15. Goldbart, R., Traitel, T., Lapidot, S. A., and Kost, J., 2002, Enzimatically controlled responsive drug delivery systems, *Polym. Adv. Technol.* 13:1006-1018.
16. Flory, P. J., 1953, Principles of polymer chemistry, Cornell University Press, New York.
17. Peppas, N. A., 1987, Hydrogels in medicine and pharmacy, CRC Press, Boca Raton, FL.
18. Rattner, B. D., Hoffman, A. S., Schoen, F. J., and Lemons, J. E., 1996, Biomaterials science: An introduction to materials in medicine, Academic Press, New York.
19. Bell, C. L., and Peppas, N. A., 1996, Water, solute and protein diffusion in physiologically responsive hydrogels of poly(methacrylic acid-g-ethylene glycol), *Biomaterials* 17(12):1203-1218.
20. Park, K., Shalaby, W. S. W., and Park, H., 1993, Biodegradable hydrogels for drug delivery, Thechnomic Publishing Company, Inc., PA.
21. Brannon-Peppas, L., and Peppas, N. A., 1989, Solute and penetrant diffusion in swellable polymers. IX. The mechanism of drug release from pH sensitive swelling-controlled systems., *J. Controlled Release* 8:267-274.
22. Annaka, M., and Tanaka, T., 1992, Multiple phases of polymer gels, *Nature* 355:430-432.
23. Firestone, B. A., and Siegel, R. A., 1988, Dynamic pH-dependent swelling properties of hydrophobic polyelectrolyte gel, *Polym. Commun.* 29:204-208.

24. Dong, L.-C., and Hoffman, A. S., 1990, Controlled enteric release of macromolecules from pH sensitive, macroporous hetrogels, *Proceedings of the International Symposium on Controlled Bioactive Materials (Proc. Int. Symp. Control. Bioact. Mater.)* **17**:325-326.

25. Kou, J. H., Fleisher, D., and Amidon, G., 1990, Modeling drug release from dynamically swelling poly(hydroxyethyl methacrylate-co-methacrylic acid) hydrogels, *J. Controlled Release* **12**:241-250.

26. Pradny, M., and Kopecek, J., 1990, Hydrogels for site-specific oral delivery. Poly(acrylic acid)-co-(butyl acrylate) crosslinked with 4,4'-bis(methacryloamino)azobenzene, *Makromol. Chem.* **191**:1887-1897.

27. Siegel, R. A., Falamarzian, M., Firestone, B. A., and Moxley, B. C., 1988, pH-Controlled release from hydrophobic/polyelectrolyte copolymer hydrogels, *J. Controlled Release* **8**(2):179-182.

28. Kono, K., Tabata, F., and Takagishi, T., 1993, pH-responsive permeability of poly(acrylic acid)-- poly(ethylenimine) complex capsule membrane, *J. Membrane Sci.* **76**(2-3):233-243.

29. Hariharan, D., and Peppas, N. A., 1993, Modelling of water transport and solute release in physiologically sensitive gels, *J. Controlled Release* **23**(2):123-135.

30. Siegel, R. A., and Firestone, B. A., 1988, pH-dependent equilibrium swelling properties of hydrophobicpoly-electrolyte copolymer gels, *Macromolecules* **21**(11):3254-3259.

31. Allcock, H. R., and Ambrosio, A. M. A., 1996, Synthesis and characterization of pH-sensitive poly(organophosphazene) hydrogels, *Biomaterials* **17**(23):2295-2302.

32. Jarvinen, K., Akerman, S., Svarfvar, B., Tarvainen, T., Viinikka, P., and Paronen, P., 1998, Drug release from pH and ionic strength responsive poly(acrylic acid) grafted poly(vinylidenefluoride) membrane bags in vitro, *Pharm. Res.* **15**(5):802-5.

33. De, S. K., and Aluru, N. R., In press, A chemo-electro-mechanical mathematical model for simulation of pH sensitive hydrogels, *Mechanics of Materials*.

34. Siegel, R. A., Johannes, I., Hunt, C. A., and Firestone, B. A., 1992, Buffer effects on swelling kinetics in polybasic gels, *Pharm. Res.* **9**(1):76-81.

35. Kim, B., and Peppas, N. A., 2003, Analysis of molecular interactions in poly(methacrylic acid-g-ethylene glycol) hydrogels, *polymer* **44**:3701-3707.

36. Wang, C., Stewart, R. J., and Kopecek, J., 1999, Hybrid hydrogels assembled from synthetic polymers and coiled-coil protein domains, *Nature* **397**:417-420.

37. Phair, R. D., 1997, Development of kinetic models in the nonlinear world of molecular cell biology, *Metabolism* **46**:1489-1495.

38. Weng, G., Bhalla, U. S., and Iyengar, R., 1999, Complexity in biological systems, *Science* **284**:92-96.

39. Barenbrug, T. M. A. O. M., Smit, J. A. M., and Bedeaux, D., 1995, Highly Swollen Gels of Semi-flexible Polyelectrolyte Chains Near the Rod Limit, *Polymer Gels and Networks* **3**(3):331-373.

40. Brannon-Peppas, L., and Peppas, N. A., 1991, Equilibrium swelling behavior of dilute ionic hydrogels in electrolytic solutions, *J. Controlled Release* **16**(3):319-329.

41. English, A. E., Tanaka, T., and Edelman, E. R., 1997, Equilibrium and non-equilibrium phase transitions in copolymer polyelectrolyte hydrogels, *J. Chem. Phys.* **107**:1645-1654.

42. Kou, J. H., Amidon, G. L., and Lee, P. I., 1988, pH-dependent swelling and solute diffusion characteristics of poly(hydroxyethyl methacrylate-co-methacrylic acid) hydrogels, *Pharm. Res.* **5**(9):592-7.

43. Ricka, J., and Tanaka, T., 1984, Swelling of ionic gels: quantitative performance of the Donnan theory, *Macromolecules* **17**:2916-2921.

44. Siegel, R. A., Firestone, B. A., Cornejo-Bravo, J., and Schwarz, B., 1991, Hydrophobic weak polybasic gels: Factors controlling swelling equilibria, in: *Polymer gels - Fundamentals and biomedical applications* (D. De Rossi, Kajiwara K., Osada Y., and Yamauchi A., eds.), Plenum Press, New York, pp. 309-317.

45. Siegel, R. A., 1993, Hydrophobic weak polyelectrolyte gels: Studies of swelling equilibria and kinetics, in: *Advances in polymer science/ responsive gels* (K. Dusek, ed.), Springer-Verlag Press, New York, pp. 233-267.
46. Vasheghani-Farahani, E., Vera, J. H., Cooper, D. G., and Weber, M. E., 1990, Swelling of ionic gels in electrolyte solutions, *Ind. Eng. Chem. Res.* **29**:554-560.
47. Yuk, S. H., and Bae, Y. H., 1999, Phase-transition polymers for drug delivery, *Crit. Rev. Therap. Drug Carr. Syst.* **16**:385-423.
48. Cobelli, C., and Foster, D. M., 1998, Compartmental models: theory and practice using the SAAM II software system, *Advances in Experimental Medicine and Biology* **445**:79-101.
49. Heatherington, A. C., Vicini, P., and Golde, H., 1998, A pharmacokinetic/pharmacodynamic comparison of SAAM II and PC/WinNonlin modeling software, *J. Pharm. Sci.* **87**(10):1255-63.
50. Kohn, M. C., 1995, Achieving credibility in risk assessment models, *Toxicol. Lett.* **79** (1-3):107-14.
51. Wolfe, R. R., 1992, Radioactive and stable isotope tracers in biomedicine: Principles and practice of kinetic analysis, Wiley-Liss, Inc., New York.
52. Yang, R. S. H., and Anderson, M. E., 1994, Pharmacokinetics, in: *Introduction to biochemical toxicology* (E. Hodgson, and Levi P., eds.), Elsevier, New York, pp. 49-73.
53. Cornejo-Bravo, J. M., Arias-Sanchez, V., Alvarez-Anguiano, A., and Siegel, R. A., 1995, Kinetics of drug release from hydrophobic polybasic gels: effect of buffer acidity*1, *J. Controlled Release* **33**(2):223-229.
54. Klumb, L. A., and Horbett, T. A., 1992, Design of insulin delivery devices based on glucose sensitive membranes, *J. Controlled Release* **18**(1):59-80.
55. Klumb, L. A., and Horbett, T. A., 1993, The effect of hydronium ion transport on the transient behavior of glucose sensitive membranes, *J. Controlled Release* **27**(2):95-114.

Smart Elastin-like Polymers

J.CARLOS RODRIGUEZ-CABELLO
Department of Condensed Matter Physics (BIOFORGE Group). E.T.S.I.I., University of Valladolid, Paseo del Cauce s/n, 47011-Valladolid, SPAIN

Elastin-like polymers are a new family of proteinaceous polymers. In these polymers converge a wide set of interesting properties that difficultly can be found together in other polymers. They are extremely biocompatible and show an acute smart and self-assembling behaviour. The increasing in complexity of the molecular design renders polymers showing combination of functionalities and complex performance. This is specially true nowadays where, taking into account their peptide nature, these polymers can be produced as recombinant proteins in genetically modified (micro)organisms. The absolute control and absence of randomness in the primary structure makes possible the realization of multifunctional polymers that can combine physical, chemical and biological functions in a desired fashion. It can be said that the molecular design is mainly limited by imagination and not by technique. This chapter is intended to show the molecular parameters that explain the smart behaviour finally observed and how the increase in complexity of the molecular designs leads to a richer behaviour of the polymer, as a way to show the enormous potential of this family in the development of advanced materials and systems for biomedicine and nanotechnology for the next decades.

1. INTRODUCTION

Smart polymers, polymers that sharply respond in a giving manner to a physical or chemical change in their environment, are an exciting field within the material development. In comparison with more conventional polymers, there are not many examples in the literature of polymers able to

Biomaterials: *From Molecules to Engineered Tissues,* edited by
N. Hasırcı and V. Hasırcı, Kluwer Academic/Plenum Publishers, 2004

respond to stimuli. However, there are examples that cover a wide range of physical and chemical behaviours and phenomena, from electroactive and light emitting polymers to polymers that reversibly segregate and dissolve in water. In all cases, they find notable applications in the fields of biology, biotechnology and biomedicine.

This chapter is devoted to the description of a new family of smart polymers, the elastin-like polymers ("ELPs"). As it will be shown, this family shows a collection of properties that places them in an excellent position among their direct competitors. In addition, they are particularly well suited to fit into the most advanced biomedical applications.

1.1   The Origin and Main Properties of Elastin-like Polymers

ELPs are non-natural polypeptides composed of repeating sequences. They have their origin in the repeating sequences found in the mammalian elastic protein, elastin. The most striking and longest sequence between cross-links in pig and cow is the undecapentapeptide $(VPGVG)_{11}$[1,2]. Along with this repeating sequence, others can be pointed out such as $(VPGG)_n$[3], $(APGVGV)_n$[4].

Initially, the monomers, oligomers and high polymers of either this repeats, others or modifications thereof have been chemically synthesised and conformationally characterized[5-7]. More recently, with the development of Molecular Biology, these and more complex ELPs have been bioproduced by using genetic engineering techniques[6,8-11]. By this way, more complex and well-defined polymers than their chemically-synthesized precursors have been obtained.

Regarding their properties, some of their main characteristics are derived from the natural protein they are based on. For example, the cross-linked matrices of these polymers retain most of the striking mechanical properties of elastin[12], i.e, an almost ideal elasticity with Young modulus, elongation at break, etc. in the range of the natural elastin and an outstanding resistance to fatigue[13,14].

A second relevant property is their extreme biocompatibility. The complete series of the ASTM recommended generic biological tests for materials and devices in contact with tissues and tissue fluids and blood demonstrate an unmatched biocompatibility[15]. In spite of polypeptide nature of these polymers, it has not been possible to obtain monoclonal antibodies against them. Apparently, the immune system just ignores these polymers because it cannot distinguish them from the natural elastin. In addition, the secondary products of their biodegradation are just simple and natural amino acids.

However, the most striking property is perhaps their acute smart nature. This property is based on a molecular transition of the polymer chain in the presence of water. This transition, the "Inverse Temperature Transition" (ITT), first described for ELPs, has become the key issue in the development of new peptide-based polymers as molecular machines and materials. The understanding of the macroscopic properties of these materials in terms of the molecular processes taking place around the ITT has established a basis for their functional and rational design. Recently, the increased knowledge in the many aspects around the ITT has allowed the systematization and compilation of these topics in a set of five axioms[16]. Moreover, the usefulness of this set of axioms is not restricted to the design of new advanced materials since they can be used, for example, to understand the relationship between folding and function in native proteins or the principles behind amphiphilic macromolecular assemblies[16]. However, the ITT is a rich phenomenon that hides many aspects, so still now many efforts are directed to unveil the basic molecular behaviour under the ITT of ELPs (see, for example, refs. 17-19).

All of the functional ELPs exhibit this reversible phase transitional behaviour[20]. In aqueous solution and below a certain transition temperature (T_t), the free polymer chains remain disordered, random coils in solution[21] that are fully hydrated, mainly by hydrophobic hydration. This hydration is characterized by the existence of ordered clathrate-like water structures surrounding the apolar moieties of the polymer[16,18,22] with a structure somehow similar to that described for crystalline gas hydrates[22,23]. On the contrary, above T_t, the chain hydrophobically folds and assembles to form a phase separated state of 63% water and 37% polymer by weight[24], in which the polymer chains adopt a dynamic, regular, non-random structure, called β-spiral, involving type II β-turns as main secondary feature, and stabilized by intra-spiral inter-turn and inter-spiral hydrophobic contacts[20]. This is the product of the ITT. In this folded and associated state, the chain loses essentially all of the ordered water structures of hydrophobic hydration[16]. During the initial stages of polymer dehydration, hydrophobic association of β-spirals takes on fibrillar form that grows to a several hundred nm particle before settling into the visible phase separated state[20,25]. This folding is completely reversible on lowering again the sample temperature below T_t[20].

2.   INCREASING COMPLEXITY IN SMART
ELASTIN-LIKE POLYMERS

2.1   Simple Thermo-responsive ELPs: Some Molecular
Considerations

The most numerous family within ELPs is that based on the pentapeptide
VPGVG (or its permutations). A wide variety of polymers have been
(bio)synthesized with a general formula (VPGX'G), where X' represents any
natural or modified amino acid[5,6,20]. All the polymers with that general
formula that can be found in the literature are functional; i.e., all show a
sharp smart behaviour. However, the achievement of functional ELPs by the
substitution of any of the other amino acids in the pentamer is not so
straightforward. For example, the first glycine cannot be substituted by any
other natural amino acid different from L-alanine[20]. This is because,
according to Urry[20], the type II β-turn formed in the folded state per
pentamer involves this glycine. The presence of bulky moieties in amino
acids with L chirality impedes the formation of the β-turn and the resulting
polymer is not functional[20]. Thus, the substitution by alanine is the only
possibility, though even in this case, the resulting polymer shows
significantly different and "out of trend" mechanical and thermal properties.

This fact is clearly seen also in the thermo-responsive behaviour of both
materials. Figure 1 shows the DSC thermograms of poly(VPGVG) and
poly(VPAVG) water solutions (see Ref. 19 for details on synthesis and
characterization of both polymers).

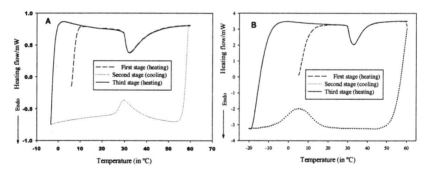

Figure 1. Typical DSC run of 125 mg mL^{-1} samples standing a heating-cooling-heating cyclic
temperature program: A) poly(VPGVG); B) poly(VPAVG). A heating rate of 8 °C min^{-1} and -
8 °C min^{-1} cooling were used in this example. Reproduced with permission from
Macromolecules **2003**, *36*, 8470-8476. Copyright 2003 Am. Chem. Soc.

The DSC patterns shown by both polymers share common features. During the first heating run, the presence of an endothermic transition is evident for both polymers. This endotherm is caused by the characteristic process of chain folding, which is accompanied by the destruction of the ordered shell of hydrophobic hydration. Obviously, this last endothermic contribution dominates this calorimetric run. T_t can be identified as the peak temperature. On the contrary, the subsequent cooling stage shows a clear exotherm for both polymers. This reflects the reverse process; i.e., the unfolding of the polymer chain and its concurrent hydrophobic hydration. However, although the main features are common for both polymers, there are significant differences that point to their divergent behaviour. For poly(VPGVG), heating T_t (T_{tH}) and cooling T_t (T_{tC}) show only marginal differences that can be attributed to the inherent thermal lags of the DSC experiment (see Fig. 1A). However, for poly(VPAVG), the difference T_{tH} - T_{tC} scores 25.6°C for this cycle run at 8 °C min^{-1} (T_{tH} = 30.7°C and T_{tC} = 6.1°C) (see Fig. 1B). This indicates that a clear hysteresis behaviour exists for poly(VPAVG); the polymer chain folds at 30.7°C but does not unfold until the temperature is undercooled down to 6.1°C. Additionally, the difference T_{tH} - T_{tC} , i.e. the degree of undercooling, for poly(VPAVG) is found to be, to a significant extent, heating/cooling rate dependent. This is specially true for T_{tC}. This difference increases with the heating/cooling rate[19], being still larger than 15°C at rates as low as 2° C min^{-1}.

The use of a model-free kinetic analysis, the Friedman´s isoconversional method[26], also revealed that the kinetics of folding-unfolding is quite different for both polymers. As expected, the kinetics of the hydration of non polar moieties for both polymers indicated a complex and multi-step process, but poly(VPAVG) showed a quite different pattern characterized by a significant decrease in the subsequent Activation Energy, as compared to poly(VPGVG). This decrease caused that the hydration-unfolding process takes place at a higher speed for poly(VPAVG) and with a simpler mechanism, which is dominated by just one limiting step, likely water diffusion[19].

The global picture taken from calorimetric and kinetic analysis clearly indicates the effect of the substitution of glycine by L-alanine in the monomer. The extra methyl group of L-alanine seems to hinder and partially block the bond rotation needed to establish or destruct the β-turns. In particular, the unfolding process of poly(VPAVG) seemed to be especially affected in a way that this unfolding only takes place after strong undercoolings. This causes that the process takes place under clear far-from-equilibrium conditions and, therefore, is strongly dominated by kinetics.

Due to the fact that, in the folded state, both polymers segregate from their solutions and spontaneously self-assemble in nano- and micro-particles, the existence of such hysteresis behaviour can be exploited in some biomedical applications such as drug delivery. Loaded particles can be prepared simply by a gentle warming of a polymer+drug solution above their corresponding T_t, which, on heating, is about 30°C for both polymers. However, while both polymers yield stable particles at body temperature causing a sustained release of the loaded drug, poly(VPAVG) particles are stable at room temperature whilst poly(VPGVG) ones are not, which clearly complicate their practical manipulation. These facts have been recently proved for the sustained release of dexamethasone phosphate by poly(VPAVG) nanoparticles[27].

2.2   pH Responsive ELPs

In all ELPs, T_t depends on the mean polarity of the polymer, being higher as the hydrophobicity decreases. This is the origin of the so called "ΔT_t mechanism"[20] and "amplified ΔT_t mechanism"[28]; i.e., if a chemical group that can be present in two different states of polarity exits in the polymer chain, and these states are reversibly convertible by the action of an external stimulus, the polymer will show two different T_t values. This T_t shift (ΔT_t) opens a working temperature window in which the polymer isothermally and reversibly switches between the folded and unfolded states following the changes in the environmental stimulus. These ΔT_t mechanisms have been exploited to obtain many elastin-like smart derivatives[20,28-30].

This mechanism is also exploited in the following model pH responding polymer; $[(VPGVG)_2\text{-}VPGEG\text{-}(VPGVG)_2]_n$. In this ELP, the γ-carboxylic group of the glutamic acid (E) suffers strong polarity changes between its protonated and deprotonated states as a consequence of pH changes around its effective pK_a.

Figure 2 shows the folded chain content as a function of T and pH for a genetically engineered polymer with the above general formula (n=15). At pH=3.5, in the protonated state, the T_t shown by the polymer is 30°C. Below this temperature the polymer is unfolded and dissolved while above it, the polymer folds and segregates from the solution. However, at pH=8.0 the rise in the polarity of the γ-carboxyl groups, as they loose their protons becoming carboxylate, is enough to increase T_t at values above 85°C, opening a working temperature window wider than 50°C. At temperatures above 30°C the polymer would fold at low pHs and unfold at neutral or basic pHs. In addition, this fact reveals the extraordinary efficiency of ELPs as compared to other pH responding polymers since this huge ΔT_t is achieved with just four E residues per 100 amino acids in the polymer backbone. This is of practical importance in using these polymers to design molecular machines and nanodevices such as nanopumps or nanovalves because just a limited number of protons are needed to trigger the two states of the system.

The emergence of an electric charge over a side chain in a given ELP due to an acid-basic equilibrium has been considered in the literature as a highly efficient way to achieve high ΔT_t. In the number of ELPs designed and studied up-to-date, the capability of the free carboxyl or amino groups of aspartic acid, glutamic acid or lysine to drive those T_t shifts is only surpassed by the ΔT_t caused by the phosphorylation of serine[20].

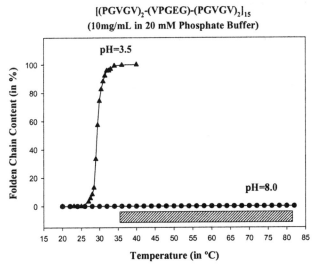

Figure 2. Turbidity temperature profiles of a model genetically engineered pH responding ELP (see Ref. 11 for details about bioproduction of this polymer). Box at the bottom represents the temperature window of working temperatures. Sample conditions are given in the plot.

2.3 Further Functionalization of Side Chains: Photochromic ELPs

The range of stimuli that can exploit the ΔT_t mechanism is not limited to those chemical reactions taking place on natural amino acid side chains. It is possible to modify certain side chains to achieve systems with extended properties. A good example of this are photo-responding ELPs, which bear photochromic side chains either coupled to functionalized side chains in the previously formed polymer (chemically or genetically engineered) or by using non natural amino acids already photochromic prior chemical polymerization.

The first example corresponds to this last kind. The polymer is an azobenzene derivative of poly(VPGVG), the copolymer poly[f_V(VPGVG), f_X(VPGXG)] (X, L-p-(phenylazo)phenylalanine; f_V and f_X, mole fractions). The p-phenylazobenzene group suffers a photo-induced cis-trans isomerization. Dark adaptation or irradiation with visible light around 420 nm induce the presence of the trans isomer, the most unpolar isomer. On the contrary, UV irradiation (at around 348 nm) causes the appearance of high quantities of the cis isomer, which is a little more polar than the trans. Although the polarity change is not high, it is enough to obtain functional polymers due to the sensitivity and efficiency of ELPs. Figure 3 shows the photo-response of one of these polymers with $f_X=0.15$. That mole fraction represents only 3 L-p-(phenylazo)phenylalanine groups per 100 amino acids in the polymer chain. In spite of the low polarity change and the exiguous presence of chromophores, it is evident (see Fig. 3A) the existence of a working temperature window at around 13°C (see Fig. 3B).

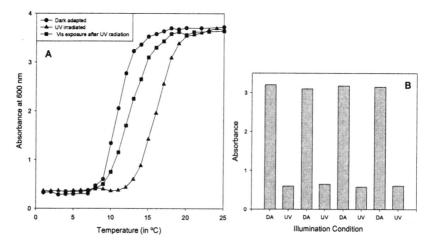

Figure 3. A), Temperature profiles of aggregation of 10 mg mL^{-1} water solutions of the photoresponsive poly[0.85(VPGVG), 0.15(VPGXG)] ((X ≡ L-p-(phenylazo)phenylalanine) under different illumination regimens. The correspondence between each profile and its illumination condition is indicated in the plot. Details on polymer synthesis and illumination conditions can be found in Ref. 30. B), Photomodulation of phase separation of 10 mg mL^{-1} aqueous samples of poly[0.85(VPGVG), 0.15(VPGXG)] at 13°C. The illumination condition prior measurement are indicated in the horizontal axis. DA, Dark adaptation; UV, UV irradiation. Reproduced with permission from Macromolecules, 2001, 34, 8072-8077. Copyright 2001 Am. Chem. Soc.

In another example, a different chromophore, a spiropyrane derivative, is attached over the free γ-carboxyl group of an E-containing ELP either chemically synthesized or genetically engineered. Figure 4 represents the photochromic reaction for this polymer (see Ref. 29). As compared to p-phenylazobenzenes, spiropyrane compounds show a photoreaction that can be driven by natural cycles of sunlight-darkness without the employment of UV sources, although UV irradiation causes the same effect as darkness but at a higher rate[31].

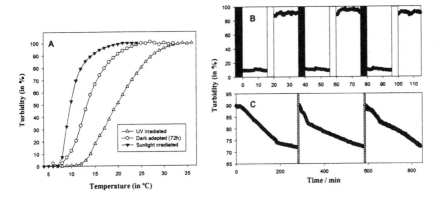

Figure 4. Photochemical reaction responsible for the photochromic behaviour of the spiropyrane-containing ELP. Reproduced with permission from Macromolecules, 2000, 33, 9480-9482. Copyright 2000 Am.Chem.Soc.

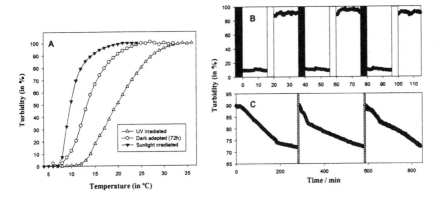

Figure 5. A), Temperature profiles of aggregation of 20 mg mL^{-1} phosphate-buffered (0.01N, pH 3.5) water solutions of the photoresponsive polymer under different illumination regimens. The correspondence between each profile and its illumination condition is indicated in the plot. Turbidity was calculated from the absorbance values obtained at 600 nm on Cary 50 UV-Vis spectrophotometer equipped with a thermostated sample chamber. B and C), Photomodulation of phase separation of 5 mg mL^{-1} aqueous samples of the photochromic polymer (T= 14°C, 0.01N phosphate buffer at pH=3.5). A) UV-sunlight cycles. Boxes in the subplot represent periods of irradiation: UV, black boxes; sunlight, white boxes. B) Darkness-sunlight cycles. Boxes in the subplot represent periods of sunlight irradiation. Reproduced with permission from Macromolecules, 2000, 33, 9480-9482. Copyright 2000 Am.Chem.Soc.

Again the difference in polarity between the spiro and merocyanine forms (see Fig.4) is enough to cause a significant T_t shift. Figure 5 shows the turbidity profiles of the polymer in the different illumination regimens (Fig. 5A) and the photomodulation of the polymer folding and unfolding (Fig. 5B and C)

The efficiency of the polymer is once more outstanding, since just 2.3 spiropyran chromophores per 100 amino acid residues in the polymer backbone were sufficient to render the clear photomodulation shown in Figs. 4 and 5.

In a different approach, it is possible to increase and further control the smart behaviour of ELPs without increasing the number of sensitive moieties. This is possible if this moiety has one state that it is able to interact with a different compound and this interaction further increases the difference in polarity between both states. This is the basis of the so called "amplified ΔT_t mechanism" and this has been proved for a p-phenylazobenzene-containing polymer of the kind shown above, poly[0.8(VPGVG), 0.2(VPGXG)], in the presence of α-cyclodextrin (αCD)[32]. αCD is able to form inclusion compounds with the trans isomer of the p-phenylazobenzene group and not with the cis due to a strong steric hindarance[32] (see Fig. 6). αCD outer shell has a relatively high polarity, which, of course, is quite more polar than the p-phenylazobenzene moiety in both the trans or cis states. The change in polarity between the dark adapted sample (trans isomer buried inside the αCD) and the UV irradiated one (cis isomer unable to form inclusion compounds) leaded to an enhanced ΔT_t (see Fig. 7).

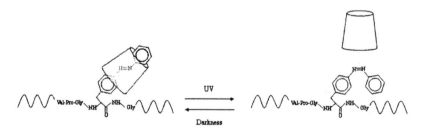

Figure 6. Schematic diagram of the proposed molecular mechanism on the interaction between the p-phenylazobenzene pendant group and the αCD. Reproduced with permission from Adv. Mater., 2002, 14, 1151-1154. Copyright 2002 Wiley-VCH.

As a result, in the αCD/poly[0.8(VPGVG), 0.2(VPGXG)] coupled photoresponsive system, αCD acts in a similar way than an amplifier acts on an electronic circuit. αCD. αCD promoted a tunable offset, gain and inversion of the photoresponse of the polymer (Fig. 7). By this way, the polymer photoresponsiveness could be shifted to room or body temperature and with a wider range of working temperatures, so as the use of precise temperature control can be avoided in most conceivable applications, being these applications in a wide range that goes from photo-operated molecular machines to macroscopic devices (photoresponsive hydrogels, membranes, etc) and nano and micro devices (phototransducer particles, photo-operated pumps, etc.). Furthermore, the amplified ΔT_t mechanism is not restricted to photoresponsive ELPs and could be exploited in some other smart ELPs responding to stimuli of a different nature. It also adds a further possibility of control, since the ability of CDs to form inclusion compounds can be controlled by different stimuli in some modified CDs[33,34].

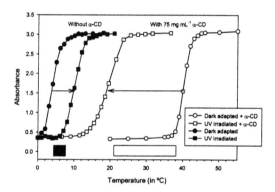

Figure 7. Temperature profiles of aggregation of a 10 mg mL⁻¹ water solutions of the photoresponsive ELP in absence and presence (75 mg mL⁻¹) of αCD under both illumination regimens. Circles represents dark adapted samples and squares UV irradiated samples. Hollow symbols, presence of αCD; filled symbols, absence of αCD. Arrows represents the sense of displacement of the turbidity profile caused by UV irradiation of the corresponding dark adapted sample. Boxes at the bottom indicate the window of working temperatures open when the system is in absence (filled box) and presence (hollow box) of αCD. Reproduced with permission from Adv. Mater., 2002, 14, 1151-1154. Copyright 2002 Wiley-VCH.

3. OUTLOOK

As a consequence of the technical advances, the limits in realizable complexity in the design of ELPs are still far from being reached. This is especially true nowadays where the incorporation of genetic engineering techniques allows the production of really complex and imaginative designs, which can be exactly defined and obtained with a total absence of randomness.

The combination of the important properties of ELPs, such as extreme biocompatibility and smart and self-assembling characteristics, with the additional power given by genetic engineering, converge in this family in a way that it is difficult to find any other group of materials possessing so huge potential. This is already starting to be exploited in smart self-assembling systems for biomedical and other technological applications. In particular, ELPs are specially suited for nanotechnological uses such as self-assembled functional nanopatterns, nanodevices (molecular machines, nanovalves, nanoactuators, etc.) and others. Their peptide nature enormously facilitates the implementation of biofunction by the integration of bioactive peptides taken from natural proteins; in this sense, some ELPs have already incorporated cell attachment sequences and other biofunctionalities. One can say that these polymers are already at the centre of gravity of nanobiotechnology.

In conclusion, in the near future, the use of these polymers and the technological concepts derived from their development will be used for sure in the most advanced designs in biomaterials, biomedicine and nanotechnology.

ACKNOWLEDGEMENTS

The work from our laboratory was supported by different grants from the Spanish Ministry of Science and Technology and the "Junta de Castilla y León".

REFERENCES

1. Sandberg, L. B., Leslie, J. G., Leach, C. T., Torres, V. L., Smith, A. R., and Smith, D. W., 1985, *Pathol. Biol.* 33: 266-274.
2. Yeh, H., Ornstein-Goldstein, N., Indik, Z., Sheppard, P., Anderson, N., Rosenbloom, J. C., Cicilia, G. C., Yoon, K., Rosenbloom, J., 1987, *Collagen Rel. Res.* 7: 235-249.
3. Sandberg, L. B., Soskel, N. T., and Leslie, J. G., 1981, *New Engl. J. Med.* 304: 566-579.
4. Indik, Z., Yeh, H., Ornstein-Goldstein, N., Sheppard, P., Anderson, N., Peltonen, L., Rosenbloom, J. C., and , Rosenbloom, J., 1987, *Proc. Natl. Acad. Sci. USA* 84:5680-5684.
5. Gowda, D. C., Parker, T. M., Harris, R. D., and Urry, D.W., 1994, Synthesis, Characterization and Medical Applications of Bioelastic Materials. In Peptides: Desing, Synthesis and Biological Activity (Basava, C. and Anantharamaiah, G. M., eds.), Birkhäuser, Boston, 81-111.
6.   Martino, M., Perri, T., and Tamburro, A. M., 2002, *Macromol. Biosci.* 2: 319–328.

7. Kurková, D., Kříž, J., Schmidt, P., Dybal, J., Rodríguez-Cabello, J. C. and Alonso, M., 2003, *Biomacromol.* **4**: 589-601.
8. McPherson, D.T., Xu, J., and Urry, D.W., 1996, *Protein Expression and Purification* **7**: 51-57.
9. Welsh, R. E., Tirrell, D. A., 2000, *Biomacromol.* **1**: 23-30.
10. Meyer, D. E., Chilkoti, A., 2002, *Biomacromol.* **3**: 357-367.
11. Girotti, A., Reguera, J., Arias, F. J., Alonso, M., Testera, A. M., and Rodríguez-Cabello J. C., 2004, *Macromolecules*, (in press).
12. Ayad, S., Humphries, M., Boot-Handford, R., and Kadler, K., Shuttleworth, A., 1994, The Extracellular Matrix Facts Book, Facts Book Series, Academic Press, San Diego,CA.
13. Urry, D. W., Luan, C. -H., Harris, C. M., Parker, T. 1997, In Protein Based Materials (McGrath, K. and Kaplan, D., eds.), Birkhäuser, Boston.
14. Di Zio, K. and Tirrell, D. A., 2003, *Macromolecules*, **36**: 1553-1558.
15. Urry, D.W., Parker, T. M., Reid, M. C., and Gowda, D. C., 1991, *J. Bioactive Comp. Polymers*, **6**:263-283.
16. Urry, D.W., 1997, *J. Phys. Chem. B*, **101**: 11007-110028.
17. Li, B., Alonso, D. O. V., Daggett, V., 2001, *J. Mol. Biol.*, **305**: 581-592
18. Rodríguez-Cabello, J. C., Alonso, M., Pérez, T., and Herguedas, M. M., 2000, *Biopolymers*, **54**: 282-288.
19. Reguera, J., Lagarón, J. M., Alonso, M., Reboto, V., Calvo, B., and Rodríguez-Cabello, J. C., 2003, *Macromolecules*, **36**: 8470-8476.
20. Urry, D.W., 1993, *Angew. Chem. Int. Ed. Engl.*, **32**: 819-841.
21. San Biagio, P.L., Madonia, F., Trapane, T.L., Urry, D.W., 1988, *Chem. Phys. Letters.*, **145**: 571-574.
22. C. Tanford, C., 1973, In *The Hydrophobic Effect: Formation of Micelles and Biological Membranes*, John Wiley & Sons.
23. Pauling, L. and Marsh, E., 1952, *Proc. Nat. Acad. Sci. USA.*, **38**: 112-116.
24. Urry, D.W., Trapane, T.L., Prasad, K.U., 1985, *Biopolymers*, **24**: 2345-2356.
25. Manno, M., Emanuele, A., Martorana, V., San Biagio, P. L., Bulone, D., Palma-Vittorelli, M. B., McPherson, D. T., Xu, J., Parker, T. M., Urry, D. W., 2001, *Biopolymers*, **59**: 51-64.
26. Friedman, H.L., 1964, *J. Polym Sci: Part C*, **6**: 183-195.
27. R. Herrero-Vanrell, R., Rincón, A., Alonso, M., Reboto, V., Molina Martinez, I., Rodríguez-Cabello, J. C. *Biomaterials* (submitted).
28. Rodríguez-Cabello, J.C., Alonso, M., Guiscardo, L., Reboto, V., Girotti, A., 2002, *Adv. Mater.*, **14**: 1151-1154.
29. Alonso, M., Reboto, V., Guiscardo, L., San Martín, A., Rodríguez-Cabello, J.C., 2000 *Macromolecules*, **33**: 9480-9482.
30. Alonso, M., Reboto, V., Guiscardo, L., Maté, V., Rodríguez-Cabello, J.C., 2001, *Macromolecules*, **34**: 8072-8077.
31. Ciardelli, F., Fabbri, D., Pieroni, 0., Fissi, A., 1989, *J. Am. Chem. Soc.*, **111**: 3470-3472.
32. Rodríguez-Cabello, J. C., Alonso, M., Guiscardo, L., Reboto, V., Girotti, A., *Adv. Mater.*, 2002, **14**: 1151-1154.
33. T. Kuwabara, T., Nakamura, A., Ueno, A., Toda, F., 1994, *J. Phys. Chem.*, **98**: 6297-6308.
34. Chokchainarong, S., Fennema,O. R., Connors, K. A., 1992, *Carbohydr. Res.*, **232**: 161-169.

Chitosan: A Versatile Biomaterial

ELVAN YILMAZ
Department of Chemistry, Faculty of Arts and Sciences, Eastern Mediterranean University, Famagusta, Northern Cyprus, Mersin 10, TURKEY

1. INTRODUCTION

Chitosan is an amino polysaccharide with a wide range of biomedical applications in addition to various other uses in food and nutrition, textile and paper industries, cosmetics, photography and water engineering.

Chitosan is a functional polymer, which is readily soluble in dilute acid solutions. In contrast to other polysaccharides that are either neutral or anionic, chitosan acts as a polycation in solution. It has complex formation and ion adsorption properties. Chitosan is a very versatile material in terms of its chemical, physical and biological properties. It is a non-toxic, mucoadhesive, biodegradable, and biocompatible polymer.

Biomedical applications of chitosan and its derivatives ranging from pharmaceutical uses to tissue engineering have been reported. It has been shown to be useful in oral, injectable, nasal, ophthalmic, transdermal, and implantable drug administration forms. It has also been found to act as a drug absorption enhancer, hypocholesterolemic agent, a wound-healing accelerator, a blood anticoagulant, and an antimicrobial agent. Its film forming property has been exploited in the design of artificial kidney membranes and artificial skin. More recent biomedical applications of chitosan include protein, gene, and vaccine delivery, owing to its microparticle and complex forming abilities together with its mucoadhesivity. Porous chitosan scaffolds have proven to be promising for use in cell transplantation and tissue regeneration.

Biomaterials: *From Molecules to Engineered Tissues,* edited by
N. Hasırcı and V. Hasırcı, Kluwer Academic/Plenum Publishers, 2004

2.   SOURCES AND PREPARATION OF CHITIN AND CHITOSAN

The precursor of chitosan is chitin, the second most abundant polysaccharide in nature after cellulose. Cellulose is mainly synthesized in plants, while chitin is synthesized mainly in lower animals. It is estimated that about 10 giga tons of chitin are produced and degraded in the biosphere each year. Chitin is found in the outer shells of crustaceans such as crabs and shrimps. These shells contain 20-50% chitin on a dry weight basis. Chitin is isolated from crab and shrimp waste available from seafood processing industries. Other potential sources of chitin production include krill, crayfish, insects, clams, oysters, jellyfish, algae and fungi. Chitin found in crustacean shells is known as α-chitin. Squid pens also contain chitin that is known as β-chitin. Chitosan is commonly produced by deacetylation of α-chitin in 40-50% aqueous alkali solution, usually sodium hydroxide, at 60–120 °C for a few hours Chitosan obtained in this way may have a degree of deacetylation up to 0.95[1].

3.   PHYSICO-CHEMICAL PROPERTIES OF CHITOSAN

Biological properties and activities of chitosan depend largely on its physicochemical properties such as the degree of deacetylation, crystallinity, molecular weight, and high charge density in solution, as well as chemical reactivity and ease of fabrication into different forms.

The chemical structure of chitin is similar to that of cellulose having acetamide ($-NH-CO-CH_3$) groups at C-2 positions instead of the hydroxyl groups (-OH) present in cellulose. Chitosan is the deacetylated form of chitin in which a fraction of acetamide groups have been replaced with amine ($-NH_2$) groups. Chitin is a polymer of β (1→ 4) linked 2-acetamido-2-deoxy-D-glucose units while chitosan is a copolymer of β (1→4) linked to 2-acetamido-2-deoxy-D-glucose   and   2-amino-2-deoxy-D-glucose   units. Chemical structure of chitosan is shown in Figure 1. The degree of deacetylation of polymers cited as chitosan changes between 40 to 100%. The presence of amino groups in chitosan structure is an advantage over cellulose since it provides it with a chemical activity suitable for a wider range of modification reactions and a unique biological functionality.

The degree of deacetylation of chitosan can be determined by elemental analysis, hydrolysis of acetamide groups, titration of free amine groups, dye adsorption, IR, UV-VIS or NMR spectroscopy[2, 3]. Metal adsorption capacity of chitosan depends largely on the degree of deacetylation. Free amine groups have an affinity for metal cations such as copper, nickel, lead, and iron.

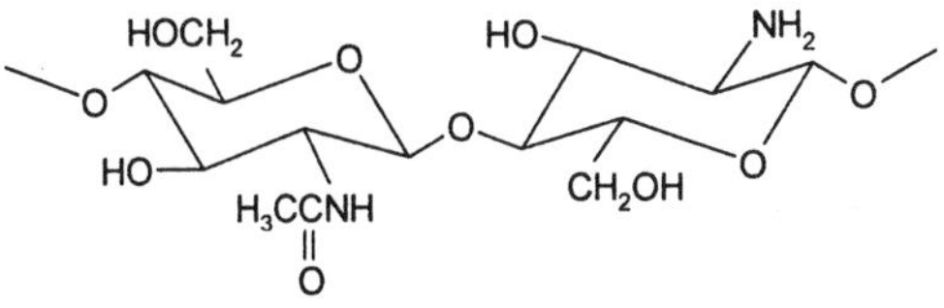

Figure 1. Chemical structure of chitosan

Chitosan is a partially crystalline polymer, which exhibits two crystalline peaks at $2\theta=10°$ and $20°$. X-ray diffraction analyses showed that the polymer chains both in anhydrous and hydrated forms have two-fold helical symmetry with extended fiber repeat of about 10.3 Å, reinforced by $O_3...O_5$ hydrogen bonds[4].

Chitosan is insoluble in water or organic solvents, but soluble in aqueous acids. It dissolves in aqueous hydrochloric, acetic, formic, oxalic and lactic acids and forms salts with these inorganic and organic acids. It acts as a polycation, positively charged polymer, with a high charge density in solution, owing to the protonation of the amine groups on the chain backbone (Fig. 2). It is a weak base with pKa value of around 6.5. Therefore, it is insoluble in neutral and alkaline media[5]. It can form complexes with negatively charged ions in solution.

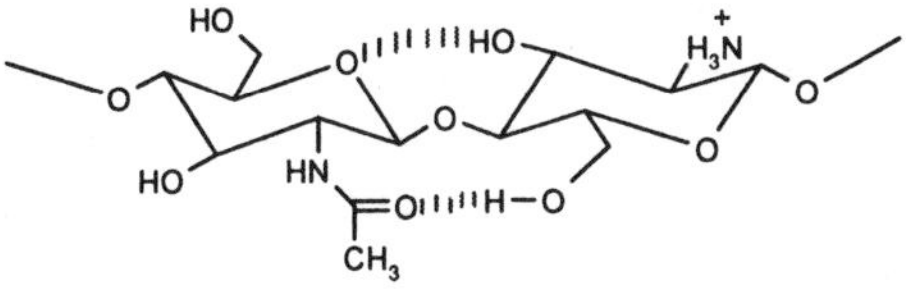

Figure 2. Chitosan in its protonated form in solution

Reactive hydroxyl and amine groups make chitosan a modifiable polymer. A variety of groups can be attached onto chitosan to provide specific functionality, alter biological properties, or modify physical properties. Examples of some modification reactions include hydrolysis, acetylation, acylation, N-phthaloylation, tosylation, alkylation, Shiff base formation,reductive alkylation, O-carboxymethylation, N-carboxyalkylation, silylation and graft copolymerisation[6].

Chitosan is available in various forms such as films, fibers, sponges, and microspheres, in addition to powders and solutions all of which are useful for biomedical applications (Fig 3). Chitosan based hydrogels whether in the microparticle or film form show pH sensitive swelling behavior, which makes them very attractive as target-specific drug delivery systems. Porous chitosan structures can be formed by freezing and lyophilizing chitosan-acetic acid solutions in suitable molds[7].

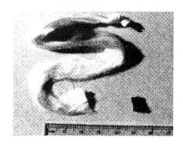

Figure 3. Optical photograph of chitosan (a) microspheres (b) fiber

4.   RELATIONSHIP OF PHYSICOCHEMICAL PROPERTIES AND BIOLOGICAL ACTIVITIES OF CHITOSAN

Favorable biological properties of chitosan include biodegradability, biocompatibility, mucoadhesivity and nontoxicity. It induces biological activities such as stimulatory effects on immune cells, local cell proliferation and integration of the implanted material with the host tissue. It is a wound-healing accelerator, a permeation enhancer, an antibacterial agent, and a hypocholesterolemic agent.

Biodegradability and biocompatibility are two very important properties that make chitosan such a useful material. It is degraded, in vivo, by enzymatic hydrolysis. The primary enzyme that acts on chitosan is lysozyme, which targets acetylated residues. The degradation products are oligosaccharides of chitosan of variable length. The degradation kinetics appears to be inversely related to the degree of crystallinity that is mainly controlled by the degree of deacetylation. Highly deacetylated forms exhibit the lowest degradation rates and may last several months in vivo. Chitosans with lower degrees of deacetylation degrade rapidly[6].

Chitosan has complex forming affinity towards negatively charged species due to its high positive charge density in solution. Its potential as a biomaterial is largely due to this complex forming ability.

Mucoadhesivity of chitosan is due to the ionic interactions between the positively charged amine group of the polymer and the negatively charged sialic acid residues in mucus. Mucoadhesive microspheres that spread on a large area of mucosa decrease the rate of clearance of the drug from mucosa. This allows a longer contact time with the absorptive epithelium in the oral and nasal cavity, and in the stomach and intestines. New bioadhesive formulations based on chitosan have been proposed for oral, nasal, ocular, buccal and vaginal-uterine therapy[8].

The cationic nature of the polymer accounts for the antibacterial activity as well. The most accepted mechanism for the antibacterial activity of chitosan is the interaction of the positive charges on the polymer backbone with the negatively charged residues at the cell surface. This causes alterations in the surface properties and the permeability of the cell membrane, which results in leakage of the intracellular material, eventually leading to death of the cells[9]. The antibacterial activity of chitosan increases with increasing degree of deacetylation. Chitosan shows antimicrobial activity against a broad spectrum of microorganisms, but its activity in its original form is limited to acidic conditions, optimum pH for highest antibacterial activity being pH 5. Above pH 6.5 it loses its solubility, cationic nature and hence antibacterial activity. Water solubility is an important factor for the application of chitosan as an antimicrobial agent. Therefore, studies are concentrated on the preparation of chitosan derivatives that can dissolve in the entire pH range. Chitosan grafted with poly(N-vinylimidazole), PNVI, and poly(4-vinylpyridine), P4VP, showed improved antibacterial activity. The results are summarized in Table 1.

Table 1. Inhibition zone diameters for chitosan and g-chitosan films prepared by solvent casting method in 3×10^3 ppm HCl solution

Sample	P. aeruginosa	E. coli	B. subtilis	S. aureus
		Inhibition zone (mm)		
Ch90[a]	10	9	10	11
G36[b]	12	13	15	14
G26[c]	14	14	17	16
G11[d]	15	16	18	17
G13[e]	17	17	23	21

[a] chitosan, 90.0%; [b] chitosan-*g*-PNVI, 82.5%; [c] chitosan-*g*-PNVI, 144.9%;

[d] chitosan-*g*-P4VP, 138.0%; [e] chitosan-*g*-P4VP, 297.0%.

Average weight of the films = 296 μg.

Chitosan acts as a permeation enhancer because it can interact with the cell membrane resulting in a structural reorganization of tight-junction associated proteins. It has been shown that the degree of deacetylation and molecular weight of chitosan affect its absorption enhancing activity.

Chitosans with high degree of deacetylation and high molecular weight have high absorption enhancing activity[10].

In vivo hydrolysis of chitosan and its derivatives by lysozyme to oligomers activate the macrophages. One of the products induced to be formed is N- acetyl-β-D-glucosamidase, which, catalyzes the production of NAG (N-acetylglucosamine), D-glucosamine and substituted glucosamines from the oligomers. NAG is a major component of dermal tissue and its presence is essential to repair of scar tissue. Glucoproteins containing NAG are one of the predominant proteins isolated during the early stages of wound healing. Hence, chitosan acts as a wound-healing accelerator[11].

Sulfonated chitosan acts as a blood anticoagulant because it is structurally similar to heparin, one of the most widely used blood anticoagulants. Conversion of position 6 into a carboxyl group in N-sulfonated chitosan gives a product with 23% of the activity of heparin. Its O-sulfonated form exhibited 45% the activity of heparin in vitro. As the sulfur content increases, anticoagulant activity increases as well[11].

Chitosan can form complex salts that bind triglycerides, fatty acids, bile acids, cholesterol and other sterols, which renders it to a hypocholesterolemic agent[12, 13]. There is a general agreement that chitosan lowers blood pressure and cholesterol according to studies with volunteers.

5.   BIOMEDICAL USES OF CHITOSAN

A broad range of biomedical applications of chitosan has been reported. It has proved to be useful for pharmaceutical applications since it can easily be fabricated into gels, microparticles and films, in addition to being mucoadhesive and a permeation enhancer. It has been proposed as a promising non-viral gene transfection agent since it can form polyelectrolyte complexes with DNA. Drugs can be covalently bonded to and carried to the target site with chitosan. Scaffolds of chitosan have proven to be useful in tissue engineering.

5.1   Drug Delivery

Chitosan has been used as a direct tableting excipient, as a binder and lubricant in wet granulated tablets, and a potential disintegrant due to its water uptake property. It has been used as an excipient for increasing the dissolution rate of poorly soluble drugs and as a stabilizing agent in emulsions. Addition of chitosan to conventional excipients such as mannitol, lactose or potato starch improved the fluidity of the powder mixtures. It is advantageous to use chitosan in the formulations of the drugs that cause stomach irritation since it has antacid and antiulcer activities[14].

Microspheres and films of chitosan have been evaluated as controlled drug release systems as a basis for new drug formulations. Poorly absorbable drugs such as peptides and proteins and vaccines are safely and effectively delivered via transmucosal delivery systems. Chitosan microspheres can be very useful in the transmucosal delivery of insulin needed in the treatment of diabetes. Mucoadhesiveness of chitosan allows adhesion of the microspheres to the cell surfaces, and protect insulin from enzymatic degradation and fast elimination. A major advantage of chitosan microspheres is that they are prepared under very mild conditions and have an excellent capacity for the entrapment of insulin, 68-93% loading efficiency, and provide a continuous release of insulin for extended periods of up to 4 days[15]. Chitosan microspheres and films can be employed in iron chelation therapy as well. Desferal, a very effective iron-chelating agent used in the treatment of thalassemia, is orally inactive. Controlled release of desferal has been achieved with chitosan membranes. This can form a basis for the development of transdermal or transmucosal formulations for desferal[16]. Deferiprone, another iron-chelating drug, is orally active but not as effective as desferal and has side effects such as causing a life-threatening decrease in white cell counts. Prolonged release of deferiprone from chitosan microspheres for to 200 days can be achieved[17]. Amount of released deferiprone changes between 16.6% and 40.1% depending upon the degree of deacetylation and degree of crosslinking of chitosan microspheres. Chitosan membranes allow much faster release of deferiprone, reaching 100% cumulative release in 6 hours[16].

Examples to stomach-targeted, colon-targeted[18], buccal and sublingual drug delivery systems based on chitosan exist in the literature[19, 20].

5.2 Gene Delivery

Non-viral vectors to transfer DNA to the target cells have received increased attention during the past few years since they do not induce immune responses or oncogenic effects like viral vectors. Several cationic polymers have been investigated for this purpose since they can form polyelectrolyte complexes with DNA. Some examples are polyethylene imine, poly(L-lysine), dendrimers, polybrene, gelatin, tetraminofullerene, poly(L-histidine)-graft-poly(L-lysine). The natural polymers are preferable over the synthetic ones as non-viral transfection agents because they are biocompatible, have low immunogenicity, and minimal cytotoxicity[21, 22].

5.3   Vaccine Delivery

Mucoadhesive and absorption enhancing chitosan is a suitable candidate for use in both oral and nasal vaccination formulations. Microparticles of chitosan can be used to deliver vaccines through the oral route, while solutions are useful for nasal vaccination. An important advantage of chitosan is that nanoparticles can be formed without using organic solvents, which may alter the immunogenicity of antigens during preparation and loading [23-25].

Furthermore, chitosan-drug conjugates have been synthesized and tested specially in the field of cancer chemotherapy [26, 27].

5.4   Tissue Engineering

Polymer scaffolds serve to support, reinforce and sometimes to organize the regenerating tissue. Some required properties of a scaffold are; the ability to release bioactive substances in a controlled manner, to be biodegradable, and to specifically interact with or mimic extracellular matrix components, and cell surface receptors. Porosity and reactive or easily modifiable groups are also needed. Chitosan having a pH dependent solubility and easy processability under mild conditions is an excellent candidate to be used as a porous scaffold in tissue engineering. Chitosan-based biomaterials have been applied in cartilage tissue engineering[28].

5.5   Biological Iron Chelator

Development of a safe and effective oral-chelating agent is a major requirement for the treatment of thalassemia (Cooley's anemia) patients who have to undergo frequent blood transfusions and experience toxic iron (III) accumulation. Chitosan flakes, powder and microspheres were recently evaluated for their ferric ion adsorption capacity from an aqueous solution of iron (III)-sorbitol-citric acid complex. Chitosan was found to be capable of removing ferric ion from solution in a pH range of 2 to 7. In vitro experiments carried out in the serum of thalassemia patients revealed that ferritin level, which is a direct measure of the amount of iron (III) stored, was decreased by 6 to 30% depending on the ferritin level of the sample[29]. Furthermore, the decrease in ferritin level was accompanied with a reduced LDH level in accordance with Muzzarelli's[12] findings. Only insignificant changes in terms of albumin, total protein, cholesterol, uric acid and glucose were observed.

6. CONCLUSION

The abundance of its precursor, the natural polymer chitin, and unique physicochemical and biological properties of chitosan make it one of the most important biomaterials. It has a high potential to find well-established applications in the biomedical field leading to commercial products on the market.

REFERENCES

1. Sanford, P.A., 1989, Chitosan: Commercial Uses and Potential Applications. In *Chitin and Chitosan:Sources, Chemistry, Biochemistry, Physical Properties and Applications* (G.Skjåk-Braek, T.Anthonsen, P.Sanford, eds.) Elsevier Applied Science, London, pp.51-69.
2. Tan,S.C., Eugene, K., Tan, T.K., and Wong, S.M., 1998, The degree of acetylation of chitosan: advocating the first derivative UV–spectrophotometry method of determination. *Talanta*, 45: 713-719.
3. Brugneretto, J., Lizardi,J., Goycoolea, F.M., Argülles-Monal, W., Desbrieres, J., and Rinaudo, M., 2001, An infrared Investigation in relation with chitin and chitosan characterization. *Polymer*, 42: 3569-3580.
4. Okuyama, K. Noguchi,K, Kanenari M, Egawa T, Osawa K, and Ogawa K, 2000, Structural diversity of chitosan and its complexes. *Carbohydrate Polymers*, 42: 237-247.
5. Anthonsen,M.W., Varum,K.M., and Smidsrod, O.,1993, Solution properties of chitosans with different degrees of N-acetylation. *Carbohydrate Polymers*, 22: 193-201.
6. Kurita, K., 2001, Controlled functionalization of the polysaccharide chitin. *Prog.Polym.Sci.*, 26: 1921-1971.
7. Madihally, S.V., and Matthew, H.W.T., 1999, Porous chitosan scaffolds for tissue engineering, *Biomaterials*, 20: 1133-1142.
8. Woodley, J., 2001, Bioadhesion-New Possibilities for Drug administration, *Clin.Pharmacokinet.* 40: 77-84.
9. Caner, H., 2002, Ph.D. Thesis *Synthesis, Characterization and Potential Applications of Poly(4_vinylpyridine) and Poly(N-vinylimidazole) Grafted Chitin/Chitosan*. Eastern Mediterranean University.
10. Bernkop-Schnürch, A., 2000, Chitosan and its derivatives: potential excipients for peroral peptide delivery systems, *Int.J.Pharm.* 194: 1-13.
11. Singh, D.K., and Ray, A.R., 2000, Biomedical applications of chitin, chitosan and their derivatives. *J.M.S.-Rev. Macromol.Chem. Phys.*, C40: 69-83.
12. Muzzarelli, R.A.A., Chitosan-based dietary foods, 1996, *Carbohydrate Polymers*, 29: 309-316.
13. Muzzarelli, R.A.A., 2000, Recent results in the oral administration of chitosan. In *Chitosan per os-from dietary supplement to drug carrier*, (R.A.A.Muzzarelli Ed.), Atec, Grottammare, Ancona pp. 3-40.
14. Felt, O., Buri, P., and Gurny, R., 1998, Chitosan: A unique polysaccharide for drug delivery. *Drug.Dev. Ind.Pharmacy*, 24: 979-993.

15. Albayrak, B., 2000, M.Sc. Thesis *Controlled Release of Insulin from Chitosan Microspheres*. Middle East Technical University.
16. Burke, A., 2002, Ph.D. Thesis *An Iron-Chelating Agent and a Controlled Release System for Iron-Chelating Drugs, Desferal and Deferiprone*. Eastern Mediterranean University.
17. Uylukcuoglu, B., 2003, M.Sc. Thesis *Chitosan Microspheres and Films Used in Controlled Release*. Middle East Technical University.
18. Hejazi,R., and Amiji, A., 2003, Chitosan-based gastrointestinal delivery systems, *J.Controlled Release*, 89: 151-165.
19. Senel, S., Kremer, M.J., Kas, S., Wertz, P.W., Hincal, A.A., and Squier, C.A., 2000, Enhancing effect of chitosan on peptide drug delivery across buccal mucosa, *Biomaterials*, 21: 2067-2071.
20. Senel, S., Ikinci, G., Kas, S., Yousefi-Rao, A., Sargun, M.F., and Hincal, A.A., 2000, Chitosan films and hydrogels of chlorhexidine gluconate for oral mucosal delivery, *Int.J. Pharm.*, 193: 197- 203.
21. Borchard, G., 2001, Chitosans for gene delivery, *Advanced Drug Delivery Reviews*, 52: 145-150.
22. Liu, W.,G., and Yao, K.D., 2002, Chitosan and its derivatives-a promising non-viral vector for gene transfection, *J.Controlled Release*, 83: 1-11.
23. Van der Lubben, I.M, Verhoef, J.C., Borchard, G., and Junginger, H.E., 2001, Chitosan for mucosal vaccination, *Adv.Drug.Delivery Reviews*, 52: 139-144.
24. van der Lubben, I.M., Verhoef, J.C., Borchard, G., and Junginger, H.E., Chitosan and its derivatives in mucosal and vaccine delivery, European J.Pharmaceutical Sciences 14 (2001) 201-207.
25. Illum, L., Jabbal-Gill, I., Hinchcliffe, M.,. Fisher, A.N, Davis, and M.S.S., 2001, Chitosan as a novel nasal delivery system for vaccines, *Adv.Drug.Delivery Reviews*, 51: 81-96.
26. Kato, Y., Onishi, H., and Machida, Y., 2000, A novel water soluble N-succinyl-chitosan-mitomycin C conjugate prepared by direct carbodiimide coupling: Physicochemical properties, Antitumor Characteristics and Systemic Retention. *STP Pharma Sci.*, 10: 133-142.
27. Onishi, H., Takahashi, H., Yoshiyasu, M., and Machida, Y., 2001, Preparation and in vitro properties of N-Succinylchitosan or carboxymethylchitin-mitomycin C conjugate microparticles with specified size. *Drug Dev. and Ind.Pharm.*, 27: 659-667.
28. Suh, J.K.F., and Matthew, H.W.T., 2000, Application of chitosan-based polysaccharide biomaterials in cartilage tissue engineering: a review. *Biomaterials*, 21: 2589-2598.
29. Burke, A., Yilmaz, E., Hasirci, N., and Yilmaz, O., 2002, Iron (III) ion removal from solution through adsorption onto chitosan. *J.Appl.Polym.Sci*, 84:1185-1192.

Bioapplication Oriented Polymers. Micro- and Nanoparticles for Drug Delivery Systems

VALERIA HARABAGIU[*], GHEORGHE FUNDUEANU[*], MARIANA PINTEALA[*], MARIETA CONSTANTIN[*] and THIERRY HAMAIDE[#]

[*]"Petru Poni" Institute of Macromolecular Chemistry, 700487 Iasi, Romania, [#]Laboratoire de Chimie et Procédés de Polymérisation, CNRS, ESCPE Lyon, 43, Bd. du 11 Novembre, 69616 Villeurbanne Cedex, France

1.   INTRODUCTION

Micro- and nanoparticles called by Paul Ehrlich "magic bullets" are one answer to the problem of drug delivery system and drug action specificity. Microparticles are homogeneous particles or monolitic microcapsules, usually spherical, with dimensions ranging between 10^{-3}-10^{-6} m. Nanoparticles and nanocapsules are defined in the same terms, with the difference that their size is situated in the range of nanometers.

These new particulate pharmaceutical formulations are generally able to carry a wide variety of drugs in a more stable and reproducible way. The reason of using micro- and nanoparticles, behind the concept, is to transport and bring the drug molecules into intimate contact with tissues and organs in a controlled way. Beside specificity, other advantages may be protection against premature inactivation, decreased toxicity and the flexibility of administration. Due to their reduced dimensions, these micro- and nanoparticles could be administered by any access way in the organism: oral, parenteral, ocular, nasal, transdermal, etc. In order to obtain the maximum desired response, the particulate carrier should simultaneously fulfill the following conditions:
-    to be biocompatible and nontoxic
-    to include enough active substance and to keep it unaltered up to the dysfunction action site

Biomaterials: From Molecules to Engineered Tissues, edited by
N. Hasırcı and V. Hasırcı, Kluwer Academic/Plenum Publishers, 2004   69

- to accumulate and to remain in the action site
- to release the drug with a convenient rate at desired site
- to display acceptable pharmaceutical characteristics regarding stability, sterilization, etc.

Micro-and nanoparticles are prepared both from natural (polysaccharides, proteins) and from synthetic polymers (polyesters, polyacrylates, etc.). From biologic point of view, the natural polymers are preferred because of their biodegradability, biocompatibility, and non-toxicity. From standpoint of chemical structure design, stability, purity, availability and low price, the synthetic polymers are preferred. The preparation of micro- and nanoparticles can be performed by a large number of methods including both those starting from monomers (polymerization and polycondensation) and those involving - usually functional - preformed polymers (suspension crosslinking, solvent evaporation, coacervation, chelatization).

2.   ε-CAPROLACTONE – DIMETHYLSILOXANE BLOCK COPOLYMERS AND THEIR USE FOR DRUG ENCAPSULATING NANOPARTICLES

Thirty years ago ε-caprolactone homopolymers (PCL) and copolymers were synthesized and evaluated as biodegradable plastics.[1,2] PCL is well known as a biocompatible and biodegradable polymer.[3] Moreover, its biodegradability can be enhanced by copolymerization.[4] PCL is a relatively hydrophobic material with low glass transition temperature, Tg = -60°C, and a crystalline structure that melts at 59-64°C, depending on the molecular weight.[4] The biodegradation of PCL takes place both *in vivo* and *in vitro* by bulk hydrolysis of the ester macromolecular chain. In living organisms the hydrolysis is followed by the solubilisation of the resulted low molecular weight fragments or small particles in the body fluids or by phagocytosis.[4-6] The overall process is rather long (2-4 years) as the hydrolysis rate of PCL chains is much lower as compared to other polyesters (polylactides, polyglycolides). A quite large amount of work was recently dedicated to the use of polycaprolactones and of their linear or crosslinked copolymers as vehicles for slow release of drugs[7-9] or as long term biodegradable-biocompatible ceramers appropriate for the repairing of skeletal tissues.[10] Nanoparticles based on amphiphilic copolymers containing poly(ethylene oxide) and PCL sequences have been proved to encapsulate various bioactive principles[11,12] or even to present a stimuli (temperature) - responsive drug release behavior.[13]

On the other hand, polysiloxanes, especially polydimethylsiloxane (PDMS), are interesting hybrid organic-inorganic polymers possessing a unique combination of properties. PDMS presents biocompatibility (physiological inertness), high gas permeability, good oxidative, thermal and UV stability. As a result of its large molar volume, low cohesive energy and density, and high chain flexibility, PDMS has a very low solubility parameter, low surface tension and it is immiscible with most organic polymers.[14,15] Its use as surface modifier in polymer alloys is limited by the tendency to migrate to the polymer-air interface where it is rejected from the material.

Different types of functional polysiloxanes[16] and siloxane copolymers[17] were synthesized and used as blend compatibilizers and surface modifiers[15] or in different pharmaceutical and biomedical applications (contact lenses, implants, transdermal penetration enhancers).[18-20]

2.1 Synthesis of PCL-PDMS Di- and Triblock Copolymers

Both PCL[5,21] and PDMS homopolymers[16] can be obtained by ring opening polymerisation of the corresponding cyclic monomers through ionic mechanisms. So far, PCL-PDMS copolymers were prepared by a two step procedure involving functional precursors.[22] We proposed the synthesis of PCL-PDMS well defined di- and triblock copolymers (Scheme 1) by coordination anionic polymerisation of CL in the presence of triethylaluminum catalyst and excess hydroxyalkyl mono- and difunctional PDMSs as chain transfer agents[23]. The methods provides copolymers with a low molecular weight polydispersity (a very clear Maldi Tof spectrum of D2 sample in Table 1 is given as an example (Fig. 1)), especially when PDMS contains the hydroxyl groups attached to the macromolecular chain through ethyl propyl ether spacer, and a good control of their molecular characteristics (CL/siloxane ratio, lengths of each block).

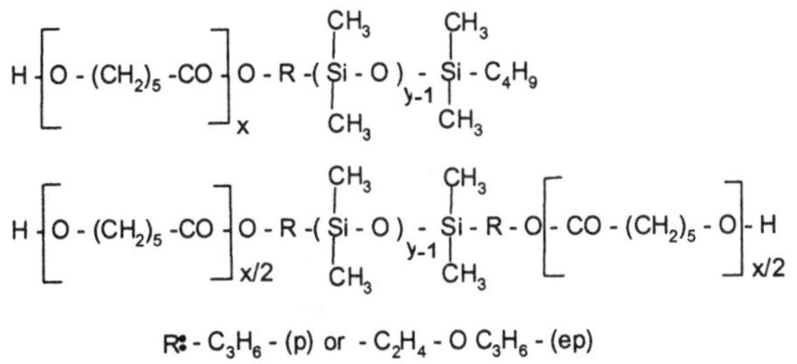

$$R = C_3H_6 - (p) \text{ or } - C_2H_4 - O\ C_3H_6 - (ep)$$

Scheme 1. PCL-PDMS di- and triblock copolymers used for the preparation of nanoparticles

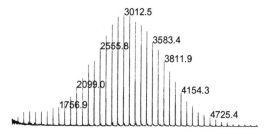

Figure 1. Maldi Tof spectrum of D2 sample in Table 1

Table 1. Synthesis[a] and characterization of CL copolymers

Code[b]	Feed			Copolymer			
	PDMS transfer agent[d]	CL/OH (molar)	Duration (min)	Mn_{PCL} (^{1}H-NMR)	SEC Mw/Mn	DSC Tm (°C)	ΔH (J/g)
D2	ep₁PMDSi	22	30	1x2500	1.08	-	-
D3	ep₁PDMS₈₀₀	28	50	1x3200	1.19	-	-
T3a	p₂PDMS₁₅₀₀	40	140	2x4800	1.38	57	50.8
T3b	ep₂PDMS₁₅₀₀	40	40	2x4700	1.18	57	50.8
T4b	ep₂PDMS₁₅₀₀	25	60	2x2850	1.11	52	46.5
T5	ep₂PDMS₃₀₀₀	45	60	2x5000	1.13	38	35.0
T6	ep₂PDMS₄₀₀₀	60	80	2x7100	1.22	35	30.0

[a] Triethylaluminum/OH: 1/10 molar ratio; temperature: 20°C
[b] D and T mean diblock and triblock PCL-PDMS copolymers
[c] D2 is a model copolymer obtained starting from 1-(2-hydroxyethoxypropyl)-1,1,3,3,3-pentamethyldisiloxane
[d] p and ep mean propylene and ethoxypropylene spacers between OH groups and the siloxane chain; 1 and 2 refer to mono- and difunctional polymers; subscript numbers after PDMS indicate the average number molecular weight of the siloxane chain

2.2   Properties of PCL-PDMS Copolymers

The advantage of the low T_g of PCL is negatively compensated in some biomedical applications by its high crystallinity and low biodegradation rate. As mentioned before, the degree of crystallinity of PCL can be reduced by copolymerisation. Siloxane – modified PCLs were previously shown to be phase-segregated semicrystalline materials.[24] They present a PDMS enriched surface as compared to the bulk composition[25] and were used to compatibilize blends of vinyl polymers.[26] Quite recently, poly(DL-lactic acid)/PCL-PDMS-PCL terpolymers were synthesized and demonstrated to support the attachment and growth of cells.[27]

The DSC analysis of PCL-PDMS copolymers shows that they are phase-segregated materials presenting two Tg values at around – 120 and –

66 °C corresponding to siloxane and caprolactone rich domains, respectively. The melting temperatures of PCL crystals, ranging between 35 and 57 °C, is slightly lower than that of PCL homopolymer, and depends on the molar ratio between the different blocks and on the molecular weight of the siloxane sequence (Table 1). The higher is the length of the last one and the siloxane/lactone molar ratio, the lower is the melting temperature and the lower is the crystallinity (as seen from the values of enthalpy of fusion).

2.3 Encapsulating Nanoparticles Based on PCL-PDMS Copolymers

Stable unloaded and loaded (with indomethacin (IMC) or vitamin E (VE)) nanoparticles of low dimensional polydispersity were prepared through nanoprecipitation in water from dilute acetone solution of PCL-PDMS copolymers, in the presence of Pluronic emulsifier (poly(ethylene oxide (PEO)-poly(propylene oxide) copolymer) (Table 2). For all the used copolymers, the dimensions of unloaded particles range between 124 and 194 nm. Particle dimensions and size distribution (PSD) are not increased by loading small amounts of IMC (10 % versus polymer weight in the nanoprecipitation feed). The nanoparticles loaded with a relatively higher amount of VE present larger dimensions and increased PSD as compared to the unloaded homologues. There is no reliable correlation between particle dimension and the molecular characteristics of CL copolymers. However, the lowest molecular weight PCL-PDMS copolymer, with a higher content of OH terminal groups, able to interact with PEO-PPO-PEO emulsifier, yields the highest dimension of the nanoparticles.

^{1}H-NMR spectra in organic solvents (Fig. 2, top) of solubilized nanoparticles show the characteristic peaks of the constituents (δ, ppm: PCL: 1.65, 2.62, 4.17; PDMS: 0.1; IMC: 3.69, 3.81, 6.40-7.50; Pluronic: 1.13, 3.10-3.90, 3.63). However, in water, the stable nanoparticles show only the peaks of Pluronic emulsifier (Fig. 2, bottom), suggesting a core-shell structure composed of a PCL-PDMS/IMC hydrophobic core and a Pluronic hydrophilic shell. The nanoparticles with a PEO-like surface are expected to have a good mobility in water and to resist protein absorption and cellular attachment.

Table 2. Size, particle size distribution (PSD) and drug loading efficiency of CL copolymer-based nanoparticles

PCL-PDMS	Unloaded particles		Loaded particles				DLE[b] (%)	
			IMC		VE		IMC	VE
	Size (nm)	PSD[a]	Size (nm)	PSD[a]	Size (nm)	PSD[a]		
D3	-	-	-	-	343	0.65	-	60.80
T3a	194	0.09	-	-	264	0.52	-	54.73
T3b	130	0.10	-	-	350	0.43	-	54.75
T4b	131	0.07	130	0.11	-	-	10.05	-
T5	130	0.14	-	-	249	0.57	-	52.80
T6	140	0.18	194	0.15	260	0.54	-	53.20

[a] PSD represents the width of the distribution as determined by dynamic light scattering measurements on a laser Coulter LS apparatus
[b] DLE represents drug loading efficiency determined by UV analysis and calculated according to DLE = [drug amount/(drug amount + polymer amount)] x 100

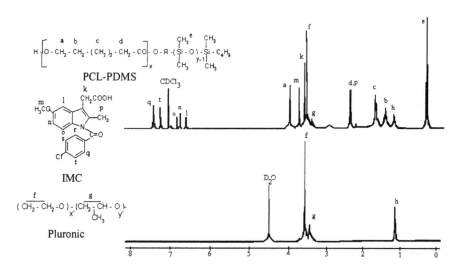

Figure 2. ^{1}H-NMR spectra of indomethacin loaded nanoparticles: solubilized in CDCl$_3$ (top) and in D$_2$O (bottom)

3. INTELLIGENT MICROSPHERES FOR CONTROLLED RELEASE OF DRUGS

While newer and more powerful drugs continue to be developed, increasing attention is being given to the methods by which these active substances are administered.

In *conventional drug delivery*, the drug concentration in the blood rises when the drug is taken, then peaks and declines. Since each drug has a therapeutic range above which it is toxic and below which is ineffective, the plasma drug concentration in a patient at a particular time depends on compliance with the prescribed routine.

The *controlled delivery devices*, which are already available commercially, can maintain the drug in the desired therapeutic range with just a single dose, localize delivery of the drug to a particular body compartment, reduce the need for follow-up care, preserve medication that is rapidly destroyed by the body, and increase patient comfort and compliance. The basic approach that drug concentration-effect relationships are significantly invariant as a function of time in man has led to the development of constant rate drug delivery systems. Nevertheless, there are a number of clinical situations where such an approach would not be sufficient. These include the delivery of insulin for patients with diabetes mellitus, antiarrhythmics for patients with heart rhythm disorders, gastric acid inhibitors for ulcer control, nitrates for patients with angina pectoris.

The field of *self-regulated delivery systems* deals with physiological functions that exhibit prominent rhythmic change such as body temperature, pH of gastric and intestinal juice, systolic and diastolic blood pressure, heart rate, renal functions. This new approach is based on polymers that show a phase transition in response to external stimuli such as temperature, pH, ionic strength, and magnetic field[30-32]. In particular, the intelligent polymer systems have a potential for applications in modulated drug delivery because these polymers not only respond to the external stimuli, as temperature or pH, but also control the release rate of drugs[33-35].

Certain polymer hydrogels, such as crosslinked structures based on poly(N-isopropyl acrylamide) (NIPAAm), possess a Lower Critical Solution Temperature (LCST) when placed in solution. When the temperature of the polymer is raised above the LCST, a phase separation occurs within the polymer accompanied by a dramatic shrinkage in volume[36,37]. It is this gel collapse phenomenon that has been utilized in the development of thermoresponsive drug delivery systems.

The biomedical and biological applications of poly(NIPAAm) gels require the derivatization of the gel without losing thermoresponsivity[38,39]. In order to satisfy the imposed requirements, like a sharp volume phase transition around the physiological pH and temperature, a series of copolymers of NIPAAm, acrylamide (AAm) and hydroxyethyl acrylate (HEA) with different molar ratios of the comonomers in the feed were synthesized and tested. The results are presented in Table 3.

Table 3. The influence of comonomers ratio on LCST (concentration of copolymer solution was 1%, w/v)

Type	NIPAAm (mM)	AAm (mM)	HEA (mM)	LCST (°C)		
				pH=7.4	pH=1.2	H_2O
P #0	10	0	0	32	-	-
P #1	10	1	0	34.5	35.1	36
P #2	10	1.5	0	36.7	37.5	38.2
P #3	10	2	0	37.2	42	43.5
P #4	10	3	0	42	46.5	47.5
P #5	10	2	1	36	38.5	40
P #6	10	3	2	39.5	42.5	43.5
P #7	10	3	1	42.8	44.5	45
P #8	10	0	1	No LCST	-	-

The LCST of the poly(NIPAAm-co-AAm-co-HEA) was shifted to a higher temperature with the increase of AAm content. In general, the LCST should increase with increasing hydrophilicity of the polymer. However, a LCST shift to a lower temperature was observed with the incorporation of the hydrophilic HEA. This is due to the formation of hydrogen bonds, which protect the NIPAAm groups from exposure to water and result in a hydrophobic contribution to the LCST.

The copolymer with a comonomer ratio of NIPAAm/AAm/HEA of 10/2/1 displays a phase transition around 36 °C in phosphate buffer, pH=7.4, and around 39 °C in pH=1.2 or water (Figure 3). However, this copolymer contains an additional crosslinkable functional group (in HEA structural units), available for microsphere synthesis by suspension reticulation method, as we previously reported[40-43].

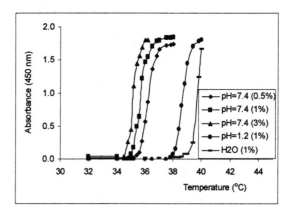

Figure 3. LCST of poly-(NIPAAm-co-AAm-co-HEA) (sample P #5, Table 3) in phosphate buffer (pH=7.4) at different concentrations, acidic buffer (pH=1.2), and distilled water

Therefore, this copolymer was solubilized in acidified water and dispersed in mineral oil at a temperature below its LCST. The droplet suspension was transformed in stable intelligent microspheres by crosslinking the OH functional groups of the HEA with glutaraldehyde (GA) (Fig. 4). The microspheres still keep the pH/thermoresponsive properties of the copolymer (Tables 4 and 5 and Fig. 5).

The swelling degree as well as water retention and porous volume decrease as the temperature increases from 4 to 45 °C (Table 4). In addition, at higher temperatures, even small molecules of the drugs are excluded from the pores of the microspheres (Table 5).

The variation of the dimensions of the pores with temperature is further exploited for the controlled release of indomethacin (Figs. 6 and 7).

Table 4. Main characteristics of poly(NIPAAm/AAm/HEA) microspheres, (P #5), (0.4 ml GA) at different temperatures

Temperature (°C)	Swelling degree (q)	Water regain (ml/g)	Porous volume (ml/g)
4	7.0	5.96	-
10	5.5	5	-
24	5.0	4.8	4.26
32	3.1	3.16	-
45	2.08	2.05	1.44

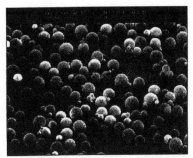 

Figure 4. Scanning electron micrographs of poly(NIPAAm-co-AAm-co-HEA) microspheres. General view (left), detail (right).

It must be underlined that the temperature has two opposite effects: decreases the dimensions of the pores of the microspheres and increases the solubility of indomethacin. Therefore, the slowest drug release was noticed at 36 °C, and not at 40 °C, as it was expected. The microspheres obtained with a lower crosslinking degree (2%) display almost a pulsatile (on-off) indomethacin release. Most drug is expelled during shrinkage and swelling. At a higher crosslinking degree, the shrinkage phenomenon is obstructed and, therefore, the drug still diffuses through the pores of the microspheres even above the LCST.

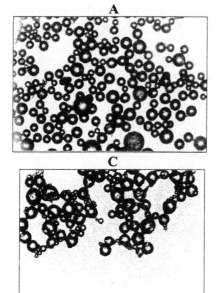

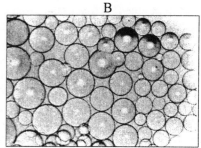

Figure 5. Optical photomicrographs of poly(NIPAAm-co-AAm-co-HEA) microspheres obtained in dried state (Panel A), swollen state in phosphate buffer, pH=7.4, under LCST (Panel B), above LCST (Panel C)

Table 5. Chromatographic behaviour of poly(NIPAAm-co-AAm-HEA) microspheres above and under LCST (The tested microspheres were crosslinked with 0.2 and 0.4 ml of glutaraldehyde; mobile phase, phosphate buffer, pH=7.4, flow rate 0.3 ml/min, injected volume 20 μl, the chromatographic procedure is described elsewhere[44])

Solute	Retention volume (ml)				
	0.2 ml GA		0.4 ml GA		
	25 °C	40 °C	25 °C		40 °C
	pH=7.4		pH=7.4	NaCl 0.1M	pH=7.4
Diclofenac	4.22	1.65	6.50	5.84	1.73
Indomethacin	4.48	1.42	7.31	7.02	1.67
Propranolol	3.69	1.45	5.69	5.42	1.80
Tetracycline	2.92	1.44	3.33	3.33	1.67
Vitamin C	2.84	1.62	2.84	2.64	1.80
Methyl-paraben	3.96	1.72	5.82	5.84	1.87
Ethyl-paraben	4.25	1.84	6.91	6.82	1.94
Propyl-paraben	4.95	1.90	8.40	8.50	1.94
Lysozyme	1.11	1.10	1.35	1.32	1.25
β-Lactoglobulin	1.11	1.10	1.35	1.25	1.25
Trypsinogen	1.11	1.10	1.35	1.25	1.25
Pepsin	1.11	1.10	1.35	1.25	1.25
Egg albumin	1.11	1.10	1.35	1.25	1.25
Blue dextran	1.11	1.10	1.22	1.25	1.25
D_2O	2.74	1.52	2.65	2.61	1.80

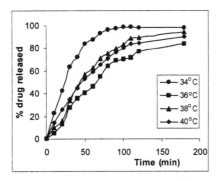

Figure 6. Effect of temperature on the release of indomethacin from microspheres crosslinked with 0.2 ml glutaraldehyde

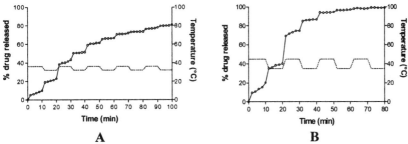

Figure 7. Effect of temperature cycling on indomethacin release from microspheres crosslinked with 0.2 (A) and 0.4(B) ml glutaraldehyde.

ACKNOWLEDGEMENTS

The work from our laboratories was supported by grants from the Romanian Ministry of Education and Research, Le Centre National de la Recherche Scientifique, France, Ministero degli Affari Esteri and Consiglio Nazionale delle Ricerche, Italy.

REFERENCES

1. Potts, J. E., Cledining, R.A., Ackart, W. B., and Niegisch, W.D. 1973, Biodegradability of synthetic polymers. *Polym. Sci. Technol.* **3**: 61.
2. Huang, S. 1985, Biodegradable polymers. In *Encyclopedia of Polymer Science and Engineering* (H. F. Mark, N. M. Bikales, C. G. Overberger, G. Menges, and J. I. Kroschwitz, eds.), John Wiley and Sons, New York, pp. 220-243.

3. Chandra, R., and Rustigi, R. 1998, Biodegradable polymers. *Prog. Polym. Sci.* **23**: 1273-1335.
4. Pitt, G., Chasalow, F. I., Hibionada, Y. N., Klimas, D. N., and Schimler, A. 1981, Aliphatic polyesters. I. The degradation of poly(lactone) in vivo. *J. Appl. Polym. Sci.* **26**: 3779-3787.
5. Pitt, C. G. 1990, Poly(ε-caprolactone) and its copolymers. In *Biodegradable Polymers as Drug Delivery Systems* (M. Chasin, and R. Langer, eds.), Marcel Dekker, New York, pp. 71-120.
6. Engelberg, I., and Kohn, J. 1991, Physico-mechanical properties of degradable polymers used in medical applications: a comparative study. *Biomaterials* **12**: 292-304.
7. Domb, J. A., Kumar N., Sheskin, T., Bentolila, A., Slager J., and Teomim, D. 2002, Biodegradable polymers as drug delivery systems. In *Polymeric Biomaterials* (S. Dumitriu, ed.), Marcel Dekker, New York, Basel, pp. 91-121.
8. Le Roy Boehm, A.-L., Zerrouk, R., and Fessi, H. 2000, Poly(ε-caprolactone) nanoparticles containing a poorly soluble pesticide: formulation and stability study, *J. Microencapsulation* **17**: 195-205.
9. Alleman, E., Gurny, R., and Doelker, E. 1993, Drug-loaded nanoparticles preparation methods and drug targeting issues. *Eur. J. Pharm. Biopharm.* **39**: 173-191.
10. Jagur-Grodzinski, J. 1999, Biomedical application of functional polymers. *React. Functional Polym.* **39**: 99-138.
11. Kim, S. Y., Shin, I. G., Lee, Y. M., Cho, C. S., and Sung, Y. K. 1998, Methoxy poly(ethylene glycol) and ε-caprolactone amphiphilic block copolymeric micelle containing indomethacin. II. Micelle formation and drug release behaviours. *J. Contr. Release* **51**: 13-22.
12. Kim, S. Y., Lee, Y. M., Shin, H. J., and Kang, J. S. 2001, Indomethacin-loaded methoxy poly(ethylene glycol)/poly(ε-caprolactone) diblock copolymeric nanospheres: pharmacokinetic characteristics of indomethacin in the normal Sprague-Dawley rats. *Biomaterials* **22**: 2049-2056.
13. Kim, S. Y., Ha, J. C., and Lee, Y. M. 2000, Poly(ethylene oxide)-poly(propylene oxide)-poly(ethylene oxide)/poly(ε-caprolactone) (PCL) amphiphilic block copolymeric nanospheres. II. Thermoresponsive drug release behaviour. *J. Contr. Release* **65**: 345-358.
14. Yilgor, I., and McGrath, J. E. 1988, Siloxane containing copolymers. *Adv. Polym. Sci.* **86**: 1-57.
15. Harabagiu, V., Pinteala, M., and Simionescu B. C. 2003, Blends and networks containing silicon-based polymers. In *Handbook of Polymer Blends and Composites* (A. K. Kulshreshtha, and C. Vasile, eds.) Rappra Technology Ltd., Shawbury, Vol. 4B, pp. 525-534.
16. Harabagiu, V., Pinteala, M., Cotzur, C., and Simionescu, B. C. 1996, Functional polysiloxanes. In *The Polymeric Materials Encyclopedia: Synthesis, Properties and Applications* (J. C. Salamone, ed.), CRC Press, Boca Raton, Fl., Vol. 4, pp 2661-2667.
17. Simionescu, B. C., Harabagiu, V., and Simionescu, C. I. 1996, Siloxane containing polymers. In *The Polymeric Materials Encyclopedia: Synthesis, Properties and Applications* (J. C. Salamone, ed.), CRC Press: Boca Raton, Fl., Vol. 10, pp. 7751-7759.
18. El-Zaim, H. S., and Heggers, J. P. 2002, Silicones for pharmaceutical and biomedical applications. In *Polymeric Biomaterials* (S. Dumitriu, ed.), Marcel Dekker, New York, Baserl, pp. 79-90.
19. Frisch, E. E. 1983, Technology of silicones in biomedical applications, In *Biomaterials in Reconstructive Surgery* (L. H. Rubin, ed.), C. V. Mosby, St. Louis, pp. 73-90.
20. Aoyagi, T., and Nagase, Y., 1995, Silicone-based polymers. In *Percutaneous penetration enhancers* (E. Smith, and H. Maibach, eds.) CRC Press, New York pp. 267.

21. Perrin, D. E., and English, J. P. 1997, Polycaprolactone. In *Handbook of Biodegradable Polymers* (A. J. Domb, J. Kost, and D. N. Weiseman, eds.), Harwood Academic Publishers, Amsterdam, pp. 63-77.

22. Yilgor, I., Steckle, Jr. W. P., Yilgor, E., Freelin, R. G., and Riffle, J. S. 1989, Novel triblock siloxane copolymers: synthesis, characterization and their use as surface modifying additives. *J. Polym. Sci.: Part A: Polym. Chem.* 27: 3673-3690.

23. Iojoiu, C., Hamaide, T., Harabagiu, V., and Simionescu, B. C. 2004, Modified poly(ε-caprolactone)s and their use for drug encapsulating nanoparticles. *J. Polym. Sci.: Part A: Polym. Chem.* 42: 689 – 700.

24. Lovinger, A. J., Han, B. J., Frank, J., Padden, Jr. J., and Mirau, P. A. 1993, Morphology and properties of polycaprolactone-polydimethylsiloxane-polycaprolactone block copolymers. *J. Polym. Sci.: Part B: Polym. Phys.* 31, 115-123.

25. Erbil, H. Y., Yaşar, B., Süzer, Ş., and Baysal, B. M. 1997, Surface characterization of the hydroxy-terminated poly(ε-caprolactone)/polydimethylsiloxane triblock copolymers by electron spectroscopy for chemical analysis and contact angle measurements. *Langmuir* 13: 5484-5493.

26. Karal, O., Hamurcu, E. E. G., and Baysal, B. M. 1998, Effect of a triblock PCL-PDMS-PCL copolymer on the properties of immiscible poly(vinyl chloride)/poly(2-ethylhexyl acrylate) blends. *Macromol. Chem. Phys.* 199: 2699-2708.

27. Kayaman-Apohan, N., Karal-Yilmaz, O., Baysal, K., and Baysal, B. M. 2001, Poly(DL-lactic acid)/triblock PCL-PDMS-PCL copolymers: synthesis, characterization and demonstration of their cell growth effect in vitro. *Polymer* 42: 4109-4116.

28. Tziampazis, E., Kohn, J., and Moghe, P. V. 2000, PEG-variant biomaterials as selectively adhesive protein templates: model surface for controlled cell adhesion and migration. *Biomaterials* 21: 511-520.

29. West, J. L., and Hubbell, J. A. 1995, Comparison of covalently and physically cross-linked polyethylene glycol-based hydrogels for the prevention of postoperative adhesions in rat model. *Biomaterials* 16: 1153-1156.

30. Peppas, L. B., and Peppas, N. A. 1991, Equilibrium swelling behavior of pH-sensitive hydrogels. *Chem. Eng. Sci.* 46: 715-722.

31. Yildiz, B., Işik, B., and Kiş, M. 2002, Thermoresponsive poly(N-isopropylacrylamide-co-acrylamide-co-hydroxyethyl methacrylate) hydrogel. *Reactive & Functional Polymers* 52: 3-10.

32. Park, T. G., and Hoffman, A.S. 1992, Synthesis and characterization of pH- and /or temperature –sensitive hydrogels. *J. Appl. Polym. Sci.* 46: 659-671.

33. Gutowska, A., Bark, J. S., Kwon, I. C., Bae, Y. H., Cha, Y. and Kim, S.W. 1997, Squeezing hydrogels for controlled oral drug delivery. *J. Contr. Release* 48: 141-148.

34. Sawahata, K., Hara, M., Yasunaga, H., and Osada, Y. 1990, Electrically controlled drug delivery system using polyelectrolyte gels. *J. Contr. Release* 14: 253-262.

35. Park, T.G. 1999, Temperature modulated protein release from pH/temperature-sensitive hydrogels. *Biomaterials* 20: 517-521.

36. Hirokawa, Y., and Tanaka, T. 1984, Volume phase transition in a nonionic gel, *J. Chem. Phys.* 81: 6379-6380.

37. Tanaka, T., Sato, E., Hirokawa, Y., Hirotsu, S., and Peetermans, J. 1985, Critical kinetics of volume phase transition of gels, *Phys. Rev. Lett.* 55: 2455-2458.

38. Yildiz, B., Isik, B., and Kis, M. 2002, Thermoresponsive poly(N-isopropylacrylamide-co-acrylamide-co-2-hydrohyethyl methacrylate) hydrogels, *Reactive & Functional Polymers* 52: 3-10.

39. Feil, H., Bae, Y.H., Feijen, J., and Kim, S.W. 1993, Effect of comonomer hydrophylicity and ionization on the lower critical solution temperature of N-isopropylacrylamide copolymers, *Macromolecules* **26**: 2496-2500.
40. Fundueanu, G., Esposito, E., Mihai, D., Carpov, A., Desbrieres, J., Rinaudo, M., Nastruzzi, C. 1998, Preparation and characterisation of Ca-alginate microspheres by a new modification method, *Int. J. Pharm.* **170**: 11-21.
41 Fundueanu, G., Constantin, M., Mihai, D., Bortolotti, F., Cortesi, R., Ascenzi, P., Menegatti, E. 2003, Pullulan-cyclodextrin microspheres. A chromatographic approach for the evaluation of the drug-cyclodextrin interactions and the determination of the drug release profiles, *J. Chromatography B* **791**: 407-419.
42. Constantin, M., Fundueanu, G., Cortesi, R., Esposito, E., Nastruzzi, C. 2003, Aminated Polysaccharide Microspheres as DNA Delivery Systems, *Drug Delivery* **10**: 1-11.
43. Fundueanu, G., Constantin, M., Dalpiaz, A., Bortolotti, F., Cortesi, R., Ascenzi, P., Menegatti, E. 2004, Preparation and characterization of starch/cyclodextrin bioadhesive microspheres as platform for nasal administration of Gabexate Mesilate (Foy®) in allergic rhinitis treatment, *Biomaterials* **25**; 159-170.
44. Fundueanu, G., Nastruzzi, C., Carpov, A., Desbrieres, J., and Rinaudo, M. 1999, Physico-chemical characterization of Ca-alginate microspheres produced with different methods, *Biomaterials* **20**: 1427-1435.

Polyurethanes in Biomedical Applications

AYER BURKE[*] and NESRIN HASIRCI[#]
*European University of Lefke, Faculty of Architecture and Engineering,Department of
Electrical and Electronic Engineering, Turkish Republic of Northern Cyprus; #Middle East
Technical University, Faculty of Arts and Sciences, Chemistry Department, Ankara 06531,
TURKEY

1.   INTRODUCTION

Polyurethanes are the most commonly used materials in the production of
blood contacting devices such as heart valves or artificial veins and arteries.
They comprise a large family of materials with the only common
characteristic of the presence of urethane linkages along the large molecular
chains. In general urethane linkages form by the reaction of isocyanates and
alcohols. During the preparation and the curing processes of polyurethanes,
besides the formation of urethane linkages, many other reactions take place
and lead to formation of various bonds such as allophanate, biuret, acylurea
or isocyanurate and these bonds may lead to further branching or
crosslinking affecting the whole physical, chemical and mechanical
properties as well as the biocompatibilities of the resulting polymers[1,2].

The synthesis of polyurethanes was first achieved in the 1930's and the
polymers obtained by the reaction of diisocyanates with glycols possessed
interesting properties as plastics and fibers. During World War II,
polyurethanes found many applications such as rigid foams, adhesives,
resins, elastomers and coatings.

Development of polyurethane elastomers[3] and flexible foams[4] based on
polyesters was achieved in 1950's and shortly thereafter they came into
commercial production. Polyurethanes developed for industrial purposes
demonstrated very good long term mechanical properties when exposed to
static or dynamic loads, and therefore, they also found various application
areas in medical technology.

Biomaterials: From Molecules to Engineered Tissues, edited by
N. Hasırcı and V. Hasırcı, Kluwer Academic/Plenum Publishers, 2004

The first generation of polyurethanes used for implant studies were industrial grade and commercially available. But it was reported that, when implanted into the muscle of dogs or when used as monocusp valvular prosthese polyester urethanes degraded rapidly[5].

This type of degradation and calcification limited the use of these polyurethanes in medical applications. On the other hand, it was reported that polyester-polyether polyurethanes demonstrated good blood compatibility, particularly with regard to long term stability and thromboresistance in intravascular replacement[6].

The initial observations showed that it was not easy to reach a general conclusion about the biocompatibility of polyurethanes since they cover a very wide range of compositions and structures in this family of polymers. Polyurethanes differ in their interactions with blood, their tendencies to calcify and their propensity toward biodegradation. This is the reason that many research groups began to search new compositions and new methods to produce new polyurethanes and to modify available ones to improve their properties to obtain materials with the desired properties for medical applications and biocompatibilities.

1.1   Types of Polyurethanes

The urethane linkages usually represent a small component in the total chain, with the greatest number of linkages contributed by the macroglycol. Therefore, many of the properties are derived from the macroglycol portions of the chain, and depending on the choice of macroglycol, the polyurethanes have been found to perform differently in varying clinical applications[7-9]

The basic classification can be given as follows:
1. Polyester-based polyurethanes undergo rapid hydrolysis when implanted in the human body, and thus are not preferred in medical applications.
2. Polyether-based polyurethanes are the polymers of choice in medical applications because they are virtually insensitive to hydrolysis, and are thus very stable in the physiological environment.
3. Polycaprolactone-based polyurethanes, due to their quick crystallization, can be used advantageously as medical, solvent-activated, pressure-sensitive adhesives.
4. Polybutadiene-based polyurethanes have been evaluated, but limited medical applications has been found to date.
5. Castor oil-based polyurethanes can be used as potting and encapsulating compounds, but due to their poor tear resistance, find limited use in medical applications.

As it can be seen polyurethanes are a very diverse family of polymers capable of exhibiting a wide range of properties varying from smooth elastomeric membranes to porous or smooth rigid bulk structures depending on their molecular compositions.

Segmented polyurethane elastomers are the ones used in the production of blood contacting devices like heart valves and artificial arteries or veins. Segmented polyurethanes are composed of soft and hard segments where hard segments are stiff blocks composed mainly of isocyanate groups while soft segments are flexible high molecular weight chains of polyether or polyesters. These segments form some domains distributed in the three dimensional structure of the polymer. Physical and mechanical properties are directly related to chemical structure of these domains as well as their ratio in the resulting polymer[10,11].

Even for a certain well defined type of polyurethane, it is possible to find contradictory results and disagreements in literature, and this might be associated with a specific application and related to various parameters such as blood flow rate, device type, time of implantation, etc.

1.2   Phase Separations of Polyurethanes

Soft and hard domains distributed in segmented polyurethanes cause phase separation and control the properties of the resultant polyurethanes. The type and the ratio of hard and soft segments are generally examined by spectrophotometric methods. Small-angle X-ray scattering was used by Garrett et al. to examine phase-separated microdomain morphology of the multiblock copolymers, synthesized from 4,4'-methylene di(p-phenyl isocyanate), poly(tetramethylene oxide), ethylenediamine, and/or 1,4-diaminocyclohexane. The hard segment content was in the range of 14 to 47 wt percent and it was reported that copolymers have relatively low overall degrees of phase separation contrary to the common notion that these copolymers are well phase separated materials. The introduction of the second diamine reduced phase separation; presumably as a consequence of disruption of hydrogen bonding in hard segment domains[12]. The phase separated morphology affects the blood compatibility properties of polyurethanes. It was claimed that hydrophilicity creates surfaces resembling the natural tissue and decreases thrombus formation. In the literature it was given that at a certain surface concentration of soft segments the number of adhered and deformed platelets on the surface of polyurethanes was minimized[13-15]. But it was also shown that the percentage of soft segments on the surface did not relate to the blood compatibility of polyurethanes[16].

2.   BIOMEDICAL APPLICATIONS

Polyurethanes, having extensive structure/property diversity, are one of the most bio- and blood-compatible materials known today. These materials played a major role in the development of many medical devices ranging from catheters to total artificial heart. One very important point is the medical purity of the materials. There should be no leachable toxic solvent, monomers, chain extender or other chemicals, which may cause toxic effects in the body[17]. Properties such as durability, elasticity, elastomer-like character, fatigue resistance, compliance, and acceptance or tolerance in the body during healing, became often associated with the chemical composition of polyurethanes. Furthermore, propensity for bulk and surface modification via hydrophilic/hydrophobic balance or by attachments of biologically active species such as anticoagulants or bio-recognizable groups are possible via chemical groups typical for polyurethane structure. These modifications are designed to mediate and enhance the acceptance and healing of the device or implant. Many innovative processing technologies are used to fabricate functional devices, feeling and often behaving like natural tissue. The hydrolytically unstable polyester polyurethanes were replaced by more resistant but oxidation-sensitive polyether polyols based polyurethanes and their copolymers containing silicone and other modifying polymeric intermediates.

The first biomedical grade polyether polyurethane was synthesized by two groups. Boretos and Pierce[18,19] introduced the biomedical application of segmented polyether polyurethanes containing hard segments of urea and soft segments of polyether linked by the urethane group. These materials sustained high modulus of elasticity, biocompatibility, resistance to flex-fatigue and excellent stability over long implantation periods.

Lyman and associates[20] based on their previous experience in the synthesis of polyurethanes for dialysis membranes, introduced a segmented polyether polyurethane which demonstrated very good thromboresistance.

Nylias, the developer of Avcothane[R], synthesized a copolymer of polyurethane and polydimethyl siloxane[21] which is blood compatible and used in the making of heart assist balloon pumps. Lelah et.al.[22] reported that, materials having higher surface soft segment concentration are more thromboresistant, but in contrast Hanson[23] reported that platelet consumption decreased as the percentage of surface carbon atoms forming hydrocarbon bonds increased.

Research on the artificial heart (carried out at Termedics, Inc., U.S.A.) has stimulated interest in the segmented polyurethane elastomers. The polymers exhibit high flexure endurance, high strength, and inherent non-thrombogenic characteristics, and therefore, have a maximum impact in medical prostheses, the most immediate and promising applications appear to be in the cardiovascular area, where chronic, nonthrombogenic interfacing with blood is of paramount importance[24].

The armory of cardiac surgeons would not be as impressive as it is without the outstanding contribution of polyurethanes in intra–aortic balloons, blood sacs for ventricular assist devices, catheters, pacemaker leads to name the most important. Results of PUs as blood conduits have still not found a niche because of the unresolved lack of long–term resistance to degradation. Breast implants covered with PU foam are part of a scientific controversy. The use of PU in contraception is limited but these materials present some interesting features. Wound dressings and scaffolds for tissue engineering could permit new developments.

On the other hand polyurethanes are also prepared in the form of microspheres as carriers for drug or bioactive substances, or can be used as liners in dentistry, coatings for metallic heart assist devices, membranes or tubes for hemodialysis systems, scaffolds for tissue engineering purposes, etc[25-27]. Table 1 shows the general applications of polyurethanes in the biomedical area.

Table 1: Current biomedical application areas of polyurethanes

Blood bags, closures, fittings	Leaflet heart valves
Blood oxygenation tubing	Mechanical heart valve coatings
Breast prostheses	Orthopedic splints, bone adhesives
Cardiac assist pump bladders, tubing, coatings	Percutaneous shunts
Catheters	Reconstructive surgery materials
Dental cavity liners	Skin dressing and tapes
Endotracheal tubes	Surgical drapes
Heart pacemaker connectors, coatings,	Suture materials
Hemodialysis tubing, membranes, connectors	Synthetic bile ducts
Lead insulators, fixation devices	Vascular grafts and patches

The human body's acceptance of synthetic polymers is highly complex, and most polymers have a tendency to form surface thrombus. A good biomaterial would have wide range of applications, including as an prosthesis device material or as coating material for components of various devices such as components of dialysis units, extracorporeal circuits, blood pressure monitors, sensors or catheters. Polyurethane elastomers are known as inherently thromboresistant. Although compatibility and non-thrombogenicity are subject to many complex factors, among which are polymer surface composition, device configuration, and blood-flow characteristics, in general the polyurethanes have performed well in numerous device configurations. Therefore, several methods have been developed for evaluating the blood interactions of polyurethanes.

3. MODIFICATIONS OF POLYURETHANES

The applications of polyurethanes are almost limitless in the medical industry. However, one should ascertain that polyurethanes are indeed the best materials to manufacture devices for specific applications. The capacity of polyurethanes to undergo modifications increase their suitability for biomedical applications.

3.1   Modifications for Blood Compatibility

When an artificial substance or an implant is exposed to blood some interactions such as protein deposition, platelet adhesion and activation, and initiation of coagulation and thrombi formation occurs. It is believed that when blood comes in contact with an artificial surface, the first reaction is the adsorption of plasma proteins on the surface. This forms a layer 100-200Å thick. It is also believed that the first adsorbed protein plays an important role in blood compatibility and adsorption of serum albumin greatly reduces, or totally prevents platelet adhesion reaction, thus passivating the surface toward the formation of mural thrombus[28]. The apparent thromboresistance of polyurethanes is thought to reside in the polyurethanes, ability to preferentially adsorb albumin. However, not all polyurethanes are equally biocompatible.

One short-term approach for the case of an implant application has been the use of a systemic anticoagulant such as heparin to prevent thrombus formation on the polymer surface, but this has serious drawbacks, in some cases it may cause hemorrhaging, and even death. To circumvent this problem, and localize the anticoagulant effect, heparin has been coated and chemically combined with the blood-contact surfaces of the polymers. This approach can be used succesfully, but after the heparin has been eluted from the coated surface, thrombotic episodes can be expected .

Some scientists prepared polyurethane and ethylene vinylacetate copolymers by adding heparin in the system during preparation, and examined the thermal stability and biological activity of the released heparin[29]. Modification by heparin grafting was studied since heparin was inexpensive and rapidly neutralized by the administration of protamine. But one drawback was that it may induce thrombocytopenia and that may lead to massive thromboembolism. Due to decrease of the amounts of clotting factors, high risk of excessive bleeding is exists in heparin applications[30]. Some scientists examined the effects of anticoagulants coated or immobilized on the surfaces of polyurethanes. The most commonly examined proteins were heparin[31], albumin[32,33] , hirudin[34] or conjugates of heparin-albumin[35]

Studies for improving blood compatibility also include the use of more biocompatible hydrophobic polymers[36], endothelial cell seeding[37], fibrinolytic enzymes[38] and self-assembly strategies[39]. Polyethylene oxide coupling is one method to increase hydrophilicty[40,41] and prevent thrombus formation. Static blood compatibility and hemocompatibility was studied in in vitro tests under a shear of blood flow by using various stearyl poly(ethylene oxide) coupled polyurethanes[42].

To modify the surfaces of polyurethanes by using negatively charged ions to resemble the endothelium layer of veins is another approach for blood compatibility. Chen et.al.[43] prepared polyurethane structures by using sulfonated or carboxylated chain extenders. and it was reported that especially for carboxylate group the degree of platelet adhesion and platelet activation was significantly reduced.

Modification with ionic functional groups such as poly(sodium vinyl sulfonate[44], propyl sulfonate[45] were also achieved and increase in biocompatibility was observed. For MDI-based polyurethanes it was reported that, at the same ionic content, sulfonate incorporation significantly reduced platelet deposition compared to carboxylate incorporated ones[46]. In contrast, it was also reported that sulfonate containing polyethylene surface had much higher platelet adhesion than the carboxylic acid containing ones[47]. Presence of charged groups decrease the contact angle of the surfaces by increasing hydrophilicities and it was shown that contact angle dropped from 70° to 43° for Biospan [TM] (Segmented poly(ether urethane urea) linear block copolymer of 4,4'-diphenylmethane diisocyanate extended with a mixture of ethylenediamine, 1,3-cyclohexanediamine and polytetra methylene oxide) when it was modified and covalently coupled with p-aminosalicylic acid and the ratio of absorption of albumin to fibrinogen increased with drug content on the surface[48].

Modifications of segmented polyurethanes with methacryloyloxyethyl phosphorylcholine by forming semi-interpenetrating polymer networks[49] or by forming alloys[50] increased the antithrombogenic property of the surfaces. Phospholipid layers[51,52] or sulfoammonium zwitterionic molecules also demonstrated lower platelet aggregation on segmented polyurethanes[53]

Use of hydrophobic materials which exhibit slow NO release is another way to create a nonthrombogenic surface since endothelial cells produce NO and it is known that this prevents platelet adhesion and activation. NO releasing polymers can be prepared by incorporating some diazeniumdiolate molecules, which donates NO onto polymeric substances[54]. Movery etal.[55] prepared solid and stable several diazeniumdiolate NO donors and incorporated them into polymers by uniform dispersion, covalent attachment or ion-pairing. These molecules spontaneously react with water and produce NO and residual di- or poly-amines. Enhanced thromboresistance compared to conventional polymers was reported and in vitro platelet adhesion decreased significantly

Polyurethanes were modified by L-cysteine by covalent immobilization and treated with S-nitroso-bovine serum albumin to supply NO in order to decrease platelet aggregation[56]. The control of release rate of NO is very important since high release may lead to toxic effects while insufficient release may not be effective to prevent platelet aggregation.

Immobilization of functionalized dextrans or biological compounds[57] or nonsteroidal antiinflammatory drugs such as salicilic acid and their derivatives[58] also increase antithrombogenic effects.

It was shown that, grafting the surfaces of vascular PU scaffolds (Estane, made of polyester type polyurethane), with biomacromolecules increased their biocompatibility. For this purpose, surfaces were aminolyzed by using

1,6-hexanediamine and the produced free amino groups were reacted with gelatin, collagen or chitosan via glutaraldehyde coupling. A drop in surface contact angle from ~83° to ~40°, and enhanced cell-material interaction and faster cell proliferation were observed on modified surfaces compared to control PU scaffolds in the carried out with human endothelial cells[59].

However upto date none of these mentioned methods have been completely successful platelet activation and coagulation and thrombi formation occurs on the interface when a polymer based medical device in come to interaction with blood in vivo. Therefore there is still and will be more intense research ion chemical and surface properties of implant materials.

3.2   Modifications for Gas Permeability

Permeability of polyurethanes is an important feature for a broad range of applications including packaging, bio-materials (e.g. for controlled release or encapsulating membranes, catheters, dialysis membranes or wound dressings, etc.), barrier materials, high performance impermeable breathable clothing and membrane separation processes. The gas permeation rate of PU membrane could be modified by controlling the ratio of hard domain to soft domain. The gas permeation property is dependent on the type, amount and molecular weight of polyols and chain extenders. The effect of chemical composition on the gas permeability might be through the degree of phase separation and the nature of chain packing.

For the PU films prepared from different isocyanate and glycol components (such as toluene diisocyanate and 1,4-butanediol hard segments; and hydroxyl terminated polybutadiene, hydroxyl terminated polybutadiene/styrene and hydroxyl terminated polybutadiene/acrylonitrile soft segments), it was reported that free volume size and fractional free-volume decreased with the increase of hard segment content. A direct relationship between the gas permeability and the free-volume has been established based on the free-volume parameters and gas diffusivity measured[60].

In vivo and in vitro stabilities of modified poly(urethaneurea) (BioSpan MS/0.4) blood sacs were reported in literature. Blood sacs were utilized primarily in left ventricular assist devices that were implanted in calves for times ranging from 5 to 160 days. In vitro cyclic testing was also conducted on similar sacs. Various analytical methods were employed to characterize the properties of the sacs after in vivo or in vitro service. The methods included ATR-FTIR spectroscopy, scanning electron microscopy and gel permeation chromatography. In general, the characteristics of implanted and in vitro cycled sacs were similar to their control sacs. It was reported that thermal and microtensile properties were unchanged after testing. The same was true for the ATR-FTIR spectra, indicating relative chemical stability for the time frames explored here. The only significant changes occurred in molecular weight and gross surface morphology. A modest increase in weight average molecular weight was observed for most implanted blood sacs, indicating some type of chain extension or branching reaction in vivo. Although the surface morphologies of implanted blood sacs were often similar to their control, sometimes limited pitting was observed on the nonblood contacting surfaces in regions that experience maximum bending during service[61].

Polyurethaneurea-polyether (PEUU) multiblock copolymers were synthesized to elicit lower permeability to water vapor and gases. For this purpose, polyisobutylene (PIB) segments were linked to the PEUU copolymer as combs. By using macromonomers with two hydroxyl sites, amphiphilic copolymers, with a polyurethaneurea-polyether multiblock backbone and polyisobutylene combs, were synthesized by condensation reaction. PIE incorporation varied between about 2 and 30%, with comb lengths ranging from around 3000 to 29 000 g/mol. Characterization of this new multiblock multicomb copolymer was performed by GPC, FTIR, solid state C-13 NMR, and Soxhlet extraction[62].

A nanocomposite approach for biomedical poly(urethane urea)s, that results in a significant reduction in gas permeability without sacrificing mechanical properties was reported. PUU/alkylammonium modified montmorillonite nanocomposites were prepared containing relatively low volume fractions of the layered silicate. X-ray diffraction experiments showed a silicate gallery spacing increase, indicating the formation of intercalated PUU/silicate structures. With higher silicate content, modulus and strength increased with no loss of ductility. Water vapor permeability was reduced by five times at 6 vol% silicate, as a result of polymer/inorganic composite formation. It was proposed that, these concurrent property enhancements are well beyond what can be generally be achieved by chemical modification of PUU polymers[63].

A series of polyurethanes with varying hard/soft segment ratio were prepared from toluene diisocyanate and polypropyleneglycol, and oxygen permeabilities were examined. It was reported that increase in the amount of soft segment increased oxygen permeation while addition of chain extender caused a drop in permeability[64].

It was reported that, for separation of gases such as O_2/N_2 and CO_2/N_2, the small addition of PDMS into polyether-based PU matrices led to both higher gas permeabilities and selectivities of O_2/N_2 and CO_2/N_2. This might be because the incorporation of PDMS leads to the phase separation in both hard segment (MDI/BD) and soft segment (polyethers) due to the difference in solubility parameter, and the dispersed PDMS phases serve to produce a more tortuous route for diffusing molecules. While the small addition of polyethers into PDMS-based PUU matrix decreased gas permeability and did not affect the selectivity of O_2/N_2. It, however, increased the selectivity of CO_2/N_2[65].

Chronic in vivo instability, however, observed on prolonged implantation, became a major roadblock for many applications. Presently, utilization of more oxidation resistant polycarbonate polyols as soft segments, in combination with antioxidants such as Vitamin E, offer materials which can endure in the body for several years. The applications cover cardiovascular devices, artificial organs, tissue replacement and augmentation, performance enhancing coatings and many others. In situ polymerized, cross-linked systems could extend this biodurability even further. The future will expand this field by revisiting chemically controlled biodegradation, in combination with a mini-version of RIM technology and minimally invasive surgical procedures, to form, in vivo, a scaffold, by delivery of reacting materials to the specific site in the body and polymerizing the mass in situ. This scaffold will provide anchor for tissue regeneration via cell attachment, proliferation, control of inflammation, and healing[66].

3.3   Modifications for Calcium Deposition

Calcification of polyurethane prosthetic heart valves is a major problem. Solidification leads to deterioration of the prosthesis and at the same time formation of tears and cracks in the prosthesis. Calcification, defined as the deposition of calcium compounds, occurs in a wide spectrum of medical devices.

The actual mechanism of calcification is not completely understood and is associated with practically all the soft implants such as bioprosthetic heart valves, polymeric blood pumps and heart valves, contact lens, etc. It is related to host metabolism as well as to the chemistry and structure of the implant. It is known that the implanted medical devices lead to the most dystrophic calcification, where the tissues are necrotic or otherwise altered in normocalcemic subjects. The glutaraldehyde (GA) treatment of tissues for crosslinking would be the major cause by deteriorating the tissue structure, while fresh tissues do not calcify. In addition, glycoaminoglycan components in the tissues are extracted during the GA treatment to provide the sites for the initial nucleation of calcium compounds. Blood components or lipids deposited would have a contribution, too. There have been a number of approaches to prevent the calcification, and aminooleic acid, after-treatment or heparin coupling, etc. is a representative one.

Decrease in calcification was reported when the bioprosthetic tissue (BT) was coupled with sulfonated PEO derivative via glutaraldehyde crosslinking. Such a decreased calcification might be explained by several effects as following; the amino end groups of NH_2-PEO-SO_3 were coupled to the residual GA groups to remove the possible contribution of GA residues for the calcification. PEO-SO_3 segments would have filled the space in the collagen matrix which was reported as nucleating sites for calcification. The enhanced blood compatibility of PEO-SO$_3$ might decrease the adhesion of blood or cellular components. This method can be a useful anticalcification treatment for implantable tissue valves and pericardium tissue patch[67].

In several reports, the complexation of calcium ions with PEO chain segments was hypothesized for the calcification of PU as in the case of PEO-polybutylene terephthalate containing PEO main chain[68].

Calcification properties of polyurethane composites made from biaxially drawn ultrahigh molecular weight polyethylene were examined by incubating the samples in calcium phosphate metastable solution. The results demonstrated that the membranes were susceptible to extrinsic calcification that was closely related to the matrix polyurethane material used. Calcification was postponed for composites compared to solution cast polyurethane membranes[69].

Polyurethane composite membranes were prepared by solution casting and then subjected to heat-compaction. It was reported that the heat compaction induced distortion of macromolecules and physical changes on microstructures and demonstrated higher affinity to calcium ions than the non-heat compacted samples[70]

Ventricles made from segmented polyurethane membranes and used in the fabrication of a totally implantable artificial heart are known to undergo biomaterial-associated calcification. As there is no effective method currently available to prevent such biomaterials from calcifying, a practical solution is to use only materials with a relatively high resistance to calcification to extend ventricular durability and ensure a longer functional life for the manufactured device. In this study, an in vitro calcification protocol was used to determine the relative resistance to calcification of six different polyurethanes, namely, Carbothane (R) PC3570A, Chronoflex (R) AR, Corethane (R) 80A, Corethane (R) 55D, Tecoflex (R) EG80A, and Tecothane(R) TT1074A. The results demonstrated that all six polyurethanes became calcified during the 60-day incubation period in the calcification solution. The degree of calcification was found to be related with the surface chemistry of the particular polyurethane, with the Tecothane TT1074A exhibiting the highest level. The Corethane 80A and 55D polymers showed a relatively low propensity to calcify. These two membranes can, therefore, be considered as the most appropriate materials for the fabrication of ventricles for a totally implantable artificial heart. In addition, since the calcification occurred primarily at the surface of the membranes, without affecting the bulk microphase structure, the issue of modifying the surface chemistry to reduce the incidence of calcification is discussed[71].

Some scientist examined the calcification and apatite formation on porous polyurethane foams. The foams were used as matrices for the production of hydroxyapatite-based calcium phosphate ceramics, which are important as bone defect fillers[72].

4.   PLASMA SURFACE MODIFICATION

Surface modification of vascular and cardiac implants as well as other medical devices by use of glow discharge plasma application is a concept studied in the last decades. The process has many consequent steps and depends on various factors including the geometrical shape of the reaction chamber, type and power of electrical discharge, type and flow rate of gas used, and properties of substrate. In glow discharge surface modifications generally radiofrequency energy is applied at low pressures.

The generated plasma contains free electrons, radicals, excited atoms and neutral particles. Constant bombardment of the substrate surface with these active particles leads to breaking some bonds of the surface molecules, and react with them by forming new bonds. Therefore, a chemical change on the surface about 30 Å occurs while the bulk structure of the substrate stays intact.

Depending on the type gas used, it becomes possible to change hydrophilicity of the surface or create a coat having completely different structure than the bulk. Most generally used gases are oxygen, nitrogen, argon, or various monomeric substances. Previous studies have shown that it is even possible to coat a very porous substrate such as activated charcoal granules · by polymerisation of hexamethyldisiloxane and create a biocompatible surface[73]. Various properties of polyurethanes prepared by using different diisocyanate compounds (such as diphenylmethane-, hexamethylene-,2,6-toluene- and 2,4-toluene-diisocyanates) and different glycols (such as polypropylene-ethyleneglycol, polypropylene glycol) were examined along with their biocompatibilities.

Plasma application by using monomers such as hydrophilic hydroxyethyl methacrylate or hydrophobic hexamethyl disiloxane changed the surface properties.

When the segmented polyurethanes prepared from toluene diisocyanate and polypropylene-ethyleneglycol were subjected to oxygen and argon plasmas it was observed that the adsorbed protein levels decreased upon treatment and the adsorption was the lowest for oxygen plasma treated samples where albumin levels were higher than fibrinogen and gammaglobulin[74].

Polyurethane elastomeric films which were prepared from toluenediisocyanate and poly(propylene-ethylene)glycol did not cause any blood cell aggregation or fibrin formation (except the one which contained high amount of glycol). Aging or accelerated aging did not demonstrate any change in mechanical properties[75,76].

For polyurethanes prepared from 2,4-2,6-toluene diisocyanate, and poly propylene glycol, it was observed that tensile strength and ultimate elongation values were both increased in the presence of chain extender, propanediol, in the structure. Oxygen permeabilities of the prepared elastomers increased as the soft segment content increased [64,77,78].

For polyurethane elastomers prepared from diphenylmethane diisocyanate (MDI) or hexamethylene diisocyanate (HDI) and poly(propylene glycol)[11], and from toluenediisocyanate (TDI) and polyol[79], without using any other ingredients such as chain extender, catalyst or solvent, it was shown that the ratio of the ingredients used initially affected the mechanical properties. An increase in the amount of MDI, HDI or TDI led to production of stiffer membranes. When MDI and HDI containing polymers were brought in contact with blood no clotting was observed even after 1 hour of contact.

Surface modification by oxygen plasma application increased surface hydrophilicity of the TDI containing membranes and the contact angle values decreased from 67 to 46 degrees. Such a surface modification is reported to affect cell attachment capability of the membranes. For Vero cells, it was observed that as the applied power increased, number of attached cells increased for 10 watt but decreased for 100 watt power applied samples. This observation indicated the hydrophobic nature of adhesion of the cells on polyurethanes[80].

The attached cell numbers were about 42-45 cells per cm^2 for the pristine membranes depending on the preparation composition. The highest attachment numbers were obtained for membranes modified with 10 watts and the cell densities were in the range of 62-70 cells per cm^2. The lowest values were observed with the samples modified with 100 watts and were in the range of 27-40 cells per cm^2. These results showed that a certain level of hydrophilicity is needed for cell attachment. Figure 1 shows the adhered Vero cells on pristine and plasma modified polyurethane membranes.

All these experiments showed that, the mechanical properties, physical forms, as well as biocompatibilities and hemocompatibilities of polyurethanes all depended on their chemical compositions (types and amounts of diisocyanate and glycol components), preparation conditions (types and amounts of solvents, chain extenders, catalysts, reaction type and temperature, curing conditions, etc), and modification parameters (by chemical molecules, irradiation by UV or gamma, or by plasma glow discharge applications).

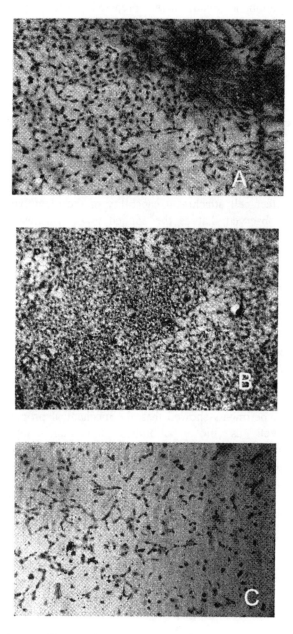

Figure 1. Attachment of Vero cells on segmented polyurethane surfaces. A) Untreated polyurethanes, B) Treated with 10 Watt oxygen plasma, C) Treated with 100 Watt oxygen plasma

5.   CONCLUSION

Polyurethanes have very high importance in the production of medical devices or systems because of their excellent mechanical properties high biocompatibilities and the availabilities in different forms changing from elastomeric membranes to porous sponges. Still an intense research is going on to synthesize new formulations, to modify the existent ones with various techniques in order to increase their bio and blood compatibilities. Most of the research is concentrated on understanding the interactions between chemical structure and cell attachment, thrombus formation or calcification under stress applications. At the moment polyurethanes are the only family of polymers used in cardiovascular applications because of their inherent hemocompatibilities.

ACKNOWLEDGEMENTS

The work was supported by research project grants of METU- AFP and Scientific and Technical Research Council of Turkey.

REFERENCES

1. Hasirci, N., 1991, Polyurethanes. In *High Performance Biomaterials: Comprehensive Guide to Medical and Pharmaceutical Application* (M. Szycher, ed), Technomic Pub.Co., Lancaster, pp.71-91.
2. Hasirci, N., 1994, Polyurethanes as biomedical materials. In *Surface Properties of Biomaterials*, (R. West and G. Batts, eds), Butterworth- Heineman Ltd., Oxford, pp.81-90.
3. Bayer, O., Muller, E., Peterson, S., Piepenbrink, H.F., and Windemuth, E., 1950, Polyurethanes VI. New Highly Elastic Synthetics, Vulcollans, *Rubber Chem. Technol.*, 23: 81
4. Bikales, M.N., 1969. Polyurethanes. In *Encyclopedia of Polymer Science and Technology*, Interscience Publishers John Wiley and Son Inc., New York, 11: 507.
5. Mirkovitch, V., Akutsu, T., and Kolff, W.J., 1962, Polyurethane Aortas In Dogs-Three-Year Results. *Trans .Am.Soc.Artif.Intern.Organs*, 8: 79.
6. Sharp, W.V., Gardener, D.L., and Anderson, G.J., 1966, A Bioelectric Polyurethane Elastomer for Intravascular Replacement. *Trans.Am.Soc.Artif.Intern.Organs*, 12: 1979.
7. Szycher, M., Poirier, V.L., and Dempsey, D.J., 1983, Development of an aliphatic biomedical-grade polyurethane elastomer. *J.Elast.Plast.*, 15: 81.
8. Szycher, M., Poirier, V.L., and Dempsey, D.J., 1983, Development and Testing of Melt-Processable Aliphatic Polyurethane Elastomers. *Trans.Soc.Biomater.*, 6: 49.
9. Ulubayram, K. and Hasirci, N., 1991, Polyurethanes: Chemistry and Properties. *Proceedings of Second Mediterranean School on Science and Technology of Advanced Polymer -Based Materials.* pp.1-4.

10. Ulubayram, K. and Hasirci, N., 1995, Preparation of Polyurethane Elastomer For Biomedical Applications. In *Proceedings of the Second National Symposium on Biomedical Science and Technology* (V.Hasirci, ed.), pp.65-66.

11. Kutay, S., Tincer, T. and Hasirci N., 1990, Polyurethanes as Biomedical Materials. *British Polymer Journal*, 23: 267-272.

12. Garrett, J.T., Runt, J., and Lin, J.S., 2000, Microphase separation of segmented poly(urethane urea) block copolymers. *Macromolecules* 33(17): 6353-6359.

13. Nojima, K., Sanui, K., Ogata, N., Yui, N., Kataoka, K., and Sakurai, Y., 1987, Material characterization of segmented polyether poly(urethane-urea-amide)s and its implication in blood compatibility. Polymer, 28: 1017-1024.

14. Takahara, A., Tashita, J.I., Kajiyama, T., Takayanagi, M., and MacKnight, W.J., 1985, Microphase separated structure, surface composition and blood compatibility of segmented poly(urethaneureas) with various soft segment components. *Polymer* 26: 987-996.

15. Takahara, A., Okkema, A.Z., Wabers, H., and Cooper, S.L., 1991, Effect of hydrophilic soft segment side chains on the surface properties and blood compatibility of segmented poly(urethaneureas). *J Biomed Mater Res*, 25: 1095-1118.

16. Lelah, M.D., Grasel, T.G., Pierce, J.A., and Cooper, S.L., 1986, Ex vivo interactions and surface property relationships of polyetherurethanes. *J Biomed. Mater. Res.*, 20: 433-468.

17. Hasirci, N., 1993, Synthesis of Polyurethane Elastomers for Biomedical Use. In *Ohio Science Workbook- Polymers* (M.R.Steward, ed.) The Ohio Academy of Sci., Columbus, Ohio, USA, pp. 38-41.

18. Boretos, J.W., and Pierce, W.S., 1968, A Polyether Polymer. *J. Biomed. Mat. Res.* 2: 121.

19. Boretos, J.W., 1973, *Concise Guide to Biomedical Polymers, Their Design Fabrication and Molding,* Charles C. Thomas-Publisher, Springfield, IL, pp.10.

20. Lyman, D.J., Loo, B.H., 1967, New Synthetic Membranes for Dialysis IV- Copolyether Urethane Membrane Systems. *J.Biomed.Mat.Res.*, 1: 17-26.

21. Nylias, E., 1970, Develpoment of Blood Compatible Elastomers Theory, Practice and In-Vivo Performance. *23rd Conference on Engineering in Medicine and Biology*, 12: 147.

22. Lelah, M.D., Lambrecht, L.K., Young, B.R., and Cooper, S.L., 1983, Physicochemical characterization and in vivo blood tolerability of cast and extruded Biomer. *J. Biomed. Mater. Res.*, 17: 1-22.

23. Hanson, S.R., Harker, L.A., Ratner, B.D., and Hoffman, A.S., 1980, In vivo evaluation of artificial surfaces with a nonhuman primate model of arterial thrombosis, *J. Lab. Clin. Med.*, 95(2): 289-296.

24. Szycher, M., and Poirier, V.L., 1984, Polyurethanes in Implantable Devices. *Plastic Technology*, pp.45.

25. Bouchemal, K., Briançon, S., Perrier, E., Fessi, H., Bonnet, I., and Zydowicz, N., 2004, Synthesis and characterization of polyurethane and poly(ether urethane) nanocapsules using a new technique of interfacial polycondensation combined to spontaneous emulsification, *Int. J. Pharmaceutics*, 269(1):89-100

26. Buma, P., Ramrattan, N.N., van Tienen T.G. and Veth R.P.H., 2004, Tissue engineering of the meniscus, *Biomaterials*, 25(9), 1523-1532.

27. Shukla, P.G., Kalidhass, B., Shah, A., and Palaskar, D.V., 2002, Preparation and characterization of microcapsules of water soluble pesticide monocrotophos using polyurethane as carrier material. *J Microencapsulation*, 19(3): 293-304.

28. Szycher, M., and Poirier, V.L., 1984, Polyurethanes in Implantable Devices. *Plastic Technology*, pp.45.

29. Huang, J.C., and Jennings, E.M., 2004, The effect of temperature on controlled release of heparin from polyurethane and ethylene vinyl acetate copolymer. *Int. J. Polym. Mater.*, 53: 69-78.
30. Han, D.K., Park, K.D., Ahn, K., Jeong, S.Y., and Kim, Y.H., 1989, Preparation and surface characterization of PEO-grafted and heparin-immobilized polyurethanes. *J. Biomed. Mater. Res.*, 23: 87-104.
31. Weerwind, P.W., van der Veen, F.H., Lindhout, T., de Jong, D.S., and Calahan, F.T., 1998, Ex vivo testing of heparin-coated extracorporeal circuits: Bovine experiments. *Int. J. Artif. Organs*, 21: 291-298.
32. Marois, Y., Chakfe, N., Guidoin, R., Duhamel, R.C., Roy, R., Marois, M., King, M.W., and Douville, Y., 1996, An albumin-coated polyester arterial graft: in vivo assessment of biocompatibility and healing characteristics. *Biomaterials*, 17: 3-14.
33. DeQueiroz, A.A., Barrak, E.R., Gil, H.A., and Higa, O.Z., 1997, Surface studies of albumin immobilized onto PE and PVC films. *J.Biomater Sci Polym Ed*, 8: 667-681.
34. Seiferd, B., Romaniuk, P., and Groth, T., 1997, Covalent immobilization of hirudin improves the haemocompatibility of polylactide-polyglycokide in vitro. *Biomaterials*, 18: 1495-1502.
35. Bos, G.W., Scharenborg, N.M., Poot, A.A., Engbers, G.H., Beugeling, T., van Aken, W.G., and Feijen, J., 1999, Blood compatibility of surfaces with immobilized albumin-heparin conjugate and effect of endothelial cell seeding on platelet adhesion. *J. Biomed. Mater. Res.*, 47: 279-291.
36. Yoda, R., 1998, Elastomers for biomedical applications, *J. Biomater. Sci. Polym. Ed.*, 9: 561-626.
37. Pasic, M., Muller Glauser, W., von Segesser, L., Odermatt, B., Lachat, M., and Turina, M., 1996, Endothelial cell seeding improves patency of synthetic vascular grafts: manual versus automatized method. *Eur. J. Cardio Thorac Surg.*, 10: 372-379.
38. Ryu, G.H., Han, D.K., Park, S., Kim, M., Kim, Y.H., and Min, B., 1995, Surface characteristics and properties of lumbrokinase immobilized polyurethane. *J.Biomed Mater Res*, 29: 403-409.
39. Ratner, B.D., 1998, Molecular design strategies for biomaterials that heal. *Macromol Symposia*, 130: 327-335.
40. Morra, M., Occhiello, E., and Garbassi, F., 1993, Surface modification of blood contacting polymers by poly(ethyleneoxide). *Clin Mater* 14: 255-265.
41. Han, D.K., Jeong, S.Y., and Kim, Y.H., 1989, Evaluation of blood compatibility of PEO grafted and heparin immobilized polyurethanes, *J.Biomed Mater Res: Appl Biomater*, 23 (A2): 211-228.
42. Wang, D.A., Ji, J., Gao, C.Y., Yu, G.H., and Feng, L.X., 2001, Surface coating of stearyl poly(ethylene oxide) coupling-polymer on polyurethane guiding catheters with poly(ether urethane) film-building additive for biomedical applications. *Biomaterials*, 22: 1549-1562.
43. Chen, K.Y., Kuo, J.F., and Chen, C.Y., 2000, Synthesis characterization and platelet adhesion studies of novel ion-containing aliphatic polyurethanes. *Biomaterials*, 21: 161-171.
44. Ito, Y., Iguchi, Y., Kashiwagi, T., and Imanishi, Y., 1991, Synthesis and nonthrombogeneity of polyetherurethaneurea film grafted with poly(sodium vinyl sulfanate). *J. Biomed. Mater. Res.*, 25: 1347-.1361.
45. Okkema, A.Z., Yu, X.H., and Cooper, S.L., 1991, Physical and blood contacting characteristics of propyl sulfonate grafted Biomer. *Biomaterials*, 12: 3-12.

46. Okkema, A.Z., and Cooper, S.L., 1991, Effect of carboxylated and/or sulfanate ion incorparation on the physical and blood-contacting properties of a polyetherurethane. *Biomaterials*, 12: 668-676.

47. Lee, J.H., Khang, G., Lee, J.W., and Lee, H.B., 1998, Platelet adhesion onto chargeable functional group gradient surfaces. *J. Biomed. Mater. Res.*, 40: 180-186.

48. Abraham, G.A., Queiroz, A.A., Roman, J.S., 2002, Immobilization of a nonsteroiddal antiinflammatory drug onto commercial segmented polyurethane surface to improve haemocompatibility properties. *Biomaterials*, 23: 1625-1638.

49. Morimoto, N., Iwasaki, Y., Nakabayashi, N., and Ishihara, K., 2002, Physical properties and blood compatibility of surface-modified segmented polyurethane by semi-interpenetrating polymer networks with a phospholipid polymer. *Biomaterials*, 23(24): 4881-4887.

50. Ishihara, K., Fujita, H., Yoneyama, T., and Iwasaki, Y., 2000, Antithrombogenic polymer alloy composed of 2-methacryloyloxyethyl phosphorylcholine polymer and segmented polyurethane. *J. Biomater. Sci. Polym. Edn.*, 11(11): 1183-1195.

51. Yoneyama, T., Sugihara, K., Ishiara, K., Iwasaki, Y., and Nakabayashi, N., 2002, The vascular prosthesis without pseudointima prepared by antithrombogenic phospholipid polymer. *Biomaterials*, 23: 1455-1459.

52. Korematsu, A., Takemoto, Y., Nakaya, T., and Inoue, H., 2002, Synthesis, characterization and platelet adhesion of segmented polyurethanes grafted phospholipid analogous vinyl monomer on surface. *Biomaterials*, 23: 263-271.

53. Zhang, J., Yuan, J., Yuan, Y., Zang, X., Shen, J., and Lin, S., 2003, Platelet adhesive resistance of segmented polyurethane film surface-grafted with vinyl benzyl sulfo monomer of ammonium zwitterions. *Biomaterials*, 24: 4223-4231.

54. Smith, D.J., Chakravarthy, D., Pulfer, S., Simmons, M.L., Hrabie, J.A., Citro, M.L., Saavedra, J.E., Davies, K.M., Hutsell, T.C., Mooradian, D.L., Hanson, S.R., and Keefer, L.K., 1996, Nitric oxide releasing polymers containing [N(O)NO]⁻ group. *J. Med. Chem.*, 39: 1148-1156.

55. Movery, K.A., Schoenfisch, M.H., Saavedra, J.E., Keefer, L.K., and Meyerhoff, M.E., 2000, Preparation and characterization of hydrophobic polymeric films that are thromboresistant via nitric oxide release. *Biomaterials*, 21: 9-21.

56. Duan, X., and Lewis, R.S., 2002, Improved haemocompatibility of cystein-modified polymers via endogenous nitric oxide. *Biomaterials*, 23: 1197-1203.

57. Logeart-Avramoglou, D., and Jozefonvicz, J., 1999, Carboxymethyl benzylamide sulfonate dextrans [CMDBS] , a family of biospecific polymers endowed with numerous biological properties: a review. *J. Biomed. Mater. Res. Appl. Biomater.*, 48(4):578-590.

58. Roman, S.J., Bujan, J., Bellon, J.M., Gallardo, A., Escudero, M.C., Jorge, E., Haro, J., Alvarez, L., Castillo, J.L., 1996, Experimental study of the antithrombogenic behavior of Dacron vascular grafts coated with hydrophilic acrylic copolymers bearing salicilic acid residues. *J. Biomed. Mater. Res.*, 32: 19-27.

59. Zhu, Y., Gao, C., He, T., and Shen J., 2004, Endothelium regeneration on luminal surface of polyurethane vascular scaffold medified with diamine and covalently grafted with gelatin, *Biomaterials*, 25:423-430.

60. Wang, Z.F., Wang, B., Yang, Y.R. and Hu C.P., 2003, Correlations between gas permeation and free-volume hole properties of polyurethane membranes, European Polymer J., 39(12):2345-2349.

61. Liu, Q., Runt, J., Felder, G., Rosenberg, G., Snyder, A.J., Weiss, W.J., Lewis, J., and Werley, T., 2000, In vivo and in vitro stability of modified poly(urethaneurea) blood sacs. *J. Biomat. Appl.*,14 (4): 349-366.

62. Weisberg, D.M., Gordon, B., Rosenberg, G., Snyder, A.J., Benesi, A., Runt, J., 2000, Synthesis and characterization of amphiphilic poly(urethaneurea)-comb-polyisobutylene copolymers. *Macromolecules*, 33 (12): 4380-4389.

63. Xu, R.J., Manias, E., Snyder, A.J., Runt, J., 2001, New biomedical poly(urethane urea) - Layered silicate nanocomposites. *Macromolecules*, 34 (2): 337-339.

64. Ulubayram, K., and Hasirci, N., 1992, Polyurethanes: Effect of Chemical Composition on Mechanical Properties and Oxygen Permeability. *Polymer*, 33(10): 2084-2088.

65. Park, H.B., Kim C.K., and Lee Y.M., Gas separation properties of polysiloxane/polyether mixed soft segment urethane urea membranes, *J. Membrane Science*, 204: 257-269.

66. Zdrahala, R.J., and Zdrahala, I.J., 1999, Biomedical applications of polyurethanes: a review of past promises, present realities, and a vibrant future. *J. Biomater. Appl.*, 14:67-90.

67- Kim, Y.H., Han, D.K., Park, D.K., and Kim, S.H., 2003, Enhanced blood compatibility of polymers grafted by sulfonated PEO via a negative cilia concept, *Biomaterials*, 24(13):2213-2223.

68. van Blitterswijk, C.A., van der Brink J., Leenders, H., Hessling, S.C., and Bakker, D., 1991, Polyactive: a bone bonding polymer effect of PEO/PBT proportion, *Trans Soc Biomater*, 14:11

69. Tang, Z.G., Teoh, S.H., McFarlane, W., Poole-Warren, L.A., and Umezu, M., 2002, In vitro calcification of UHMWPE/PU composite membrane, *Materials Sci. Eng.*, C-20:149-152.

70. Tang, Z.G., Teoh, S.H., McFarlane, W., Warren, L.P., and Umezu, M., 2003, Compression- induced changes on physical structures and calcification of the aromatic polyether polyurethane composite. *J. Biomater. Sci. Polym. Ed.*, 14(10): 1117-1133.

71. Yang, M., Zhang, Z., Hahn, C., King, M.W., and Guidoin, R., 1999, Assessing the resistance to calcification of polyurethane membranes used in the manufacture of ventricles for a totally implantable artificial heart. *J. Biomed. Mater. Res.*, 48(5): 648-659.

72. Miao, X., Hu, Y., Liu, J., and Wong, A.P., Porous calcium phosphate ceramics prepared by coating polyurethane foams with calcium phosphate cements, *Materials Letters*, 58:397-402.

73. Hasirci, N., 1987, Surface modification of charcoal by glow-discharge: the effect on blood cells. *J. Appl. Polym. Sci.*, 34: 2457-2468.

74. Kayirhan, N., Denizli, A., Hasirci, N., 2001, Adsorption of Blood Proteins on Glow-discharge Modified Polyurethane Membranes. *J. Appl. Polym. Sci.*, 81: 1322-1332.

75. Hasirci, N., and Burke, A., 1987, A novel polyurethane film for biomedical use. *J. Bioactive and Compatible Polymers*, 2: 131-141.

76. Burke, A., Hasirci, V.N.and Hasirci, N., 1988, Polyurethane Membranes, *J. Bioactive and Compatible Polymers*, 3: 232-242.

77. Ulubayram, K., and Hasirci, N., 1991, Polyurethanes: Chemistry and Properties, *Procedings of Second Mediterranean School on Science and Technology of Advanced Polymer -Based Materials* , pp: 1-4.

78. Ulubayram, K., and Hasirci, N., 1993, Properties of plasma modified polyurethane surfaces. *J. Colloids and Surfaces: Biointeractions*, 1: 261-269.

79. Ozdemir, Y., and Hasirci, N., 2002, Surface modification of polyurethane membranes. *Technology and Health Care*, 10: 316-319.

80. Ozdemir, Y., Serbetci K., and Hasirci, N., 2002, Oxygen Plasma Modification of Polyurethane Membranes. *J. Mat. Sci, Materials in Medicine, JMS: MIM*, 13: 1147-1152.

Chemical Durability of Alumina and Selected Glasses in Simulated Body Fluid

Effect of Composition and Surface Abrasion

MURAT BENGISU[*] and ELVAN YILMAZ[**]

*Department of Industrial Engineering, Eastern Mediterannean University, Famagusta TRNC via Mersin 10, Turkey; **Department of Chemistry, Eastern Mediterranean University, Famagusta TRNC via Mersin 10, Turkey*

1. INTRODUCTION

Biocompatible ceramics, glasses, and glass-ceramics are gaining increased importance in biotechnological applications. Ceramic materials can be bioinert, bioactive, bioresorbable, and biotoxic[1]. The chemical durability of bioceramics is of concern for many biotechnological applications such as bone implants, bone defect fillings, and drug delivery. Crystalline ceramics such as Al_2O_3 and partially stabilized ZrO_2 (PSZ) and various glasses based on SiO_2 have been used for a range of biomedical applications. Al_2O_3 and PSZ are known to be nearly bioinert materials. Examples of their use include dental implants, ball and socket hip replacements, knee prostheses, bone screws, and segmental bone replacements[2]. These materials can be used in the dense form where bone attachment is of mechanical nature, for example by bone growth into surface irregularities. A second possibility is the use of porous inert implants where bone can grow into porosities and a better mechanical attachment occurs. Bioactive glasses join the bone by chemical attachment, which is usually stronger than mechanical attachment. It is commonly accepted that the prerequisite for glasses and glass-ceramics to chemically attach to bone is the formation of a hydroxycarbonate apatite (HCA) layer[3]. 45S5 Bioglass® is the first commercialized glass with such features. This SiO_2-based bioglass, whose composition is shown in Table 1, is well studied and has been in use as implant material for various bone parts and filler material for dental applications[2,4]. S53P4 is a bioactive glass with similar composition to 45S5 (Table 1). This glass has been used for clinical trials with various target applications such as interpositional graft in the repair of septal perforations, filling material in the treatment of frontal sinusitis, and bone-graft material for spinal surgery[5-7].

A borate-based glass was also proposed for biomedical applications, especially as a coating on orthopaedic titanium alloys[8] since borate-based glasses were shown to bond more favourably to titanium than silica-based glasses[9].

The aim of the present study is to compare the in vitro behaviour of alternative bioglasses in simulated body fluid (SBF) with each other and with reference materials and assess the effect of surface abrasion on the dissolution rate. Such comparative data is not widely available although it is valuable for choosing the most suitable material for a specific application. The effect of surface abrasion on the dissolution rate is interesting both for modelling the corrosion behaviour and to simulate the effect of possible wear on the glass surface to the service in vivo, for example in hip joints or dental applications.

Table 1 . Nominal chemical compositions and biocompatibility of studied materials

Oxide	Al_2O_3	Microscope slide glass	45S5	S53P4	H6b
	Amount present (mol%)				
SiO_2	-	72,2	46,13	53,85	6,5
Na_2O	-	14,3	24,35	22,65	15,0
CaO	-	6,4	26,91	21,78	35,0
K_2O	-	1,2	-	-	-
B_2O_3	-	-	-	-	41,5
P_2O_5	-	-	2,60	1,72	1,0
Al_2O_3	99	1,2	-	-	1,0
MgO	-	4,3	-	-	-
Other	1	0,4	-	-	-
Biocompatibility	Bioinert	Bioinert	Bioactive	Bioactive	Bioresorbable [i]

[i] Present study

2.   EXPERIMENTAL

2.1   Materials and Sample Preparation

Ten mm long pieces were cut from Al_2O_3 bars with 6.25 mm wide square cross-sectioned bars (McMaster-Carr, USA). Microscope glass slides (Isolab, Germany) were cut to smaller pieces of $26 \times 10 \times 1$ mm^3 while some were used in the as-received conditions and dimensions.

Bioactive glass batches were prepared by mixing suitable amounts of oxide powders or corresponding sources that yield the desired amount of oxides when heat-treated. After mixing, glass samples were prepared by heating the batches in alumina crucibles for an hour at 1350°C in the case of 45S5 and S53P4 compositions and at 1300°C in the case of H6b. Molten glass was poured into preheated graphite molds. These were then annealed at 500°C (which is below the crystallization temperature) for an hour and cut to smaller pieces. All cutting was performed using a diamond wafering blade (Metkon, Bursa, Turkey) in mineral oil. All samples were abraded dry using 320 and 600 grit SiC paper. Four samples per composition were prepared in the case of bioactive glasses and six samples per composition were prepared from the remaining materials. Half the number of each set of samples was annealed at 500°C for two hours. The remaining samples were exposed to SBF without annealing.

2.2   Chemical Durability Experiments

SBF (500 mL) was prepared by mixing 0.153 g $MgCl_2$, 3.939 g NaCl, 0.187 g KCl, 0.139 g $CaCl_2$, 0.071 g Na_2HPO_4, 0.177 g $NaHCO_3$, 0.0355 g Na_2SO_4, and 3.0275 tris(hydroxyl methyl) aminomethane. Solution pH was adjusted to pH 7.1 by adding 0.1 M HCl. Samples were immersed in 400 mL SBF in pyrex tubes with caps. The test tubes were placed in an oven with a stable temperature of 37°C. Samples were weighed at various intervals up to 52 days and the pH of SBF in each tube was measured with a digital pH-meter (Philips PW9421). Before weighing, samples were visually analyzed to monitor any macroscopic changes on the surface and then carefully dried using soft tissue paper. A digital balance (Ohaus, USA) with a precision of 1×10^{-5} g was employed to track weight changes.

2.3   Infrared Spectroscopy

Fourier Transform Infrared Spectroscopy (FTIR) was used to monitor chemical changes at the glass surfaces. KBr-pelletized powder samples and a

Mattson Satellite 5000 FTIR Spectrophotometer device were used. Two types of samples were prepared for FTIR analysis. One type involved samples actually employed for the chemical durability experiments, removed from the oven but kept in their test solution, for a total of 120 days. The surface layers were scraped off from such samples. Another set of samples was prepared specifically for FTIR studies by breaking and grinding the glasses into powder form and keeping them in SBF for various durations.

3.   RESULTS AND DISCUSSION

The specific weight reduction, calculated by dividing the weight reduction (as compared to the initial weight) to the whole part surface in contact with the liquid medium, as a function of time, is shown in Figure 1. Al_2O_3 was used as a reference material in this study due to its nearly inert behaviour as a bioceramic[2]. Similarly, microscope slide glass is a reference material since it is not bioactive, as opposed to the remaining glass types used in this study. As expected, Al_2O_3 did not lose any weight (Fig.1) within the precision level of the instrument used. Negligible pH fluctuations were observed in the test tubes containing alumina samples which can be attributed to some ion exchange from the test tubes. No significant difference was observed in terms of weight change or pH values between abraded and abraded + annealed alumina samples.

Microscope slide glass was tested under three conditions. Some of these glasses were in the as-received condition. This is a condition where the glass surface is polished to a low surface irregularity in order to provide suitable transparency. Abraded samples had a higher irregularity than the as-received samples. Figure 2a shows the difference in the specific weight reduction of these samples and Figure 2b shows the corresponding pH values. Note that these glasses are also very resistant to corrosion and nearly bioinert (see Fig.1 for comparison with other glasses). The higher corrosion rate of abraded samples compared to as-received and abraded + annealed ones can be attributed to the increased surface area by the irregularities and the increased number of surface defects (scratches and microcracks) as a direct result of the abrasion process.

pH values increased with increasing glass dissolution. This can be explained by ion exchange between Na^+, K^+, and Ca^{2+} from the glass and H^+ or H_3O^+ from the solution[2,4]. 45S5 glass exhibited one order of magnitude faster dissolution compared to microscope slide glass (Figs.1 and 3a). 45S5 is a well-known bioactive glass that forms HCA in SBF. In these glasses, abraded samples demonstrated slower weight reduction than abraded + annealed samples (Fig. 3a). This behaviour is in contrast to microscope slide glass. The slower weight reduction in the abraded samples may be attributed to faster HCA layer formation than annealed samples. pH values showed a good agreement with weight reduction (Fig. 3b).

S53P4 samples initially lost some weight but both abraded and abraded + annealed samples started to gain weight after 4 days of immersion in SBF (Fig. 4a). Again, pH values agreed well with weight changes (Fig. 4b).

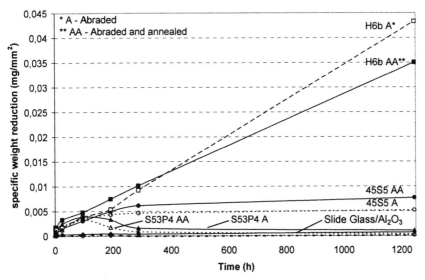

Figure 1. Specific weight reduction of various glasses in simulated body fluid (SBF) Al₂O₃ (as a reference material) with time.

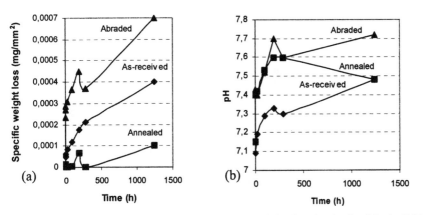

Figure 2. (a) Specific weight reduction of microscope slide glass in simulated body fluid (SBF) with time and (b) corresponding pH values.

In both of 45S5 and S53P4 glasses, a SiO_2-rich amorphous layer depleted in alkali and alkali-earth ions forms on the surface. Migration of Ca^{2+} and PO_3^{-} ions to the surface leads to a CaO-P_2O_5-rich film on top of the silica-rich layer which subsequently crystallizes and forms the HCA layer[2,4].

The specific weight reduction and corresponding pH change involving H6b glasses are shown in Figures 5a and 5b. H6b glass, as opposed to 45S5 and S53P4, has a borate-based composition and therefore a different glass structure. The fastest dissolution rate among the materials studied occurred in H6b samples. This can be attributed to the presence of B_2O_3 in the composition. Trigonal B atoms are very electrophilic and readily attacked by water. However, tetrahedral B atoms are coordinatively saturated and kinetically stable towards hydrolysis or dissolution in aqueous solutions, as is the case for borosilicate glasses. A nearly constant dissolution rate of 5×10^{-7} mg/(mm^2 min) was observed in H6b regardless of the annealing treatment. The pH increases almost linearly like the dissolution behaviour (Fig. 5b). Samples which were abraded exhibited higher specific weight reduction compared to abraded + annealed ones. The same argument used for slide glass is also valid here. H6b was shown to convert to hydroxyapatite during incubation in SBF[8]. H6b exhibits a linear dissolution behaviour with a dissolution rate two orders of magnitude higher than slide glass and approximately one order of magnitude higher than 45S5. Since H6b glass had an average density of 2.36 g/mL, it is calculated that it would take about 33 days for a coating with a 10 μm thickness to dissolve completely. Based on these dissolution properties, it can be argued that H6b is a bioresorbable glass. This behaviour may be preferred for certain applications over the bioactive behaviour of 45S5 and S53P4 glasses.

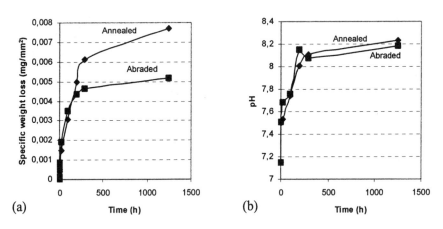

Figure 3. (a) Specific weight reduction of 45S5 glass in SBF with time and (b) corresponding pH values.

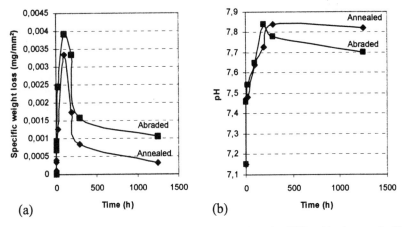

Figure 4. (a) Specific weight reduction of S53P4 glass in SBF with time and (b) corresponding pH values.

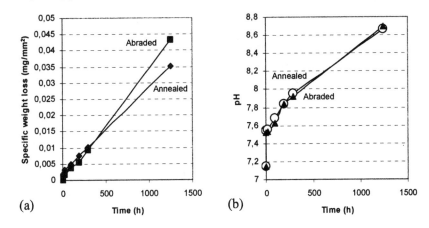

Figure 5. (a) Specific weight reduction of H6b glass in SBF with time and (b) corresponding pH values.

FTIR spectra of the silica rich glass powders studied, namely microscope slide glass, 45S5 and S53P4, bear common characteristic features. Each spectrum (Figs. 6a and f) exhibits a broad absorption band at 3420-3440 cm^{-1} due to the stretching vibrations of [O-H] groups, a sharp band at 1620-1640 cm^{-1} representing molecular water, a strong, broad band at 1030-1040 cm^{-1} with a shoulder at 930-940 cm^{-1} due to the antisymmetric stretching of bridging oxygens of [Si-O-Si] and vibrations of nonbridging oxygens respectively. The bands in the 730-760 cm^{-1} are due to symmetric stretching of [Si-O-Si] bonds. Bending vibration of [Si-O-Si] bonds are observed at 640-690 cm^{-1} region.

The effect of corrosion on the glass powders after 24 days of contact with SBF, can be followed from the FTIR spectra (Figs. 6b, d, g). 45S5 and S53P4 behave similarly. The [O-H] stretching vibrations in the 3420 – 3440 cm^{-1} region become stronger indicating the formation of [Si-OH] bonds. Leaching of the modifier ions and break up of the silica network result in a sharper [Si-O-Si] absorption which has been shifted to longer wave numbers, 1074 and 1089 cm^{-1}, in 45S5 and S53P4 respectively. A shoulder around 1100 cm^{-1} is now clearly visible. Migration of carbonate and phosphate ions from the solution onto the glass surface is indicated by the appearance of two new peaks at 1410 – 1420 cm^{-1} and 603 cm^{-1} due to ionic carbonate and phosphate absorptions respectively (Fig. 6g). No such changes could be observed in the spectrum of the microscope slide glass surface since this glass is bioinert (Fig. 6b).

The FTIR spectra of the layers scraped from the surfaces of the glass samples 45S5 (Fig. 6h) and S53P4 after 120 days of immersion in SBF are almost identical with each other. They are characterized by a very strong, sharp, intense absorption band at 1039 cm^{-1} of phosphate and silica absorption band. Two sharp bands at 603 and 566 cm^{-1} emerge, belonging to the phosphate groups. Absorptions in the 1500 –1400 cm^{-1} region and 872 cm^{-1} are due to the presence of the carbonate ion. It can be concluded that migration of these ions onto the glass surface occurs to form a carbonated HCA layer.

Borate-rich H6b glass exhibits characteristic [B-O] stretching bands belonging to BO_3 and BO_4 groups at 1404 and 1025 cm^{-1} respectively (Fig. 6c). Upon exposure to SBF, the [O-H] absorption band at 3420-3440 cm^{-1} shows a remarkable increase in absorption and a new band at 1635 cm^{-1} appears (Fig. 6d) indicating the inclusion of water molecules in the structure, during the ion exchange process. The FTIR spectrum of the layer scraped from the surface of the H6b glass sample that was immersed in SBF solution for 120 days, exhibits strong [O-H] bands at 3440 and 1653 cm^{-1} due to hydration. Absorption bands at 1436 with a shoulder at 1318 cm^{-1} and the band at 1028 cm^{-1} of the borates are available (Fig.6d and e). A carbonate peak at 873 cm^{-1} and a phosphate peak at 603 cm^{-1} have emerged after SBF contact, indicating the initiation of a HCA layer. Other peaks belonging to carbonate and phosphate peaks in the 1300-1500 cm^{-1} region might have been masked by the [B-O] stretching.

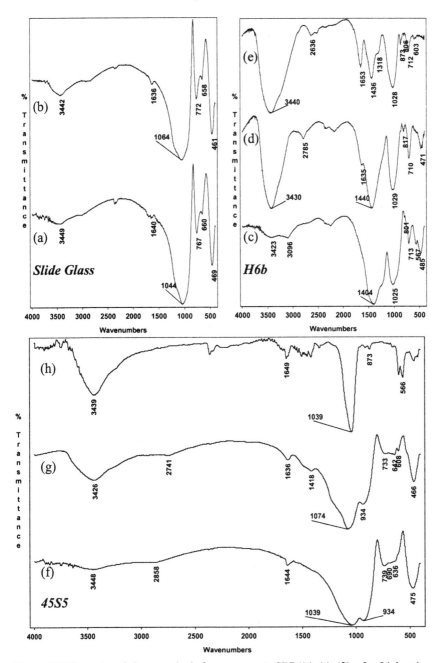

Figure 6. FTIR spectra of glass samples before exposure to SBF ((a), (c), (f)); after 24 days in SBF in powder form ((b), (d), (g)), and after 120 days, from the surface layer ((e) and (h)).

4.   CONCLUSIONS

The following conclusions are deduced from the present study.
1. Annealing of abraded slide glass and H6b glass reduces the dissolution rate in SBF.
2. Borate-based H6b glass corrodes linearly and at a faster rate than silicate-based 45S5 and S53P4 glasses. Based on the calculated dissolution rate, H6b glass exhibits bioresorbable behaviour.
3. S53P4 exhibited weight gain in SBF after 8 days of exposure while other glass types did not gain any weight during the course of the experiments.
4. According to FTIR studies, carbonate and phosphate ions migrate from solution onto the surfaces of all glass types studied except for slide glass, when exposed to SBF.

REFERENCES

1.   Dubok, V.A., 2000, Bioceramics-Yesterday, Today, Tomorrow. *Powder Metallurgy and Metal Ceramics*, 39: 381-393.
2.   Hench, L.L., 1991, Bioceramics: From Concept to Clinic. *J. Am. Ceram. Soc.* 74: 1487-1510.
3.   Peltola, T., Jokinen, M., Rahiala, H., Levanen, E., Rosenhol, J.B., Kangasniemi, I., and Yli-Urpo, A., 1999, Calcium Phosphate Formation on Porous Sol-Gel Derived SiO_2 and $CaO-P_2O_5-SiO_2$ Substrates in Vitro. *J. Biomed. Mater. Res.* 44: 12-21.
4.   Jones, J.R., Sepulveda, P., and Hench, L.L., 2001, Dose-Dependent Behavior of Bioactive Glass Dissolution. *J. Biomed. Mater. Res. (Appl. Biomater.)* 58: 720-726
5.   Peltola, M.J., Suonpaa, J.T.K., Andersson, O.H., Maattanen, H.S., Aitasalo, K.M.J., Yli-Urpo, A., and Laippala, P.J., 2000, In-Vitro Model for Frontal Sinus Obliteration with Bioactive Glass S53P4. *J. Biomed. Mater. Res. (Appl. Biomater.)* 53: 161-166
6.   Lindfors, N.C., Tallroth, K., and Aho, A.J., 2002, Bioactive Glass as Bone-Graft Substitute for Posterior Spinal Fusion in Rabbit. *J. Biomed. Mater. Res. (Appl. Biomater.)* 63: 237-244
7.   Stoor, P., Soderling, E., and Grenman, R., 2001, Bioactive Glass S53P4 in Repair of Septal Perforations and its Interactions with the Respiratory Infection-Associated Microorganisms Heamophilus Influenza and Streptococcus Pneumaniae. *J. Biomed. Mater. Res. (Appl. Biomater.)* 58: 113-120
8.   Brown, R.F., Teitelbaum, H.K., Adams, N., and Brow, R.K., April 2002, In Vitro Assessment of a Novel Borate-Based Bioactive Glass. MRS Spring Meeting, San Fransisco CA.
9.   Brow, R.K., Saha, S.K., and Goldstein, J.I., 1993, Interfacial Reactions Between Titanium and Borate Glass. *Proc. Mat. Res. Soc.* 314: 77-81

Degradable Phosphazene Polymers and Blends for Biomedical Applications

MARIO CARENZA*, SILVANO LORA*, and LUCA FAMBRI#
*Istituto per la Sintesi Organica e la Fotoreattività, CNR, v.le dell'Università, 2- 35020
Legnaro-Padova, Italy; #Dipartimento di Ingegneria dei Materiali e Tecnologie Industriali,
Università di Trento, via Mesiano, Trento, Italy

1. INTRODUCTION

In the biomaterials science, bioresorbable polymeric materials are of great importance in the case of short-term applications that need a temporary presence of implant. It has been known for many years that polymer scaffolds are useful materials for initial cell attachment and subsequent tissue formation both in vitro and in vivo. A number of requirements should be met for a proper use of polymer scaffolds, such as biodegradability with a controllable degradation rate, structure, porosity and surface properties to allow cells to be seeded for a successful growth and differentiation. Mechanical properties comparable to those of the natural extracellular matrices are also desirable.

Aliphatic polyesters, such as poly(lactic acid), PLLA, poly(glycolic acid), PGA, and poly(lactide-co-glycolide), PLGA, represent an important class of biodegradable polymers approved by FDA for clinical use[1-3].

In recent years attention has been addressed to polyorganophosphazenes, POPs, which are inorganic polymers with alternating nitrogen and phosphorous atoms along the polymer chain and two organic side groups bonded to each phosphorous atom. POPs with amino acid esters or other organic groups as substituents are biocompatible and, therefore, can be an excellent

Biomaterials: From Molecules to Engineered Tissues, edited by
N. Hasırcı and V. Hasırcı, Kluwer Academic/Plenum Publishers, 2004

platform for tissue engineering applications due to their hydrolytic instability, non-toxic degradation products, ease of fabrication and matrix permeability[4, 5].

In this short survey, recent literature reports on the synthesis of POPs and degradation of their homo- and copolymers as well as blends of POPs and poly(α-hydroxyesters), including some preliminary results obtained in our laboratories, are presented.

2.   SYNTHESIS OF POLYPHOSPHAZENES

2.1   Methods of Synthesis

Even though there are several methods for the preparation of POPs[6, 7], the most popular one is based on ring opening polymerization at a temperature of 250°C in glass vials sealed under vacuum of the cyclic trimer hexaclorocyclotriphosphazene, $(NPCl_2)_3$, that yields the linear poly(dichlorophosphazene), $(NPCl_2)_n$. Since the latter polymer is very reactive due to fact that the polar P-Cl bond causes a high hydrolytic instability, replacement of chlorine atoms with nucleophilic substituents, such as alcohol, phenol, primary and secondary amine groups, gives rise to a wide variety of stable, high molecular weight POPs. The overall synthesis pathway for these polymers is summarized in the Scheme 1.

More than 700 different derivatives have been claimed to be synthesized and some of them can be used in biomedicine due to their biocompatibility[8]. In particular, POPs with amino acid ester or imidazolyl groups as substituents show hydrolytic instability giving rise to low molecular weight degradation products, such as alcohol, amino acid, phosphoric acid and ammonia.

2.2   Synthesis of Poly[bis(ethyl alanato) phosphazene]

We used the general method to prepare the poly[bis(ethyl alanato) phosphazene], PAlaP, from $(NPCl_2)_n$, L-alanine ethyl ester hydrochloride and triethylamine, as shown in the Scheme 2.

The reaction was carried out in toluene solution at room temperature and under nitrogen atmosphere. The insoluble chloride was removed by centrifugation and PAlaP was obtained by precipitation in n-heptane.

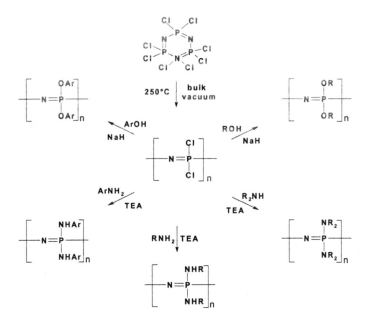

Scheme 1

$[NPCl_2]_n$ + 4n $N(CH_2CH_3)_3$ + 2n $CH_3CH(NH_2)CO_2CH_2CH_3 \cdot HCl$

       Triethylamine   L-alanine ethyl ester
                           hydrochloride

$[NP(NHCHCH_3CO_2CH_2CH_3)_2]_n$ + 4n $N(CH_2CH_3)_3 \cdot HCl \downarrow$

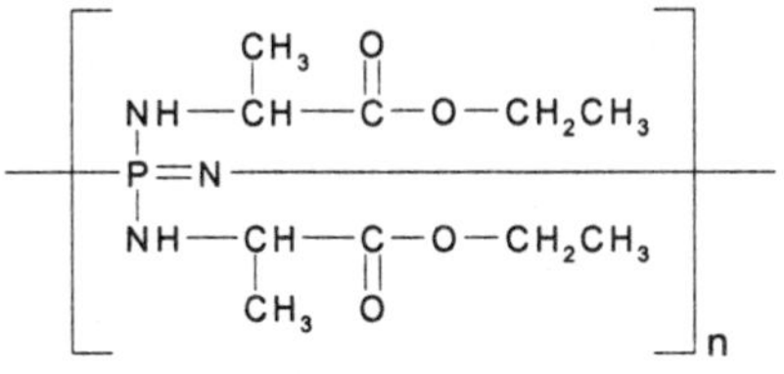

Poly[bis(ethyl alanato) phosphazene]

Scheme 2

    The elemental analysis (%) for C, H, N was 43.29, 7.19, and 15.06, respectively, while chlorine was found to be absent (% calculated values were 43.30, 7.20, 15.16 and 0.0, respectively). The intrinsic viscosity of PAlaP in THF at 25°C was: $[\eta] = 2.16$ dl/g.

3.   DEGRADATION OF POLYPHOSPHAZENES

3.1   Degradation of Polyphosphazene Homo- and Co-polymers

The main clinical application of biodegradable polyphosphazenes appears to be related to drug release[8]. Several degradation pathways of amino acid substituted POPs have been reported by Allcock *et al*[4]:

   a) Hydrolytic scission of the P-N bond of the amino acid group;

   b) Hydrolytic scission of the ester bond to yield carboxylic acid groups;

   c) Attack on the skeleton nitrogen by the carbonyl groups of the substituent units;

   d) Hydrolysis of the P-Cl bond to give P-OH, if Cl atoms are still present as residual of the polymer synthesis.

The proposed degradation mechanisms give rise, through the formation of a phosphazane intermediate, to the chain cleavage and the production of amino acid, phosphate, ammonia and alcohol. *In vitro* degradation studies demonstrated that the rate of degradation depends on both the nature of the substituents and the structure of amino acids. Moreover, the hydrolytic degradation rate can be controlled by the percent ratio of the side groups.

As an example, Crommen *et al*[9] studied polyphosphazenes having amino acid esters as substituents, *i.e.* glycine, $[NP(GlyOEt)_2]_n$, alanine, $[NP(AlaOEt)_2]_n$, and glycine + phenylalanine, $[NP(GlyOEt)_x(PheOEt)_y]_n$, with $x + y = 2$. As it can be seen in Fig. 1, the mass loss is almost the same for glycine and alanine-substituted phosphazenes. On the contrary, the introduction of phenylalanine side group gives rise to a much slower mass loss, probably due to the more hydrophobic nature of this derivative and the higher glass transition temperature of the copolymer [10].

Laurencin *et al*[5] reported the degradation rate of poly[(imidazolyl) (methylphenoxy)] phosphazene] (Pol. **2-4**) and poly[(ethyl glycinato) (methylphenoxy) phosphazene] (Pol. **5-8**), with methylphenoxy, imidazolyl and glycinato groups in different proportions, in a 0.1 M sodium phosphate buffer at 37°C and pH = 7.4. Figure 2 shows that the rate of degradation increases with increasing of imidazolyl (Fig. 2a) and glycinato (Fig. 2b) side groups. Again, this behaviour can be explained with the fact that the hydrophilic imidazolyl and glycinato groups increase the water uptake and hence the cleavage rate of the polymer chains.

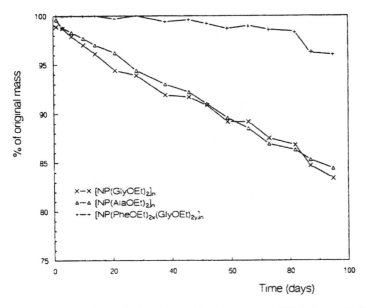

Figure 1. Percent mass loss of poly[amino acid ethyl ester] at 37°C (with permission from Elsevier, ref. 9).

3.2 Degradation of Polyphosphazene Blends

Polymers of lactic and glycolic acid and their copolymers are widely used in biomedical applications. However, the use of these polymers as scaffold materials for organ regeneration is limited in some cases because the products of their hydrolytic degradation at high concentration, i.e. lactic and glycolic acid, can be toxic for cells. Moreover, it was observed in vivo that acidity accumulation induces an inflammatory reaction or even tissue necrosis. Ibim *et al*[11] and Ambrosio *et al*[12] have designed biodegradable polyphosphazene/poly(α-hydroxyesther) blends whose degradation products are less acid than those of the poly(α–hydroxyesther) alone. In their studies, the degradation characteristics of the blends of poly(lactide-glycolide) (50:50 PLGA) and poly[(ethyl glycinato)$_x$(p-methylphenoxy)$_y$phophazene]s, with x + y = 2 and the ratio x:y = 1:3, 1:1 and 3:1, respectively, were qualitatively and quantitatively determined. The results indicate that blends degrade at an intermediate rate with respect to that of the parent polymers.

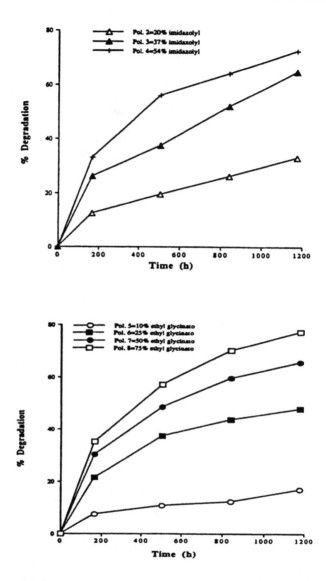

Figure 2. Percent degradation of imidazolyl-substituted polyphosphazenes (polymer 2-4) and ethyl glycinato-substituted polyphosphazenes (polymers 5-8) (with permission from Wiley, ref. 5).

Recently, Qiu and Zhu[13] carried out a thorough study of the rate of degradation rate of blends of poly[ethyl glycinato)phosphazene], PGP, with PLGA as reported in Fig. 3. The results show that the extent of mass lost as a function of time can be easily tuned by regulation of the PLGA content.

In our laboratories, blends of poly[bis(ethyl alanato) phosphazene], PAlaP, with some poly(α-hydroxyesters) were prepared in form of films by casting from solvent and assayed for degradation tests. At scheduled times up to 60 days, the water uptake and the mass loss of the films incubated at 37°C and pH = 7.4 were determined from the wet and dried samples, respectively.

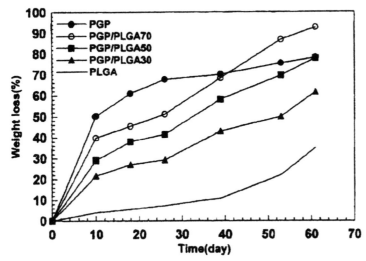

Figure 3. Percent weight loss of PGP, PLGA and blends PGP/PLGA at 37°C (with permission from Wiley, ref.13).

As an example, the mass change dependence as a function of time for PAlaP, PLGA 75-25 and the blend 50:50 (w/w) PAlaP/PLGA 75-25 are in the Figures 4, 5 and 6, respectively.

In the case of PAlaP (see Fig. 4), the polymer initially absorbs water up to a value of 4-5% and then undergoes the hydrolysis process. In the first 2-3 weeks the mass loss could be attributed to the release of small molecules such as alcohol and aminoacid derived from the lateral substituents with the formation of OH and COOH groups, according to the degradation pathways a) and b) reported above. The subsequent mass loss arises from the progressive fragmentation of the main polymer chain (pathway c).

Figure 5 shows that the absorption of water in the case of PLGA takes place quickly in the first 10 days, due to the high hydrophilicity and hydrolysability of the glycolic component, and the consequent formation of the hygroscopic OH and COOH groups.

On the contrary, the subsequent slow increase after 30 days can be due to the formation of the same small groups of the less hydrolysable lactic component. The loss mass regularly decreases in the whole period of experiment reaching the value of 15% after 60 days.

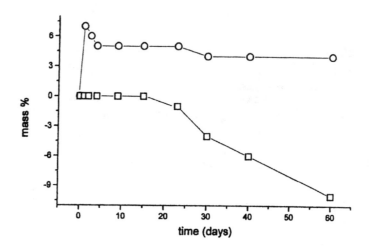

Figure 4. Water uptake (o) and mass loss (□) of PAlaP.

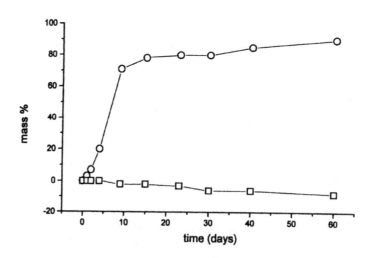

Figure 5. Water uptake (o) and mass loss (□) of PLGA 75-25.

Finally, when the blend PAlaP/PLGA is concerned (see Fig. 6), the water uptake in the whole interval of the experiment is comprised between those of PAlaP and PLGA since the glycolic component in the blend is halved with respect to the pristine PLGA. The mass loss after 60 days is about 12% with respect to the initial mass.

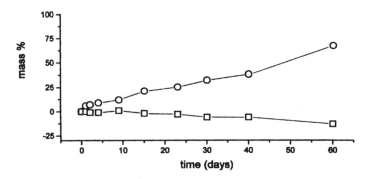

Figure 6. Water uptake (o) and mass loss (□) of blend PAlaP/PLGA 75-25 (50:50, w/w).

In summary, our experiments show that water uptake is in the order: PLGA > PAlaP/PLGA > PAlaP, while the loss mass is a little higher than that of the two parent polymers (9-10%).

4.   CONCLUSIONS

Polyphosphazenes having amino acid ester side groups are hydrolytically labile polymers that can degrade to non toxic products, such as alcohol, amino acid, phosphoric acid and ammonia. Their blends with poly(α-hydroxyesters) give rise to degradation kinetics and mass loss rate that can be easily modulated by a proper choice of the amino acid side groups and/or the composition of the blend. These materials are promising candidates for drug delivery systems as well as scaffolds in tissue engineering.

ACKNOWLEDGEMENTS

This work was supported by the Ministero dell'Istruzione, dell'Università e della Ricerca (PNR 2001-2003, FIRB Art. 8).

REFERENCES

1. Thomson, C., Wake, M.C., Yaszemski, M.J., and Mikos, A.G., 1995, Biodegradable polymer scaffolds to regenerate organs. *Adv. Polym. Sci.* 122: 245-274.
2. Kim, B.-S., and Mooney, D.J., 1998, Development of biocompatible synthetic extracellular matrices for tissue engineering. *Trends Biotechnol.* 16: 224-230.
3. Fambri, L., Migliaresi, C., Kesenci, K., Piskin, E., 2002, Biodegradabile Polymers. In *Integrated Biomaterials Science* (R. Barbucci, ed.), Kluwer Plenum, New York, pp. 119-187.
4. Allcock, H.R., Fuller, T.J., Mack, D.P., Matsumura, K., and Smeltz, K.M., 1977, Synthesis of poly[(amino acid alkyl ester)phosphazenes]. *Macromolecules* 10: 824-830.
5. Laurencin, C.T., Norman, M.E., Elgendy, H.M., El-Amin, S.F., Allcock, H.R., Pucher, S.R., and Ambrosio, A.A., 1993, Use of polyphosphazenes for skeletal tissue regeneration. *J. Biomed. Mater. Res.* 27: 963-973.
6. Neilson, R.H., and Wisian-Neilson, P., 1988, Poly(alkyl/arylphosphazenes) and their precursors. *Chem. Rev.* 88: 541-562.
7. Sulkowski, W.W., 2003, Some Aspects of Synthesis and Investigation of Poly(diorganophosphazene)s. In *Phosphazenes – A Worldwide Insight* (M. Gleria and R. DeJaeger, eds.), Nova Science, New York, pp. 69-106.
8. Lakshmi, S., Katti, D.S., Laurencin, C.T., 2003, Biodegradable polyphosphazenes for drug delivery applications. *Adv. Drug Deliv. Rev.* 55: 467-482.
9. Crommen, J., Vandorpe, J., and Schacht, E., 1993, Degradable polyphosphazenes for biomedical applications. *J. Controlled Release* 24: 167-180.
10. Crommen, J.H.L., Schacht, E.H., and Mense, E.H.G.,1992, Biodegradable polymers. I. Synthesis of hydrolysis-sensitive poly[(organo)phosphazenes]. *Biomaterials* 13: 511-520.
11. Ibim, S.E.M., Ambrosio, A.M.A., Kwon, M.S., El-Amin, S.F., Allcock, H.R., and Laurencin, C.T., 1997, Novel polyphosphazene/poly(lactide-*co*-glycolide) blends: miscibility and degradation studies. *Biomaterials* 18: 1565-1569.
12. Ambrosio, A.M.A., Allcock, H.R., Katti, D.S., and Laurencin, C.T., 2002, Degradable polyphosphazene/poly(α-hydroxyester) blends: degradation studies. *Biomaterials* 23: 1667-1672.
13. Qiu, L.Y., and Zhu, K.J., 2000, Novel blends of poly[bis(glycine ethyl ester)phosphazene] and polyesters or polyanhydrides: compatibility and degradation characteristics in vitro. *Polym. Int.* 49: 1283-1288.

Molecularly Imprinted Polymers

KEZBAN ULUBAYRAM
Department of Basic Pharmaceutical Sciences, Faculty of Pharmacy, Hacettepe University, Sıhhiye 06100 Ankara, TURKEY

1.   INTRODUCTION

Molecularly imprinted polymers (MIPs)[1-4] are highly stable polymeric molds that possess selective molecular recognition properties for various kinds of molecules. MIPs consist of highly crosslinked polymers that are synthesized in the presence of a template (imprint) molecule. After removal of template, a cavity is left, which retains affinity and selectivity for the template. Some of MIPs (such as antibody mimics), under optimized conditions, have high selectivities and affinity constants comparable with naturally occuring recognition systems such as monoclonal antibodies and receptors[2]. In addition, their unique stability is superior to that demonstrated by natural biomolecules; and they are robust and inexpensive. The simplicity of their preparation and the ease of adaptation to different practical applications make them very useful for chemical, pharmaceutical, and biotechnological industries.

The imprinted polymers have been used as stationary phases for chromatographic separation of molecules (e.g. enantiomeric or molecules with minor structural differences), as selective sorption elements for molecular and ionic separations, as recognition elements in sensors and immunoassays and as catalysts[2,3,5-10]. Molecular imprinting has enlarging applications in the field of biomaterials. They can be used for drug release matrices and alter the surface biocompatibility of medical devices[11].

The concept of molecular imprinting and applications of MIPs are introduced in this chapter.

Biomaterials: *From Molecules to Engineered Tissues,* edited by
N. Hasırcı and V. Hasırcı, Kluwer Academic/Plenum Publishers, 2004

2.   MOLECULAR IMPRINTING

The theories of molecular recognition have been considered as the origin of molecular imprinting. Molecular recognition is the fundamental property of many biological systems such as enzyme-substrate, antibody-antigen and hormone-receptor. In an enzyme or antibody (host), functional moieties of amino acid residues are precisely placed at the binding sites complementary to the target molecule (guest) and often bind them strongly through non-covalent interactions such as hydrogen bonds, ionic and hydrophobic interactions. The binding sites of the host and guest molecules complement each other in size, shape, and chemical functionality. Antibody-antigen interactions, as well as guest binding by membrane-receptor, occur essentially in the same manner. A number of amino acid residues of antibodies (or receptors) are oriented complementarily to the functional groups of the target antigens (or hormones). All these biological molecules show high selectivity and high binding activity toward the target guest molecule. Therefore, such biological recognition elements are essential for a wide range of bio-analytical procedures. However, a large portion of the biological molecules is protein and these molecules are unstable at high temperatures, in organic solvents, under serious pH conditions, etc. Many scientists started to mimic nature in order to develop synthetic recognition elements to reduce the complexity of biological systems and instability of biomolecules.

Cram, Lehn, and Pedersen (Nobel Laureates in 1987) established the basic principles of "accurate molecular recognition" in the 1960s and 1970s[12]. A multitude of synthetic or semisynthetic systems such as crown ethers[13], cyclodextrins[14], and cyclophanes[15] have been developed as recognition elements in analytical applications for separation and isolation processes. Yet, low yields in production, complicated organic synthesis, high cost and long times required for preparation of host compounds (synthetic recognition systems) compelled scientists to search for new approaches.

This new technology, called "Molecular Imprinting"[16-18] seemed to be promising to overcome all those difficulties. This technique leads to highly stable synthetic polymers that possess selective molecular recognition properties. They are known as "Molecularly Imprinted Polymers" because there are specific recognition sites within the polymer matrix that are complementary to the analyte in the shape and positioning of functional groups.

Molecular imprinting has been succesfully applied to the recognition of small organic molecules, inorganic ions, and even biological macromolecules. Some applications can be seen in Table 1.

Table 1. Various compounds used in molecular imprinting

COMPOUNDS	EXAMPLE
Drugs	Naproxen[7], propranolol[8], nortriptyline[19], theophylline[2], chlorphenamine[20]
Hormones	Cortisol[21]
Steroids	Cholesterol[22]
Pesticides	Atrazine, bentazone[23]
Proteins	Urease, ribonuclease A[24]
Amino acids	Tyrosine, aspartic acid, phenylalanine, tryptophan, histidine[25]
Carbohydrates	Galactose[26], glucose[27]
Nucleotide bases	Adenine[28]

2.1   General Principles of Molecular Imprinting Process

The molecular imprinting method is quite simple and easy to perform. Neither precise molecular design nor complicated organic synthesises are necessary. All required are functional monomer(s), a template (imprint molecule), porogen and crosslinking agent. Monomer(s) having one or more functional groups can specially interact with template molecule. The following steps are usually involved in the general approach of this technique (Fig. 1). The first step is association of the desired template molecule with a solution of monomer(s) by either non-covalent or reversible covalent interactions in the presence of a porogen to form monomer-template complex (or covalent adduct) (a through b in Fig.1). In this stage, pre-organization of the binding sites is achieved by assembling the functional monomers around the template. In order to obtain specific and homogenous binding sites, it is crucial to maintain stable monomer- template complexes during imprinting process. Then, this complex is polymerised with a high amount of crosslinker monomer in order to produce rigid polymer (b through c). In the last step (c through d), removal of the template leaves behind cavities possessing the shape and arrangement of the functional groups corresponding to the structure of the template. That is MIPs have an induced molecular memory that recognizes the template selectively. Polymer has a macroporous structure allowing the template diffusion into and out of polymer matrix. Most of MIPs are synthesized by free radical polymerisation of functional unsaturated vinylic, acrylic, and methacrylic monomers with an excess of crosslinking di- or tri- unsaturated vinylic, acrylic, and methacrylic monomers, resulting in porous organic polymer matrix.

The molecular recognition properties of molecularly imprinted polymers can be determined by the ability of resolution of enantiomers and discrimination between structurally related compounds such as theopylline and caffeine[2].

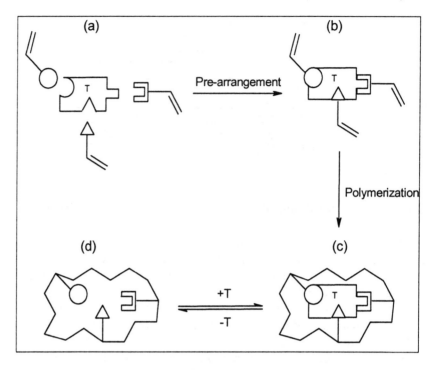

Figure 1. A scheme of molecular imprinting process:
  (a) orientation of three different functional groups of monomer(s) to the template (T) molecule;
  (b) formation of template-monomer complex (or covalent adduct);
  (c) formation of rigid polymer matrix;
  (d) removal of template leaving a functionalised cavity in polymer.

2. 2   Classification of Molecular Imprinting

Selective recognition of substrates by MIPs is critically dependent upon the nature of the binding interactions operating in the system. Molecular imprinting process may be classified in two ways with respect to the interaction mechanism in prepolymerisation: (1) the pre-organized system, developed by Wulff et al.[29], is based on reversible covalent bond formation (e.g., boronate esters, imines, and ketals) between the monomer and template and (2) the self-assembly system, developed by Mosbach et al.[30], depends on non-covalent interactions (such as ionic, or hydrophobic interactions, hydrogen bonding, and metal coordination) between the template and monomer precursors.

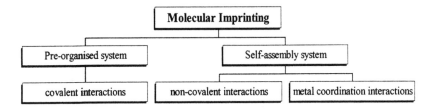

Figure 2. Classification of molecular imprinting process according to interaction mechanism during prepolymerisation, between monomer and template.

2.2.1 Molecular Imprinting with Covalent Interactions

The pre-organized molecular imprinting approach involves formation of strong, reversible, covalent linkage, which connects functional monomers with template before polymerisation. The linkages available at present are boronic acid esters, acetals, ketals, Schiff bases, disulfide bonds etc. Then this covalent adduct is copolymerised with a crosslinker in a suitable porogen to produce rigid macroporous polymer. The template molecule is fixed within the polymer matrix by means of this covalent linkage. Later on, the covalent linkage is cleaved and the template is removed from the polymer. The same chemical bonds in the initial template-monomer complex are reformed during any subsequent binding of template to the imprinted polymer matrix, since covalent imprinting method produces an exactly fitted recognition site.

Wulff[29] et al. introduced first covalent imprinting by using boronic acid group in 1972. In their work[31], two molecules of 4-vinylphenylboronic acid (structure 1, template containing monomer) were covalently bound by diester linkages to the template (Fig. 3). Phenyl-α-D-mannopyranoside (structure 2) acted as the template. After polymerisation of the template monomer with a high amount of crosslinker, the boronic acid ester was cleaved, and then the phenyl-α-D-mannopyranoside was removed. Removal of the template from the resulting polymer by water or methanol leaves a polymer cavity containing boronic acid residues in the exact spatial arrangement necessary to rebind the template. The accuracy of the systems was tested by the ability of the polymer to resolve the racemate of the template; phenyl-α-D,L-mannopyranoside. The selectivity of separation was extremely high and α values (separation factor) between 3.5 and 6.0 were obtained[1,16]. A large number of different templates have been used in covalent imprinting protocol[16,32].

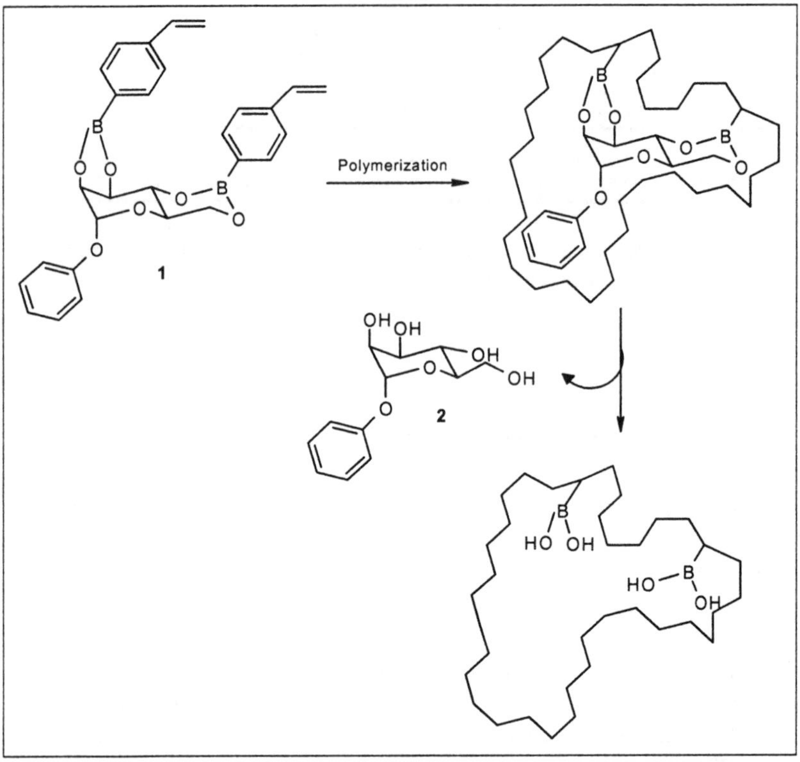

Figure 3. Covalent approach to molecular imprinting of phenyl-α-D-mannopyranoside[16, 31].

2.2.2 Molecular Imprinting with Non-Covalent Interactions

The self-assembly molecular imprinting approach involves non-covalent adduct between a functional monomer(s) and template molecule produced from non-covalent interactions such as hydrogen bonds and ionic, hydrophobic interactions and metal coordination[33]. Although these interactions are individually weak relative to covalent bonds, the cumulative effect of many such interactions may be very significant. In addition to these strong polar interactions, other types of interactions may also come into play, such as van der Waals interactions, charge-transfer interactions, and π-π interactions. These interactions give rise to three-dimensional geometry of the site, and contribute to the overall site quality[34]. In self-assembly imprinting mechanism, the functional monomers form spontaneous interactions with the template to produce one or more complexes. Once polymerisation has been initiated, the polymer chains begin to grow, leading to the formation of new self-assembly complexes.

Development of non-covalent approach was introduced by the group of Mosbach[30] in early 1980s and significantly broadened the scope of molecular imprinting. Their methodology for preparation of MIPs that are selective for dansyl-L-phenylalanine is illustrated in Fig. 4. A combination of methacrylic acid and vinylpyridine monomers prearranges with the amino acid derivative in acetonitrile solution before polymerisation. Following crosslinking with ethyleneglycol dimethacrylate by free-radical polymerisation, a rigid polymer is produced that, after extensive washing to remove the imprint molecule, retains recognition sites specific for dansyl-L-phenylalanine[35].

Substrate recognition and rebinding by imprinted polymers depend on the nature of the interaction involved. For the molecularly imprinted polymer to be useful, fast desorption and rebinding kinetics under mild conditions are as important as the substrate selectivity. This means that selection of appropriate binding interactions is very critical while designing molecularly imprinted materials. Different molecular interactions offer varying levels of specificity and reversibility. Covalent binding during the imprinting procedure has the advantage of stability and high functional group specificity (exact stoichiometry), however, it is slow in guest binding and release. In general, non-covalent interactions are more useful due to their versatility, faster kinetics and the requirement of milder conditions for their formation and breakage. Although the non-covalent imprinting system is simple, various complexes between the functional monomers and template were formed during the initial stages of polymerisation. In turn those complexes create different binding sites with a range of affinities.

Recently, combination of covalent and non-covalent approaches has been used in molecular imprinting[36,37]. Template-functional monomer complex is formed with covalent interactions, but the guest rebinding employed non-covalent interactions. Thus disadvantage of slow release kinetics of covalent imprinting is eliminated.

3. FACTORS THAT INFLUENCE THE RECOGNITION PROPERTIES OF MIPs

The affinity and selectivity of the imprinted polymers towards the template molecule are dependent on several factors such as template shape, the number of template interaction sites, concentration of functional monomer-template, polymerisation temperature and pressure etc. Here we briefly discuss the factors such as choice of monomer, crosslinker, solvent that affect the recognition properties of the MIPs. The roles of the other factors have been described in the literature[38,39].

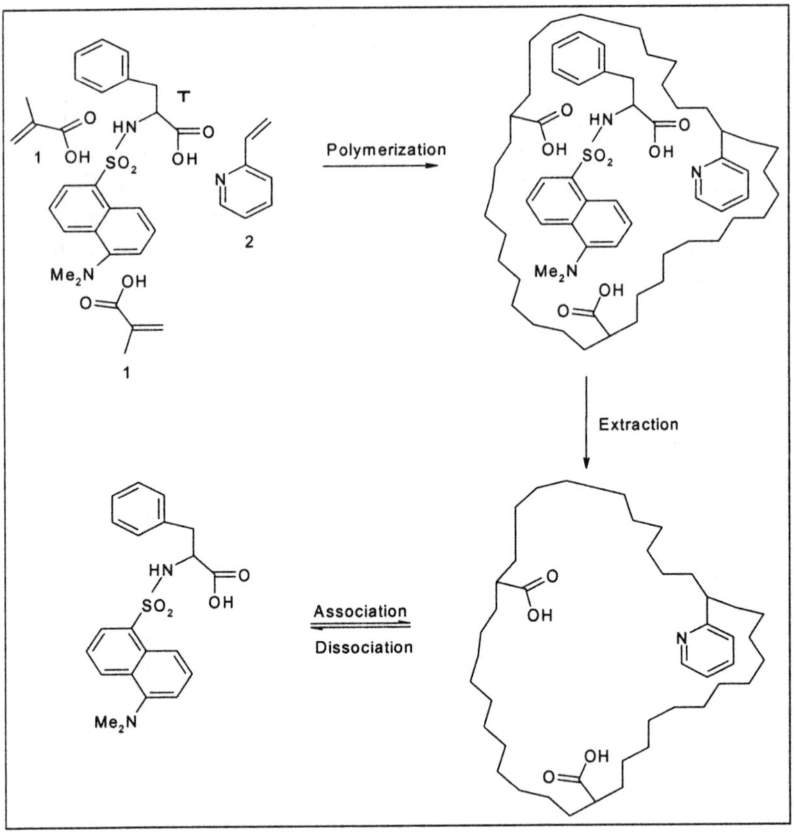

Figure 4. Imprinting of dansyl phenylalanine (T) using both MAA (1) and 2- vinylpyridine (2-VP, 2) as functional monomers[35].

3.1   Functional Monomers

Complementing functional groups and strength of interactions between monomer-template are the basis for the choice of functional monomer. For the templates containing acid groups, basic functional monomers are preferably chosen. For example; 2- or 4-vinylpyridine (VP) is particularly well suited for the imprinting of the carboxylic acid templates. Vice versa, methacrylic acid (MAA) is frequently used for basic templates. As a functional monomer, MAA has been used in a large variety of applications, because, carboxylic acid group acts as a hydrogen bond donor and acceptor. Typical functional monomers used in self-assembly systems are carboxylic acids (acrylic acid, methacrylic acid, vinylbenzoic acid), sulphonic acids (acrylamido-methylpropane sulphonic acid), and heteroaromatic weak bases (vinylpyridine, vinylimidazole). Furthermore, combinations of two or more functional monomers, giving terpolymers or higher may have better recognition ability than that of the recognitions observed for the corresponding copolymers[40].

3.2 Crosslinkers

Crosslinkers are as important as functional monomers in molecular imprinting. They affect the chemical and physical properties of the polymer matrix since a high degree of crosslinking (70-90 %) is necessary for achieving specificity. The required solubility of crosslinker in pre-polymer solution and compatibility with the functional monomer in terms of their reactivity reduces the possible choices. Ethyleneglycol dimethacrylate (EDMA) is the most commonly used crosslinker for the methacrylate based systems since it provides mechanical and thermal stability, good wettability in most rebinding media, and rapid mass transfer with good recognition properties. Except for the trimethylolpropane trimethacrylate (TRIM), none of the methacrylate- based crosslinkers provide similar recognition properties. In the imprinting of peptides, TRIM has been used to prepare resins possessing a higher sample load capacity and a better performance than similar resins prepared using EDMA as crosslinker[6].

3.3 Solvents

Solvents have a crucial role in the imprinting process since they may either enhance or destabilize the specific interaction between the template and monomer. Solvent that is used as porogen, determines the strength of non-covalent interactions in addition to its influence on the polymer morphology. One of the roles is to provide porous structures to imprinted polymers and promote their rates of guest binding as porogen in polymerisation. The choice of solvent is more critical in non-covalent imprinting procedures. Since all non-covalent forces are influenced by the properties of solvent, non-polar solvents normally lead to a good recognition. The use of more polar solvent will inevitably weaken the interaction forces formed between the template and monomer, resulting in poorer recognition. Chloroform is one of the most widely used solvents since it satisfactorily dissolves many monomers and templates and hardly suppresses hydrogen bonding. Other best imprinting solvents with very low dielectric constant are toluene and dichloromethane.

4.   PROPERTIES OF IMPRINTED POLYMERS

Molecularly imprinted polymers offer considerable advantages over biomolecules as recognition systems in terms of their chemical and physical stability. Due to their highly cross-linked polymeric nature, these materials are resistant to mechanical stresses, high pressures and temperatures, also to acids, bases and organic solvents. Polymers withstand exposure to temperatures of up to 150 ^{0}C for 24 hours without loss of affinity for the template. Temperatures above this point induce rapid loss in binding capacity and resultant mass loss. Furthermore, the polymers can be used repeatedly, in excess of 100 times during a period of years without loss of the "memory effect". MIPs can be prepared in a variety of configurations, which provide considerable advantages in practical applications. MIPs can be synthesized as macroporous block polymers, which are subsequently ground to particles with a diameter of about 25 μm. They are used directly as separation media in chromatography[41] or in batch binding assays[42]. For chromatographic conditions, spherical polymer beads have been developed by suspension polymerisation to acquire homogenous particles[43]. Imprinted polymer can be synthesized in situ within a high performance liquid chromatography column[44] and capillary columns for use in capillary electrophoresis[45] or in a vial for batch use[46]. Other configurations are thin films or membranes as the recognition element in a sensor and transducer surface[47].

5.   APPLICATIONS OF IMPRINTED POLYMERS

MIPs were originally designed as artificial antibodies. In the last decade, besides recognition studies, applications of MIPs have been expanded to the areas including separations and isolations (chiral, substrate-selective separation), antibody/receptor binding mimics, enzyme mimics/catalysis, biosensor-like devices and materials with specific functions.

5.1   Imprinted Polymers as Tailor-made Separation
       Materials

Molecularly imprinted polymeric materials have been used extensively as stationary phases for separation in high performance liquid chromatography (HPLC). The most widely used method has been bulk polymerisation followed by grinding, sieving and packing into HPLC columns.

Separation of molecules with closely related structures, drugs, different steroids, herbicides, sugars, amino acids and derivatives, metal ions, microorganisms, and toxins have been achieved[32,33]. Another important area for application of MIPs is the separation of chiral compounds such as peptides and drugs. There are 500 drugs on the market which are optically active, about 90% of these are administered as racemic mixtures[17]. Separation of enantiomers is required according to their pharmacokinetic and toxicological profiling. The selectivities of MIPs are in many cases comparable to those of commercially available chiral stationary phases. For example, a separation factor (α) of 17.8 was found for the separation of the two enantiomers of a dipeptide on poly(methacrylic acid-co-EDMA) imprinted with one of the enantiomers[48]. In addition to HPLC, the use of molecularly imprinted materials has been expanded to capillary electrophoresis[49] and thin layer chromatography[50]. MIPs have been investigated as highly selective sorbents for solid phase extraction (SPE) in order to concentrate and clean up samples prior to analysis. Highly selective sorbents to be used in SPE towards a large number of analytes of environmental[51] and pharmaceutical[52] interest, can be prepared by molecular imprinting.

Sales of chromatography columns are estimated to approach $500 million/year and membrane separation market to grow to more than $1 billion/year in the USA. MIPs are expected to take a 1-3% portion of this market[53]. Some pharmaceutical companies are already using MIPs on small scale for drug separation and purification.

5.2   Imprinted Polymers as Antibody Mimics

MIPs are essentially plastic antibodies. The first example of imprinted polymer as antibody in immunoassay protocols was reported in 1993[2]. In this study, MIPs against theopylline and against diazepam were used to develop an assay for the determination of these drugs in human serum. Later, this assay was further improved in organic solvent, buffer systems and plasma systems for cyclosporin, (s)-proprananol, morphine and many other analytes[3]. Superior characteristics of MIPs in comparison to antibodies in conventional immunoassays are observed with respect to chemical, mechanical and thermal stabilities.

5.3   Sensors Involving Imprinted Polymers

Specific recognition phenomena play a key role in sensor technology. Many sensors for environmental monitoring, biomedical and food analysis rely on biomolecules, such as enzymes or receptors as the specific recognition elements. Because of the poor chemical and physical stability of biomolecules, artificial receptors are gaining increasing interest. Molecular imprinting is an alternative way to produce a synthetic recognition system in sensor devices[9,54]. MIPs sensors have been developed for many different target species including gaseous substances (anesthetics, respiratory by-products, inflammables, toxics, and nerve gases); metabolites (glucose, urea, hormones, steroids, drugs of abuse); ions (H^+, Na^+, Ca^{2+}, heavy metals); toxic organic vapors (benzene, toluene); and proteins and microorganisms (viruses, bacteria, parasites)[9]. Until now, the most convincing demonstration of the usefulness of a "real" biomimetic sensor based on molecular imprints is an optical-fiber-like device in which a fluorescent amino acid derivative (dansyl-L-phenylalanine) binds to the polymer particles, resulting in fluorescent signals that vary as a function of the concentration of the derivative. Chiral selectivity was shown by using the corresponding D-enantiomer as a control[35].

5.4   Imprinted Polymers as Artificial Enzymes

One of the most important challenges for the application of molecularly imprinted polymers is their use as enzyme mimics. Catalytic MIPs have been produced by three approaches, based on the template used: i) substrate analogues, ii) transition state analogues and, iii) product analogues. The common approach has been the use of transition state analogues in the imprinting protocol, thus stabilizing the reaction transition and enhancing the rate of product formation. It was reported that a 10^3-10^4 fold increased rate of hydrolysis was obtained by preparing the imprinted polymer against phosphonic ester as transition state analogue for alkaline ester hydrolysis[4,10].

6.   CONCLUSION AND FUTURE DIRECTIONS

MIPs are easily prepared in large amounts, robust, inexpensive, and in many cases, possess an affinity and specificity that is suitable for industrial applications. However, this technique has some fundamental drawbacks unresolved yet. While natural antibodies can function in water, this technique is inapplicable to water-soluble substances. MIP utility should be extended to recognition of large biomolecules as it was realized for proteins. The recent research areas include reduction of binding site heterogeneity in non-covalent approaches, improvement of mass transfer, especially in non-covalent imprinting, development of new and better binding sites in imprinting, improvement in manufacturing procedures, better recognition in aqueous systems. Certainly, with the developments in the above fields, molecular imprinting will be widely used in the near future in sciences, industry, and daily life.

ACKNOWLEDGEMENT

I would like to thank Yeliz Tunç for her assistance in preparation of the figures given in the text.

REFERENCES

1. Wulff, G., 1995, Molecular imprinting in cross-linked materials with the aid of molecular templates - A way towards artificial antibodies. *Angew. Chem. Int. Ed. Engl.* 34: 1812-1832.
2. Vlatakis, G., Andersson, L.I., Müller, R. and Mosbach, K., 1993, Drug assay using antibody mimics made by molecular imprinting. *Nature* 36: 645-647.
3. Andersson, L. I., 2001, Application of molecularly imprinted polymers in competitive ligand binding assays for analysis of biological samples. *In Molecularly Imprinted Polymers* (B. Sellergren, ed.) Elsevier Science, Amsterdam, pp. 341-353.
4. Sellergren, B., and Shea, K.J., 1994, Enantioselective ester hydrolysis catalyzed by imprinted polymers. *Tetrahedron Asymmetry* 5: 1403-1406.
5. Baggiani, C., Giraudi, G., Giovannoli, C., Trotta, F., and Vanni, A., 2000, Chromatographic characterization of molecularly imprinted polymers binding the herbicide 2,4,5-trichlorophenoxyacetic acis. *J. Chromatogr. A* 883: 119-126.
6. Kempe, M., 1996, Antibody-Mimicking Polymers as Chiral Stationary Phases in HPLC. *Anal. Chem.* 68: 1948-1953.
7. Hoginaka, J., and Sanbe, H., 2001, Uniformly sized molecularly imprinted polymer for (s)-naproxen. Retention and molecular recognition properties in aqueous mobile phase. *J. Chromatogr. A* 913: 141-146.
8. Suedee, R., Srichuna, T., and Martin, G.P., 2000, Evaluation of matrices containing molecularly imprinted polymers in the enantioselective-controlled delivery of β-blockers. *J. Cont. Rel.* 66: 135-147.
9. Kugimiya, A., and Takeuchi, T., 2001, Surface plasmon resonance sensors using molecularly imprinted polymer for detection of sialic acid. *Biosensors & Bioelectronics* 16: 1059-1062.
10. Robinson, D.K., and Mosbach, K., 1989, Molecular imprinting of a transition state analogue leads to a polymer exhibiting esterolytic activity. *J. Chem. Soc. Chem. Commun.* 649: 969-970.
11. Bures, P., Huang, Y., Oral. E. and Peppas, N.A., 2001, Surface modifications and molecular imprinting of polymers in medical and pharmaceutical applications. *J. Cont. Rel.* 72: 25-33.
12. Lehn, J.M., 1988, Supramolecular Chemistry-Scope and Perspectives: Molecules, Supermolecules, and Molecular Devices (Nobel Lecture). *Angew. Chem. Int. Ed. Engl* 27: 89-112.
13. Breslow, R., 1986, *Advanced in Enzymology and Related Areas of Molecular Biology* Vol.58 (A. Meister, ed), Wiley, pp.1-60.

14. Schneider, H-J, 1991, Mechanisms of Molecular Recognition-Investigations with Organic Host-Guest Complexes. Angew Chem. 103, 1419, *Angew. Chem. Int. Ed. Engl.* 30: 1417-1436.
15. Wenz, G., 1994, Cyclodextrins as building blocks for supramolecular structures and functional units. Angew Chem., 106, 851, *Angew. Chem. Int. Ed. Engl.* 33: 803-822.
16. Wulff, G. and Biffis, A., 2001, Molecularly imprinting with covalent or stoichiometric non-covalent interactions. In Molecularly Imprinted Polymers (B. Sellergren, ed). Elsevier Science, Amsterdam, pp. 71-111.
17. Mosbach, K., 1994, Molecular Imprinting. *Trends Biochem. Sci.* 19: 9-14.
18. Komiyama, M., Takeuchi, T., Mukawa, T. and Asanuma, H., 2003, Fundamentals of Molecular Imprinting. In Molecular Imprinting, (M. Komiyama, T. Takeuchi, T. Mukawa, and H. Asanuma, eds.) Wiley, Weinheim. pp.9-19.
19. Khasawneh, A.M., Vallano, P.T., and Remcho, V.T., 2001, Affinity screening by packed capillary high performance liquid chromatography using molecular imprinted sorbents. *J. Chromatogr. A.* 922: 87-97.
20. Chen, W., Liu, F., Zhang, X., Li, A.K., and Tong, S., 2001, The specificity of a chlorphenamine imprinted polymer and its application. *Talanta* 55: 29-34.
21. Baggiani, C., Giraudi, G., Trotta, F., Giovannoli, C., and Vanni, A., 2000, Chromatographic characterization of a molecular imprinted polymer binding cortisol. *Talanta* 51: 71-75.
22. Sreenivasan, K., 1998, Effect of the type of monomers of molecularly imprinted polymers on the interaction with steroids. *J. Appl. Polym. Sci.* 68: 1863-1866.
23. Baggiani, C., Trotta, F., Giraudi, G., Giovannoli, C., and Vanni, A., 1999, A molecularly imprinted polymer for the pesticide bentazone. *Anal. Commun.* 36: 263-266.
24. Kempe, M., Glad, M., and Mosbach, K., 1995, An Approach Towards Surface Imprinting Using the Enzyme Ribonuclease A, *J. Mol. Recogn.* 8: 35-39.
25. Liao, Y., Wang, W., and Wang, B., 1998, Enantioselective polymeric transporters for tryptophan, phenylalanine and histidine prepared using molecular imprinting techniques. *Bioorg. Chem.* 26: 309-322.
26. Nilsson, K. G. I., Sakaguuchi, K., Gemeiner, P., and Mosbach, K., 1995, Molecular imprinting of acetylated carbohydrate derivatives into methacrylic polymers. *J. Chromatogr. A* 707: 199-203.
27. Wizeman, W. J., and Kofinas, P., 2001, Molecularly imprinted polymer hydrogels displaying isomerically resolved glucose binding. *Biomaterials* 22: 1485-1491.
28. Spivak, D.A., and Shea, J. K., 1998, Binding of nucleotide bases by imprinted polymers. *Macromolecules* 31: 2160-2165.
29. Wulff, G., and Sarhan , A., 1972, Use of polymers with enzyme analogue structures for the resolution of enantiomers. *Angew. Chem. Int. Ed. Engl.* 11: 341-344.
30. Arshady, R., and Mosbach, K., 1981, Synthesis of substrate selective polymers by host-guest polymerization. *Makromol. Chem.* 182: 687-692.
31. Wulff, G., Vesper, W., Grobe-Einsler, R. and Sarhan, A., 1977, Enzyme-analague built polymers, 4) On the synthesis of polymers containing chiral cavities and their use for the resolution of racemates. *Makromol. Chem.* 178: 2799-2817.
32. Komiyama, M., Takeuchi, T., Mukawa, T. and Asanuma, H. (eds), 2003, *Molecular Imprinting*, Wiley, Weinheim.
33. Bartsch, R.A. and Maeda, M. (eds), 1998, *Molecular and Ionic Recognition with Imprinted Polymer*, ACS-Symposium Series 703, Oxford Univ. Press, Washington, DC.
34. Mosbach, K., Haupt, K., Liu, X. C., Cormack, P. A. G. and Ramström, O., 1998, "Molecular Imprinting: Status artis et quo vadere ? *In Molecular and Ionic Recognition*

with Imprinted Polymers. (R. A. Bartsch and M. Maeda, Eds). ACS-Symposium Series 703, Oxford University Press, Washington, DC. pp. 29-48.

35. Kriz, D., Ramström, O., Svensson, A. and Mosbach, K., 1995, A Biomimetic Sensor Based on a Molecularly Imprinted Polymer as a Recognition Element Combined with Fiber-Optic Detection. *Anal. Chem.* 67: 2142-2144.

36. Whitcombe, M.S., Rodriquez, M.E., Villar, P. and Vulfson, E.N., 1995, A new method for the introduction of recognition site functionality into polymers prepared by molecular imprinting: Synthesis and characterization of polymeric receptors for cholesterol. *J. Am. Chem. Soc.* 117: 7105-7111.

37. Sellergren, B. and Andersson, L, 1990, Molecular recognition in macroporous polymers prepared by a substrate analogue imprinting strategy. *J. Org. Chem.* 55: 3381-3383.

38. Nicholls, I.A. and Andersson, H.S., 2001, Thermodynamic principles underlying molecularly imprinted polymer formulation and ligand recognition. *In Molecularly Imprinted Polymers* (Sellergren, B. ed). Elsevier Science, Amsterdam, pp. 60-70.

39. Sellergren, B., 2001, The non-covalent approach to molecular imprinting. *In Molecularly Imprinted Polymers* (B. Sellergren, ed). Elsevier Science, Amsterdam, pp. 114-184.

40. Moring, S. E., Wong, S. O., and Strobaugh, J. F., 2002, Target specific sample preparation from aqueous extracts with molecular imprinted polymers. *J. Pharm. Biomed. Anal.* 27: 719-728.

41. Ji, Z., and Xiwen, H., 1999, Study of the nature of recognition in molecularly imprinted polymer selective for 2-aminopyridine. *Anal. Chim. Act.* 381: 85-91.

42. Kempe, M., and Mosbach, K., 1991, Binding studies on substrate and enantio-selective molecularly imprinted polymers. *Anal. Lett.* 24: 1137-1145.

43. Baggiani, C, Trotta, F., Giraudi, G., Giovannoli, C., and Vanni, A., 1999, Chromatographic characterization of a molecularly imprinted polymer binding theophylline in aqueous buffers. *J. Chromatogr. A* 786: 23-29.

44. Matsui, J., Kato, T., Takeuchi, T., Suzuki, M., Yokoyama, K., Tamiya, E., and Karube, I., 1993, Molecular recognition in continuous polymer rods prepared by a molecular imprinting technique. *Anal. Chem.* 65: 2223-2224.

45. Brüggemann, O., Freitag, R., Whitcombe, M. J., and Vulfson, E. N., 1997, Comparison of polymer coatings of capillaries for capillary electrophoresis with respect to their applicability to molecular imprinting and electrochromatography. *J. Chromatogr. A* 781:43-53.

46. Takeuchi, T., Fukuma, D., and Matsui, I., 1999, Combinatorial molecular imprinting: An approach to synthetic polymer receptors. *Anal. Chem.* 71:285-290.

47. Hedborg, E., Winquist, F., Lundström, I., Andersson, L.I., and Mosbach, K., 1993, Some studies of molecularly-imprinted polymer membranes in combination with field-effect devices. *Sensors and Actuators A* 37-38:796-799.

48. Ramström, O.,Nicholls, I.A., and Mosbach, K., 1994, Synthetic peptide receptor mimics: highly stereoselective recognition in non-covalent molecularly imprinted polymers. *Tetrahedron:Assymmetry* 5:649-656.

49. Nilsson, K., Lindell, K., Norrlow, O., and Sellergen, B., 1994, Imprinted polymers as antibody mimetics and new affinity gels for selective separations in capillary electrophoresis. *J.Chromatogr. A* 680: 57-61.

50. Kriz, O., Kriz, C. B., Andersson, L., and Mosbach, K., 1994, Thin layer chromatography based on molecular imprinting technique. *Anal. Chem.* 66: 2636-2639.

51. Matsui, J., Fujiwara, K., Ugata, S., and Takeuchi, T., 2000, Solid-phase extraction with a dibutylmelamine-imprinted polymer as triazine herbicide-selective sorbent. *J. Chromatogr. A* 889: 25-31.

52. Sellergen, B., 1994, Direct drug determination by selective sample enrichment on an imprinted polymer. *Anal. Chem.* 66: 1578-1582.
53. Piletsky, S.A., Alcock, S., and Turner, A.P.F., 2001, Molecular Imprinting: at the edge of the third millennium. *TRENDS in Biotechnology* 19: 9-12.
54. Piletsky, S.A., Piletskaya, E.V., Sergeyeva, T.A., Panasyuk, T.L., and El'skaya, A.V., 1999, Molecularly imprinted self-assembled films with specificity to cholesterol. *Sensors and Actuators B* 66: 216-220.

Simulating Hydroxyapatite Binding of Bone-Seeking Bisphosphonates

JENNIFER E.I. WRIGHT[*], LIYAN ZHAO[*], PHILLIP CHOI[*] and
HASAN ULUDAG[*,#]
[*]Department of Chemical and Materials Engineering, Faculty of Engineering, [#]Faculty of
Pharmacy and Pharmaceutical Sciences, University of Alberta, Edmonton, Alberta,
CANADA

1.   INTRODUCTION

An ideal therapeutic agent for bone diseases is expected to restrict
its pharmacological activity to bone sites, with minimal effects at non-
skeletal sites. This requires the therapeutic agent to be localized to
bones after systemic administration with minimal distribution to other
sites. This is feasible if the therapeutic agent exhibits a high
affinity to bone tissue alone, so that it is bound and retained in
bones. Bone is distinguished from the rest of the body by the
presence of a hydroxyapatite, (HA; $Ca_{10}(PO_4)_6(OH)_2$) mineral, which is
not present in other tissues under normal circumstances. Therefore,
molecules with high affinity to HA are expected to be specifically
delivered to bones after systemic administration. Although HA is
known for its binding capability to a wide variety of molecules (the
basis for HA affinity chromatography), most therapeutic agents
intended for bone diseases do not have a particular affinity to HA. This
is because of a lack of a high enough specific affinity, and competitive
binding from the ions, organic and proteinous molecules present in
the physiological milieu. An exceptional class of molecules is
bisphosphonates (BPs), which exhibit a strong affinity to HA under
physiological conditions. Systemic administration of BPs typically
results in 20-50% deposition of the molecules at bone tissues, with

Biomaterials: From Molecules to Engineered Tissues, edited by
N. Hasırcı and V. Hasırcı, Kluwer Academic/Plenum Publishers, 2004   139

minimal accumulation at other sites (usually excreted in urine). BPs are structurally analogous to the endogenous inorganic phosphate, pyrophosphate (Figure 1). Replacement of the central oxygen atom in the pyrophosphate with a carbon in BPs allows one to incorporate additional functionalities to the BPs. Depending on the structural details, BPs exhibit a variety of pharmacological activities at bone tissues, the most significant of which is the inhibition of osteoclast-mediated bone resorption[1]. An active research program is currently pursued by all major pharmaceutical companies in order to design bone-seeking BPs with tailored pharmacological activities.

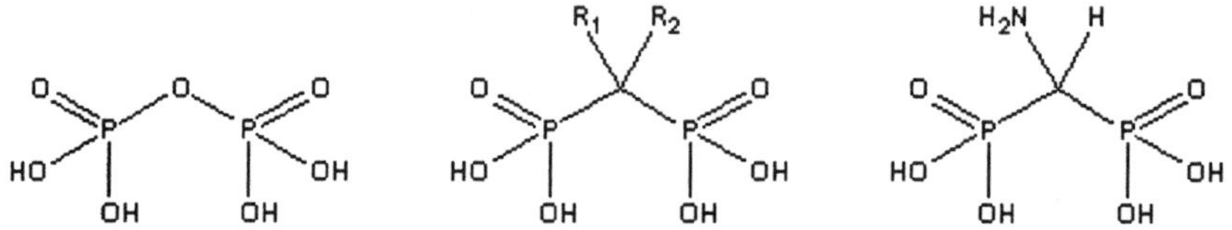

Figure 1. Structure of pyrophosphate (left), a general bisphosphonate (middle) and the aminoBP (right) used for the simulations in this study.

In addition, BPs are being currently pursued as carriers of drugs to bone tissue. The basic premise of this approach is to formulate a specific drug of interest (which exhibits no particular bone affinity) with a bone-seeking BP, so that the complex will be deposited to bone after systemic administration[2]. This approach has been applied to several classes of drugs, including radio-pharmaceuticals, small organic molecules (e.g.,estrogen, glucocorticoids, anti-inflammatory agents), as well as the macromolecules such as proteins. Proteins in particular are attractive because of the availability of recombinant technology derived molecules (e.g., cytokines) that exhibit strong activities on cellular systems but cannot be used due to the same actions exhibited on non-skeletal sites[3]. A research program in the authors' lab has been steadily establishing the foundation for utilizing BPs to deliver proteins to bones[4]-[7].First, an aqueous-based chemistry was developed to link amine-containing BPs to a model protein (Bovine Serum Albumin, BSA). BP conjugate of BSA was shown to exhibit a mineral affinity in proportion to the degree of BP derivatization[4]. The conjugates were observed to exhibit a high bone affinity, as evident by ~ 10-fold increased retention of the conjugates when they were directly injected into an osseous site (tibia)[5].

Upon systemic injection, the conjugates were also found to be bone-seeking; up to 7-fold increased bone deposition was observed depending on the bone site, post-injection time and the route of administration[6]. While these studies demonstrated the proof of principle for the targeting approach, we are pursuing parallel studies to better understand the factor's contribution to the bone affinity of the conjugates[7,8].

A critical issue in the design of protein-BP conjugates is understanding the factors that influence the bone affinity. The bone affinity is expected to primarily depend on the interaction between the bone mineral HA and the BP moiety of the conjugate. We have recently begun to employ molecular simulation studies to elucidate factors that contribute to BP binding to HA. Molecular simulation studies are viewed as a relatively quick means of shedding insights into the mode of interactions, as compared to experimental approaches, which necessitate synthesis and characterization of numerous BPs. In this communication, we report our preliminary studies on setting up the appropriate system to obtain binding energies between a BP and the biological HA.

2. MATERIALS AND METHODS

All simulations were performed on a Silicon Graphics Workstation using a molecular modeling package Cerius[2] version 4.2, from Acceryls Inc. (formerly Molecular Simulations Inc.) (San Diego, CA). The Universal 1.02 Force Field[9] was used in all calculations. This particular force field was chosen since it is an excellent general-purpose force field and was used previously in a study simulating the interactions of phosphocitrate with HA[10]. Determining which force-field to use is important in the reliability of any molecular modelling exercise. If the relative binding energies are dominated by the well-understood electrostatic interactions, the simulation results will not be strongly dependent on the force-field chosen[11]. In studies involving interactions of molecules with HA surfaces, the Consistent Valence Force Field (CVFF) has been also used by others[11,12]. This force field is suitable for peptides, proteins, and a wide range of organic systems and is primarily intended for studies of structure and binding energies, although it predicts vibrational frequencies and conformational energies reasonably well. The molecular and atomic charges of HA surfaces and BP molecules used in this study were determined using the charge equilibration[13] method of Cerius[2].

The temperature for simulations was typically set at 300 K (room temperature), but some studies employed 1000 K to overcome the relatively high barriers of rotation for single bonds[14]. This allowed collection of representative conformational distributions in a reasonable period of time. A dielectric constant of 1.0 was used[10]. Although a distance dependent dielectric of 2.0*r could be used to further diminish the effect of finite crystal thickness[11], we chose to restrict our simulations to a constant dielectric constant. This distance dependent dielectric constant is also expected to better model the effects of solvent and counter-ions in solution. Previous calculations using the 1.0 dielectric constant gave substantially larger binding energies for individual surfaces, but had no effect on identifying the most strongly binding surfaces[11].

The Cerius[2] software contains a built-in crystal structure of HA (Figure 2), with a space group of $P6_3/m$ and unit cell dimensions of a=b=9.42 Å, c=6.88 Å and $\alpha=\beta=90°$, $\gamma=120°$.

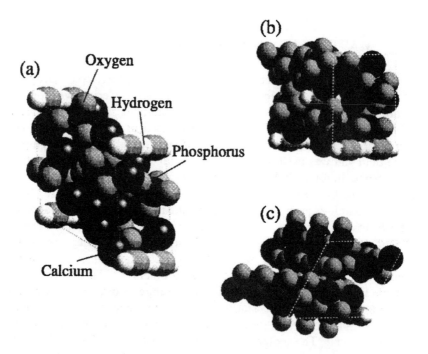

Figure 2. (a) HA crystal structure; (b) (100) surface; (c) (001) surface

This is the crystal structure of biological HA, considered to have an effective hexagonal symmetry, and is disordered compared to the monoclinic form (space group $P2_1/b$ and unit cell dimensions of a=9.42 Å, b=18.84 Å, c=6.88 Å) which was used by others[11]. Using $P6_3/m$ unit cell, surfaces were generated by cleaving the HA crystal along (100), (010), and (001) planes. A thickness of 20 Å was typically used for cleaving surfaces, but the thickness was also varied for one set of studies. A BP was then manually positioned near the HA surface within the electrostatic interaction range. 1-Amino-1,1-diphosphonate methane (aminoBP) was used for all simulations (Figure 1). The conjugate gradient method of Cerius[2] for the energy optimization procedure was applied. During minimization, the calcium atoms and phosphate groups of the HA surface were fixed while the -OH groups and aminoBP were mobile. The total simulation time was set to 500 ps.

3.   RESULTS AND DISCUSSION

3.1   Binding Energy and Effect of Crystal Thickness

The difference between the energy when a BP is adsorbed on the surface and when it is separated beyond an interaction distance is taken as the binding energy[10]. To determine the binding energy of aminoBP on the HA surface, the energy of the system was minimized before simulations with the molecule at various distances from the surface. As the aminoBP moved closer to the surface, the total energy decreased (Figure 3a). After the aminoBP was moved further than 100 Å, the energy of the system did not vary significantly, indicating that at this distance the molecules are separated beyond the interaction distance. For all binding energy calculations in this study, 1000 Å was taken as the separation distance for no interaction. When the aminoBP was placed within the HA crystal surface, the aminoBP, appeared to be visually distorted, causing a significant increase in the energy of the system (-92 kcal mol^{-1}).

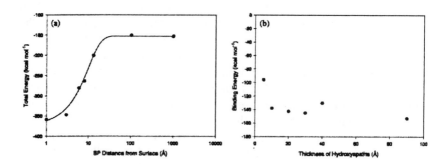

Figure 3. (a) Change in energy of the system as BP is moved away from the HA surface; thickness of HA: 10 Å; surface plane: (100); surface area: 9.42 x 6.88 Å. (b) Change in binding energy at different thicknesses of (100) HA surface with surface area of 9.42 x 6.88 Å.

A negative value for the binding energy indicates that an energetically favorable interaction has been established. This binding energy is affected by bond and non-bond interactions[11]. Bond contributions reflect differences in the conformation of the interacting molecule (aminoBP) and is expected to represent only a small fraction of the relative binding energy. Non-bond energies (van der Waals and electrostatic contributions) are expected to dominate the overall binding and are due largely to interactions between the ligand and crystal surface.

In a previous paper, the relative binding energies of the most strongly bound peptide conformers docked on HA surfaces were dominated by electrostatic contributions, by a factor of 10 as compared to other contributing terms[11]. The internal energies were relatively small, suggesting that the chosen peptides did not have to distort greatly to dock onto these surfaces. The most strongly bound conformations of a peptide docked onto the HA surfaces were those with the most charged atoms. Weakly bound conformations were characterized by charged groups not making contact with the HA surface. The electrostatic energy gained by placing the charged side-chains in contact with the surface more than compensated for any energy required to distort the molecule. In this regard, there is good agreement between our results and the simulations involving peptide binding to HA surfaces.

A critical issue in determining the binding energy is the thickness of the HA surface. A thicker surface is likely to account for a more representative BP interaction at the expense of computational time. Previous simulation studies have determined the appropriate crystal thickness by increasing the thickness until the change in binding energy differed by <0.4

kJ/mol[11]. A crystal thickness of 14 Å or thicker was found to be adequate, when the adsorbing species were $H_2PO_4^-$ or HCO_3^-. This binding energy was calculated by using electrically neutral crystals. It was reported that some negatively charged species can bind to the negatively charged HA[15]. Simulation studies of phosphocitrate adsorption onto HA have been carried out with crystal surfaces having total electric charge equal to 0 as well as -5.4[10]. The binding energy of negatively-charged phosphocitrate to the charged HA was diminished in strength but still substantial compared to the neutral HA.

The optimal thickness of the HA surface used in these simulations was determined by increasing the thickness until there was no significant change in binding energy. Figure 3b shows the calculated binding energy with different thicknesses of the HA crystal. The binding energy decreased and leveled off between thicknesses of 10 to 20 Å. A thickness of 20 Å was chosen for further simulations. Model crystal planes were constructed to be >3-times larger than the longest dimension of the interacting molecule. This was carried out to minimize electrostatic edge effects that may influence the preferred calculated conformation of the molecule.

3.2   Effect of Initial Location and Orientation of BP on Binding Energy

The calculated binding energy, although expected to represent the most energetically favorable state, may depend on the initial conditions of the HA-aminoBP system. During the simulations, the molecule may settle in a local minimum as opposed to a global minimum, causing the binding energy calculated to be a 'local minimum', rather than the 'true minimum'. To explore the influence of initial conditions, aminoBP was arbitrarily placed in the vicinity of the (001) surface, and rotated 90 degrees in the x, y and z axis and allowed to interact with the HA surface. The (001) surface has been described previously to be the most favorable growing surface of HA[12]. Following the simulations, these different orientations of the BP molecule gave different final energies of the system as well as different binding energies (Figure 4). The orientation with the amine group of BP closest to the surface produced the highest energy, while the orientations with the phosphonate groups closer to the surface produced lower energies, indicating that the phosphate groups adsorb better on the surface than the amine group of the chosen BP. This preferred orientation was consistent with the preferred binding groups of phosphocitrate on the HA surface[10], as well as phosphate groups of phosphopeptides facing the HA surface directly[11].

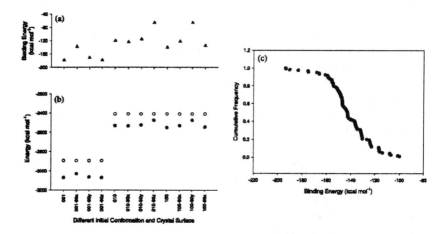

Figure 4. Effect of initial orientation of aminoBP with respect to HA surface on the binding energy. An arbitrary initial orientation was used on 3 surfaces, after which the molecule was rotated 90° in the x, y, and z directions, (a) Effect of different initial orientation of aminoBP and different HA surfaces on binding energy and (b) total energy of the system when BP is at a distance of 1 Å (•) and 1000 Å (o) from the HA surface, (c) Effect of initial location of BP on the energy of the system; thickness of HA: 20 Å; surface plane: (001); surface area; 18.84 Å x 13.76 Å.

Effect of orientation was further explored on (100) and (010) surfaces. The (100) surface was examined due to its importance for the HA crystal morphology and experimental evidence implicating this face as the binding site for anionic species, such as small molecules, polymers and anionically modified cell surfaces[10]. The (010) surface of HA was also examined, although this surface is identical to the (100) surface in a hexagonal crystal such as HA[11]. The overall energy of the HA-BP system on the (001) surface is lower than that of the (100) and (010) surfaces (Figure 4), indicating the former to be the primary binding surface. The binding energy of the BP on the later surfaces was relatively similar. Note the orientation of the aminoBP had a significant impact on the binding energy on all surfaces. Phosphocitrate has been shown to bind most favorably to the (100) surface of HA[10], unlike our results that indicate the (001) surface to be the most favorable for aminoBP binding. This suggests that the most favorable binding surface might be ligand dependent.

Finally the effect of the starting location of aminoBP on binding energy was determined. This was carried out by moving the aminoBP systematically over the HA surface and carrying out the simulations. Figure 4c shows the distribution of the average minimum energy for each initial location following the simulations. The binding energy was clearly different depending on the starting conditions. The average binding energy ($\sim$ -142 kJ mol^{-1}) was associated with a coefficient of variation of 12%. This variation is likely to represent relative orientations of charged groups on the surface and the aminoBP. However, the variation in starting location seemed to result in as much variability, as the starting orientation of the aminoBP.

4. CONCLUSION

Molecular modeling was undertaken to investigate the binding of BPs to HA. It was shown that the overall energy of the system decreases when BPs adsorb on the surface, leading to calculations of binding energy. All binding energies were negative indicating that an energetically favorable interaction was established between the model BP (aminoBP) and HA surfaces. The binding energy was found to depend on the initial conditions of the BP molecule. Further studies will examine the effect of different BPs on the calculated binding energy. These studies will be used to examine the design of BP conjugates, providing insight into the choice of appropriate BPs for high HA affinity, hence can be used in the design of bone-specific delivery systems.

ACKNOWLEDGEMENTS

This research was supported by the Canadian Institutes of Health Research (CIHR) and the Whitaker Foundation.

REFERENCES

1.   Rogers, M.J., Gordon, S., Bendford, H.L., Coxon, F.P., Luckman, S.P., Montkonen, J., and Frith, J.C., 2000, *Cancer* 88: 2961-2978.
2.   Uludag, H., 2002, *Curr. Pharm. Design* 8: 1929-1944.
3.   Gittens, S., and Uludag, H., 2001, *J. Drug Targeting* 9: 407-429.
4.   Uludag, H., Kousinioris, N., Gao, T., and Kantoci, D., 2000, *Biotech. Prog.* 16: 258-267.
5.   Uludag, H., Gao, T.J., Wohl, G., Zernicke, R.F., and Kantoci, D., 2000, *Biotech. Prog.* 16: 1115-1118.
6.   Uludag, H., and Yang, J., 2002, *Biotech. Prog.* 18: 604-611.
7.   Gittens, S., Matyas, J.R., Zernicke, R., and Uludag, H., 2003, *Pharm. Res.* 20: 978-987.
8.   Gittens, S.A., Kitov, P.I., Matyas, J.R., Loebenberg, R., and Uludag, H., Submitted to *Pharm. Res.*
9.   Rappe, A.K., Casewit, C.J., Colwell, K.S., Goddard-III, W.A., and Skiff, W.M., 1992, *J. Am. Chem. Soc.* 114: 10024-10035.
10. Wierzbicki, A., and Cheung, H.S., 2000, *J. Mol. Struct. (Theochem)* 529: 73-82.
11. Huq, N.L., Cross, K.J., and Reynolds, E.G., 2000, *J. Mol. Model.* 6: 35-47.
12. Neves, M., Gano, L., Pereira, N., Costa, M.C., Costa, M.R., Chandia, M., Rosado, M., and Fausto, R., 2002, *Nucl. Med. Bio.* 29: 329-338.
13. Rappe, A.K., and Goddard III, W.A., 1991, *J. Phys. Chem.* 95:3358-3363.
14. Kitov, P.I., Shimizu, H., Homans, S.W., and Bundle,D.R., 2002, *J. Am. Chem. Soc.* 125: 3284-3294.
15. Barroug, A., Lernoux, E., Lemaitre, J., and Rouxhet, P.J., 1998, *J. Colloid Interface Sci.* 208: 147-152.

Biochips: Focusing on Surfaces and Surface Modification

ERHAN PİŞKİN and BORA GARİPCAN
Hacettepe University-Center of Bioengineering and Bioengineering Division, and TÜBİTAK: Center of Excellence-BİYOMÜH, Beytepe, Ankara, Turkey

1. INTRODUCTION

Thousands of genes and their products (i.e., proteins) in a given living organism function in a complicated and orchestrated way that creates the mystery of life. In "genomics", briefly genes of organisms, their functions and activities are investigated. Genomics is naturally linked to "proteomics" – large scale study of all proteins encoded by an organism's genome. Genomics' combination with proteomics resulted with a fascinating growth in the biomedical research and in the field of large scale and high throughput biology.[1,2]

Traditional methods in molecular biology generally work on a "one gene in one experiment" basis, which means that the throughput is very limited and the "whole picture" of gene function is hard to obtain. Biochips, or the so-called "microarrays" are "Labs on Chips" and allow simultaneous analysis of multiple samples. "Macrochips" contain sample spot sizes of about 300 microns or larger and can be easily imaged by existing gel and blot scanners. The sample spot sizes in a "microchip" are typically less than 200 microns in diameter. Thousands of probe ("ligand") molecules (such as ssDNA, proteins, oligopeptides) are immobilized on microarray surfaces as schematically depicted in Figure 1.

Biomaterials: *From Molecules to Engineered Tissues,* edited by
N. Hasirci and V. Hasirci, Kluwer Academic/Plenum Publishers, 2004

Figure 1. ssDNA molecules on microarray surfaces.

DNA biochip technology uses the microscopic arrays of DNA molecules for biomedical analysis such as gene expression analysis/profiling and polymorphism or mutation detection/mapping, DNA sequencing, and gene discovery[3,4]. Protein microarrays are used for analysing gene function, regulation and a variety of other applications including study of enzyme-substrate, DNA-protein and protein-protein interactions. It is obvious that proteins are more difficult to handle than DNA. Proteins are chemically and structurally much more complex and heterogeneous than nucleic acids. In contrast to DNA, proteins easily lose their 3D-structure and biochemical activity due to denaturation, dehydration or oxidation. The state of proteins, such as post-translational modifications, partnership with other proteins, protein subcellular localization, and reversible covalent modifications (e.g. phosphorylation) make them difficult to study. Recently, there have been considerable achievements in preparing microarrays containing over 100 proteins and even an entire proteome[2,5-9].

2.   BIOCHIP TECHNOLOGY

The major steps of biochip technology include the manufacture of microarrays, immobilization labeling of the ligand molecules, interaction of immobilized ligands with the target molecules and the subsequent analysis of these interactions[4].

There are two main approaches for manufacturing DNA arrays. In the first approach (in situ synthesis) oligonucleotide probes are synthesized directly on the surface using photolithographic techniques and modified ODN synthesis chemistry (Figure 2A)[10,11]. This state-of-the-art methodology provides the capability of producing and interrogating hundreds of thousands of microarrayed elements per assay and allows wide flexibility in design of custom arrays without preliminary efforts in synthesis and maintenance of a vast library of modified ODNs[12]. On the other hand in situ synthesis does not allow quality control and purification of generated features, and the less efficient phosphoramidite monomer coupling gives poor yields for longer ODN probes.

In the second approach, pre-synthesized ODNs containing end group modifications that react with functional groups on the solid support are immobilized onto the support material surfaces by using several robotic printing techniques, such as "microspotting" and "ink-jetting" (Figures 2B and C). Using conventional techniques such as polymerase chain reaction and biochemical synthesis, strands of identified DNA are made and purified. A variety of probes are available from commercial sources, many of which also offer custom production services. The ability to purify full-length products before arraying is a major advantage of immobilization[13]. Protein libraries are also available or can be custom made as probes for the production of protein chips by a similar immobilization strategy discussed above.

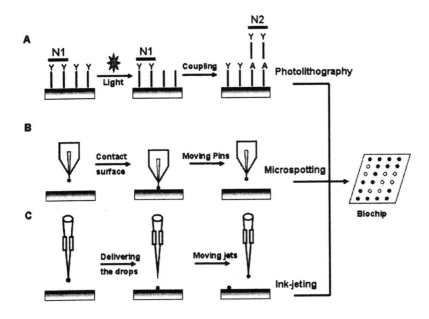

Figure 2. Putting probes onto microarray surfaces: (A) Photolithography; (B) microspotting; (C) ink-jetting.

Several supports like glass, silicon, gold, and polymeric membranes are used as solid substrates for the immobilization of the oligodeoxynucleotide (ODN) or polypeptides in microarray fabrication. An effective immobilization technique should provide the followings: (i) a high density of the probe molecules (e.g. DNA) on the surface of the microarray; (ii) efficient interaction reactions (e.g. hybridization in the case of DNA arrays) for capturing target molecules; (iii) low background; (iv) stability; (v) reproducibility; and (vi) sensitivity[14].

An immobilization reaction should have several characteristics. Firstly, the reaction should occur rapidly and therefore allow the use of low concentrations of reagents for immobilization. The chemistry should require little, if any, post-synthetic modification of ligands before immobilization to maximize the number of compounds that can be generated by solution or solid-phase synthesis and minimize the cost of these reagents. The immobilization process should occur selectively in the presence of common functional groups, including amines, thiols, carboxylic acids, and alcohols, to ensure that ligands are immobilized in an oriented and homogeneous manner. Correct orientation of the ligand molecules on the surface, and use of a spacer arm are important and critical, and make the ligand available for the target (Figure 3). Surface density of the ligand should be optimized. Low density surface coverage will yield a correspondingly low hybridization signal rate. High surface densities may result in steric interference between the covalently immobilized ligand molecules, impending access to the target molecules. Finally, the reaction should have well-behaved kinetics and be easily monitored with conventional spectroscopic methods to control the density of ligands on the chip[15].

A variety of synthetic chemical approaches have been used to immobilize DNA molecules to glass and other surfaces including photoactivatable chemistries[12,16], hydrogel-based chemistries[17], polylysine-mediated surfaces, dendritic linkers[18], and homo-or hetero-bifunctional cross-linkers[19-23].

There are several techniques to follow (measure) the interaction of the probe molecules (immobilized onto the array surface) and the target molecules (in the sample). In most of the biochips the measuring device, the so-called "readers", is based on measurement of fluorescence. Note that a fluorescent label needs to be attached to the target molecules. This is a time consuming step and needs expensive labeled substances and detection equipment. Unlike DNA, proteins are much more sensitive to their physiological environment and can easily be degraded by physical or chemical effects. Therefore, marker binding may inversely affect the natural activity of a protein[1,24].

These problems cause increasing interest in label-free monitoring methods such as Surface Plasmon Resonance (SPR) biosensors and Ellipsometry/SPR in DNA and protein chip technology[24,25]. In general SPR has gained attention as a label-free method for real-time detection of the binding of biological molecules onto functionalized surfaces[25]. SPR technology is expected to play an important role especially in proteomics research because it does not require labelling of the protein target[26-28].

3. SUBSTRATES FOR BIOCHIPS

3.1 Glass

Glass is one of the most preferred solid supports for DNA and protein microarray technology. Table 1 exemplifies some commercially available glass slides. Primarily its low cost and low intrinsic fluorescence, transparency, resistance to high temperature and a relatively homogeneous chemical surface makes it a popular support. In addition, glass offers a number of practical advantages over porous membranes and gel pads. As liquids cannot penetrate the surface of the support, target nucleic acids have direct access to the probe without internal diffusion[29]. Microscope slides are commonly used in laboratories because they are easy to handle and adaptable to automatic readers.

Table 1. Some commercially available glass slides

Erie Scientific Co., Porstmouth, NH, USA, (Untreated)
Knittel Glaser, Germany, (Untreated)
Corning Life Sciences, Corning, USA, (Untreated)
AminoPrep, Sigma, USA, (Aminopropylsilane coated).
CMT-GAPS, Corning, Corning, USA, (Aminopropylsilane coated)
SilanePrep, Sigma, Germany, (Aminoalkylsilane derivatized)
Cell Associates, Houston, TX, USA, (Silanated, amine)
PolyPrep, Sigma, USA, (Polylysine-coated)
Schleicher and Schuell, Dassel, Germany, (Nylon coated)
Menzel-Glaser, Braunschweig, Germany, (Untreated and isothiocynate modified)
TeleChem International Inc., Sunnyvale, CA, USA (Aldehyde modified)
SurModics, Inc., Eden Prairie, MN, USA (Amino modified)
Orchid Bioscience, Inc., Princeton, NJ, USA (Mercaptosilane derivatized)

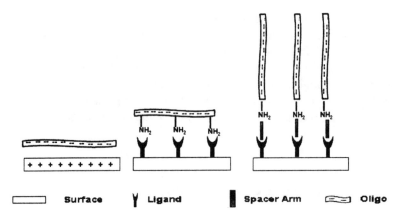

Figure 3. Probe ODN immobilization onto microarray surfaces

A number of chemical modification strategies have been employed by researchers for the attachment of oligonucleotides to glass surfaces. There are advantages and disadvantages of these approaches. Ligands (e.g., ODNs) can be immobilized onto glass surface by hydrophobic interaction. Glass surfaces can also be coated with polycations (e.g. polylysine) that allows ionic interaction between the negatively charged ODNs, which is a noncovalent immobilization approach. However, in both cases, due to the relatively weak bonds between the ligand molecules and the support, these microarrays often are susceptible to detachment of the ligand molecules from the surface when exposed to stringent hybridization conditions (high salt and/or high temperatures). These substrates have resulted in experimental inconsistencies that beget inconclusive data interpretation.

Covalent coupling chemistries for the immobilization of the ligand molecules onto glass slides are preferred, because they provide stable and reproducible immobilization[14]. Here, at the first step glass slide surfaces are activated to form hydroxyl groups. Several solutions have been applied to do this, as given in Table 2. The most successful, and widely used one is the Piranha solution.

Table 2. Solutions to form hydroxyl groups on glass slides

25% NH_4OH
KOH solution pH: 11
NaOH solution (10%)
H_2O_2 (30%) and H_2SO_4 (70%) (Piranha Solution)
1 N HNO_3
HCl (saturated)-H_2O (3:1)

At the second step, glass surface is modified usually with various type of silane compounds (Table 3) via the surface OH groups. Glass surfaces can be modified by silane chemistry to introduce specific functions such as amino, epoxy, carboxylic acid, isocyanate, sulfhydryl, and aldehyde, which are then used for the immobilization of ODNs carrying different functional groups on their 5'-ends (Table 4)[3].

Table 3. Examples of silane compounds

Silane Compound	Functional Group Created
3-Aminopropyltrimethoxysilane	amino groups
p-Aminophenyltrimethoxysilane	amino groups
Diethoxymethylaminopropylsilane	amino groups
Triethoxyaminopropylsilane	amino groups
Glycidyloxopropyltrimethoxysialne	epoxy
3-Mercaptopropyltrimethoxysilane	sulfhydryl groups
3-Isocyanatepropyltrimethoxysilane	isocyanate groups

Table 4. Modified glass surfaces and suitable ODNs with functional end groups

Surface	ODN
Amino carrying surfaces	Carboxylated-ODN
	Phosphorylated-ODN
Isocyanate, epoxy or aldehyde carrying surfaces	Aminated-ODN
Aminosilane via hetero bifunctional cross linker or mercaptosilane modified	Thiol modified-ODN
	Bisulfide modified-ODN
Immobilized streptavidin carrying surfaces	Biotin modified-ODN

Electrophilic glass coatings such as epoxysilane[30] or phenylisothiocyanate[31] have been prepared by direct silylation, but chemical instability of the reactive molecular monolayer can lead to variation in performance. In general, glass surfaces with strongly electrophilic groups have poor shelf-life and are difficult to control. Glass supports coated with milder electrophiles such as aliphatic aldehydes or carboxylic acids have also been used for preparation of microarrays[3]. Although these supports are more stable, reaction of alkylamine-modified ODNs with aldehyde-modified supports require post-arraying reduction of the unstable Schiff base linkages with sodium borohydride. Use of carboxylic acid supports requires use of carbodiimide coupling agents. More commonly, amine-modified slides are activated with homo- or hetero-bifunctional coupling agents to provide an electrophilic surface just prior to arraying[32,33]. Thiol-modified or disulfide-modified oligonucleotides have also been grafted onto aminosilane via a heterobifunctional crosslinker[32] or on 3-mercaptopropylsilane. Note that use of coupling agents is difficult for researchers since the conditions are difficult to reproduce and reagents are hazardous.

Phosphite-triester chemistry is another approach for directly attaching the first nucleotide to a glass plate by a covalent bond between hydroxyl group of the glass and the phosphate group of the protected deoxyribonucleotide. This covalent bond is similar in structure and strength to the phosphodiester bonds within the DNA molecule[34].

Modification of nucleic acids by silane compunds allow molecules to be attached covalently to unmodified glass surface directly. Different procedures developed to covalently conjugate an active silyl moiety on the oligonucleotides or cDNA in solutions to form a new class of modified nucleic acid, namely silanized nucleic acids. The silanized oligonucleotides and cDNA were shown to immobilize readily on glass slides upon deposition. The immobilization is fast and chips produced with this method gave strong hybridization signals with negligible background[35].

Dendritic linker systems attach to silica surfaces, enabling the highly efficient immobilization of amino-modified nucleic acids. These surfaces contain a thin layer of covalently immobilized and cross-linked polyamidoamine (PAMAM) starburst dendrimers[18]. PAMAM dendrimers, initially developed by Tomalia et al. in the early 1980s[36], provide a high density of terminal amino groups at the outer sphere. Dendrimer surface containing non-crosslinked PAMAM moieties were prepared for microarray applications. The investigation of the surfaces for microarray production revealed a significantly increased binding capacity, highly homogeneous oligomer spots as well as a remarkable stability against regeneration procedures. Moreover, the dendrimer-based DNA microarrays were used to reliably discriminate SNPs[37].

A method of conjugating polyethylenimine onto glass microarray surfaces was also prepared for the production of spotted microarrays. The method involves cyanuric chloride, a multifunctional cross-linking agent that provides residual electrophilic activity for reaction with deposited ODNs[38].

Glass is also the most popular platform for protein (e.g. antibody) arrays[39,44]. However, it must be thoroughly cleaned with 1:1 methanol and hydrochloric acid[44] and then chemically modified to facilitate covalent binding[42]. The latter involves coupling the protein molecules to the glass, which is vital for high-affinity probe–substrate interactions. Silanization is a common coupling method. Silanized slides (containing amine attachment sites) can be obtained commercially or can be individually silanized using 2% (3-mercaptopropyl) trimethoxysilane in toluene[44]. Antibodies also bind to silylated glass following treatment with an aldehyde-containing silane reagent[41]. They can also bind to glass slides coated with an agarose[43] or polylysine[40] film.

Surface modification of glass slide surfaces by ozonization and further ligand immobilization studies by common ozon generators and also AFM are under investigation by the authors' group as an ongoing research project named as "Preparation and Use of Single and Multichannel Ellipsometer Glass Slides Carrying Oligonucleotide and Oligopeptide Probes by Classical Ozonization or by AFM" As seen in Figure 4, OH groups on the modified glass slides are silanized, and then converted into ozonide by ozonization, and these unstable group to different functional groups. ODNs with different functional groups and amino acids as pseudo specific ligands for protein chips are prepared at the last step by using several activating agents described above.

3.2   Silicon Surfaces

The use of silicon wafers as a solid support for biochips has several additional advantages compared to the commonly employed microscope slides. Silicon wafers have less surface roughness, which increase uniformity of probe deposition with a higher density and smaller size of DNA spots. The flat surface allows high-density microarrays to be analyzed with confocal laser-based scanners. Silicon surfaces provide a better signal-to-noise ratio because silicon wafers show less background fluorescence and the dark, non-transparent surface absorbs excitation light. Finally, silicon wafer technology is readily available which certainly facilitate the fabrication of the chip-based devices[20].

Unoxidized crystalline silicon offers advantages as a substrate for immobilization because of its high purity, highly organized and defined crystalline structure, robustness, and its ubiquitous use in microelectronics industry[45].

Native silicon surfaces react with air under ambient conditions to form a thin surface layer of silicon oxide. This oxidized silicon surface is chemically similar to glass and suffers from some of the same drawbacks, namely inhomogeneity and variability in the relative number of Si-O-Si and Si-OH linkages. This inhomogeneity can lead to difficulties in the reproducibility and homogeneity of the subsequent ligand immobilized surfaces[46,47]. Recently, chemical pathways for direct functionalization of silicon substrates without an oxide layer has opened up new possibilities for highly controlled ligand immobilization which involves direct carbon–silicon bonds, and have resulted in methyl, chlorine, ester, or acid terminated substrates[45].

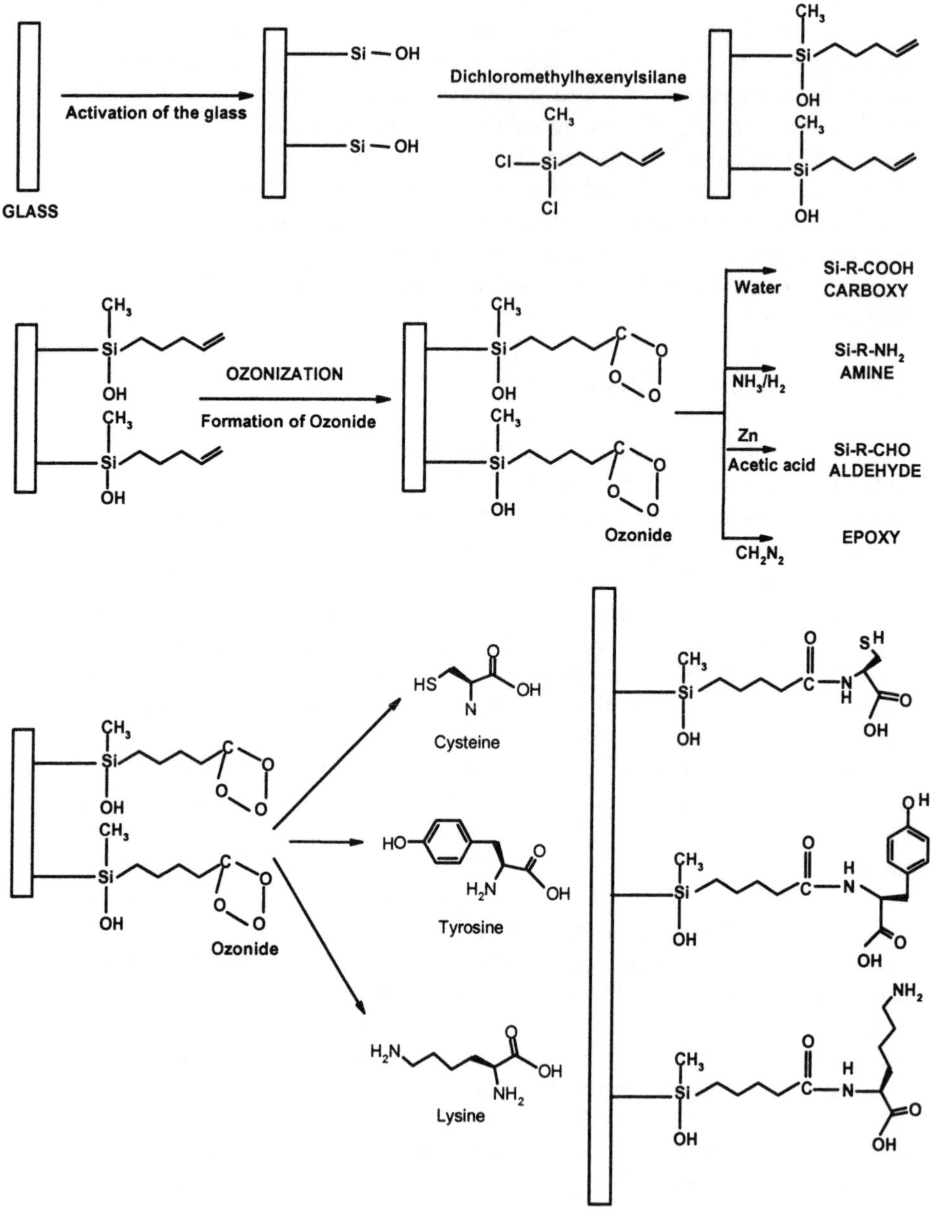

Figure 4. Surface modification of glass slides by ozonization and ligand immobilization.

3.3   Gold Slides

Gold-coated glass slides are used in SPR microarray applications. Several immobilization chemistries that are suitable for SPR applications have been reported, to immobilize the oligonucleotides and oligopeptides. Alkanethiol modification, hydrogel coating, silane derivatization, polypeptides and polymer films are applied to functionalize gold surfaces to create DNA and protein microarrays[48].

Classical immobilization of biotin modified-DNA onto gold surface by using dithiocompounds, used also by the authors group, is schematically given in Figure 5. Here, a dithio-compound (e.g. 3,3'-dithiodipropionic acid) is first reacted with the gold surface, then by carbodimide activation of the carboxylic acid groups, avidin (or streptavidin) is attached. Finally biotin-carrying DNA is immobilized specifically with the binding reaction of avidin with biotin.

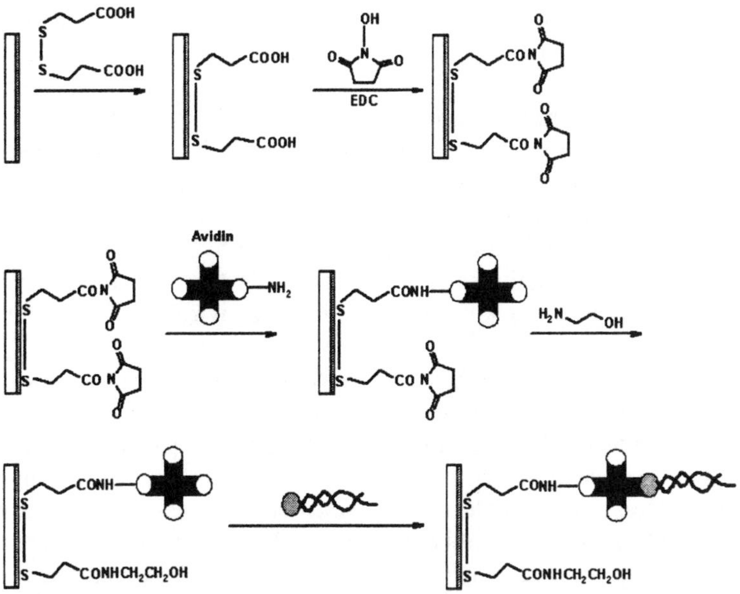

Figure 5. Immobilization of DNA onto gold surfaces via modification by 3,3'-dithiodipropionic acid.

In the mid-1980s, the spontaneous adsorption of organosulfur compounds on gold was a widespread method for the preparation of self-assembled monolayers (SAMs)[49-52]. If the alkyl chain is long enough, these architectures are very stable and show a molecular orientation nearly perpendicular to the surface[50,53-55]. Moreover, the alkyl chains can be terminated by reactive head groups that are capable of supporting the immobilization of biomolecules via covalent chemical coupling, electrostatic physiosorption or supramolecular interactions[56].

A more reliable immobilization method is covalent attachment of the biological ligand to the metal surface via a linker layer. The linker layer serves as a functionalized structure for further modification of the surface, as well as creating a barrier to prevent proteins and other ligands from coming into contact with the metal. This has been conventionally accomplished by first forming a self-assembled monolayer of long-chain alkanethiols with suitable reactive groups on one end of the molecule and a gold-complexing thiol on the other[57]. Both chemisorbed and physiosorbed alkanethiol molecules were involved during monolayer formation, with the overall formation kinetics determined by the relative solubility of the alkanethiol in the solvent (ethanol or heptane) [58]. The recognition element can then be attached directly to the alkanethiol monolayer. The physisorption and chemisorption of protein molecules onto self-assembled monolayers of carboxylate-terminated mercaptoalkanoic acids on gold surfaces were studied by SPR.

A flexible hydrogel matrix composed of carboxylmethylated dextran chains (100–200 nm) forms a porous three-dimensional linking matrix. It increases the number of immobilization sites as compared with a bare metal film, thereby increasing the response of the sensor surface. It is possible to immobilize various molecules to this dextran layer by conventional chemical methods. This hydrophilic surface provides little tendency for nonspecific adsorption of proteins as well as a more natural environment for antibody–antigen association.

The thiol-bonded monolayer described above is relatively stable under controlled conditions but can be destabilized by chemicals and other factors in the environment[59,60]. Alternative methods, employing a diverse range of reagents, have been described in the literature for the purpose of antibody immobilization. For example, organic silica compounds have been used to attach small amounts of antibodies to a gold surface for sensitive SPR immunosensing[61]. The surface was initially functionalized using g-aminopropylethoxysilane followed by a reaction of the surface amino groups with glutaraldehyde and coupling of either anti-human serum albumin or monoclonal antiatrazine antibodies.

Recombinant proteins produced by genetic engineering are crucial for fabrication of protein chips. Genetically modified proteins have specific affinity tags on the N- or C-terminus. Several affinity ligands, such as nitrilotriacetic acid (NTA), have been used to easily purify recombinant proteins. NTA has high affinity for histidine (His)-tagged proteins in the presence of nickel. NTA-containing lipid monolayers were transferred to a gold layer deposited on a glass slide and His-tagged synthetic peptides were immobilized directionally[62]. Approximately 6000 yeast proteins were expressed as His-tag fusions, spotted onto Ni-NTA functionalized slides, and 80% were found to be active[63,64].

There has been research on SPR imaging which has the potential to be used in high-throughput detection of DNA-protein and protein interactions.[65] However, the technology to fabricate SPR based protein or DNA chips in a versatile manner was still lacking[26].

3.4   Polymeric Supports

Latex beads[66], polymeric membranes such as nylon and polystyrene have been explored as surfaces for immobilization of DNA Cyanuric chloride-modified ODNs have been coupled to amine-modified nylon beads, but these electrophilic ODNs required a post-DNA synthesis coupling step and degraded on drying[67]. ODNs bearing aliphatic aldehydes have been used for coupling to hydrazide-modified latex beads, but the electrophilic aldehyde groups were generated after DNA synthesis using periodate oxidation of a vicinal diol[66]. In addition, aliphatic aldehydes undergo side reactions and a borohydride reduction step was required to obtain stable linkages[13].

Due to the porous structure of filter membranes, relatively large amounts of DNA can be applied to such material, resulting in strong signal intensities and a good dynamic range. This effect is partly offset, however, by the higher background, produced mainly by the very same structural characteristic-the large surface area per spot-responsible for the high loading capacity. Filter arrays can be re-used frequently, because the DNA sticks to the nylon surface, after some initial losses of up to 50%[68]. On the other hand, much DNA is required for filter arrays production, since spot sizes cannot be reduced to a level possible with glass or other non-porous media[33].

4.   CONCLUSION

Finally despite the lack of an ideal universal surface or immobilization approach, existing methods, those discussed above are more than adequate for many applications[69].

ACKNOWLEDGEMENTS

Erhan Pişkin has been supported by Turkish Academy of Sciences (TÜBA) as a full member.

REFERENCES

1. Eing, A., and Vaupel, M., 2002, Imaging Ellipsometry in Biotechnology. *Scientific Series*, pp 1-10.
2. Zhu, H., and Snyder, M., 2003, Protein chip technology, *Curr. Opin. in Chem. Biol.,* 7, 55-63.
3. Zammatteo, N., Jeanmart, L., Hamels, S., Courtois, S., Loutte, P., Hevest, L., and Remacle, J., 2000, Comprasion between Different Strategies of Covalnet Attachment of DNA to Glass Surfaces to Build DNA Microarray. *Anal. Biochem.,* 248, 143-150.
4. Dolan, P. L., Wu, Y., Ista, K. L., Metzenberg, L. R., Nelson, M.A., and Lopez, G.P., 2001, Robust and efficient synthetic method for forming DNA microarrays. *Nucleic Acids Res.* , 21, e107.
5. Habb, BB., 2001, Advances in protein microarray technology for protein expression and interaction profiling. *Curr. Opin. in Drug Disc.,* 4, 116-123.
6. Cahill, DJ., 2001, Protein and antibody arrays and thier medical applications. *J. of Immunol. Methods,* 250, 81-91.
7. Stoll, D., Templin, MF., Schrenk, M., Traub, PC, Vohringer, CF., and Joos, TO., 2002, Protein Microarray Technology. *Front Biosci.,* 7:c, 13-c32.
8. Zhu, H., and Snyder, M., 2002, Omic approaches for unravelling signalling networks. *Curr. Opin. in Cell. Biol.* , 14, 173-179.
9. Lueking, A., Horn, M., Eickhoff, H., Bussow, K., Lehrach, H., and Walter, G., 1999, Protein microarrays for gene expression and antibody screening. *Anal. Biochem.,* 270, 103-111.
10. McGall, G.H., Barone, A.D., Diggelmann, M., Fodor., S.P.A., Gentalen, and E., Ngo., N, 1997, The efficiency of light directed synthesis of DNA arrays on glass substrates,. *J. Am. Chem. Soc.,* 119, 5081-5090.
11. Southern, E.M., Case-Green, S.C., Elder, J.K., Johnson, M., Mir, K.U., Wang, L., and Williams, J.C., 1994, Arrays of complementary olionucleotides for analysing the hybridization behaviour of nucleic acids. *Nucleic Acids Res.,* 22, 1368-1373.
12. Fodor, S. P. A. 1997, DNA SEQUENCING: Massively Parallel Genomics. *Science,* 277, 393-395.
13. Podyminogin, M.A., Lukhtanov, E.A., and Reed, M.W., 2001, Attachment of benzaldehyde-modified oligodeoxynucleotide probes to semicarbazide-coated glass. *Nucleic Acids Res.,* 29, 5090-5098.
14. Charles, P.T., Vora, G.J., Andreadis,J.D., Fortney, A.J., Meador, C.E., Dulsey, C.S., and Stenger, D.A., 2003, Fabrication and Surface Characterization of DNA Microarrays Using Amine- and Thiol-Terminated Oligonucleotide Probes. *Langmuir,* 19, 1586-159.

15. Houseman, B. T., Gawalt, E. S., and Mrksich, M., 2003, Maleimide-Functionalized Self-Assembled Monolayers for the Preparation of Peptide and Carbohydrate Biochips. *Langmuir*, 19, 1522-1531.

16. Lipshutz, R. J., Fodor, S. P. A., Gingeras, T. R., and Lockhart, D. J., 1999, High density synthetic oligonucleotide arrays. *Nat. Genet.*, 21, 20-24

17. Guschin, D., Yershov, G., Zaslavsky, A., Gemmell, A., Shick, V., Proudnikov, D., Arenkov, P., and Mirzabekov, A. 1997, Manual manufacturing of oligonucleotide, DNA, and protein microchips. *Anal. Biochem.* 250, 203-211.

18. Benters, R., Niemeyer, C. M., and Wohrle, D. 2001, Dendrimer-activated solid supports for nucleic acid- and protein-microarrays. *Chem.Biol. Chem.*, 2, 686-694.

19. Andreadis, J. D.; and Chrisey, L. A. 2000, Use of immobilized PCR primers to generate covalently immobilized DNAs for *in vitro* transcription/translation reactions. *Nucleic Acids Res.*, 28, e5.

20. Lenigk, R., Carles, M., Ip, N. Y., and Sucher, N. J., 2001, Surface Characterization of a Silicon-Chip-Based DNA Microarray. *Langmuir*, 17, 2497-2501.

21. Chrisey, L. A., Lee, G. U., and O'Ferrall, E., 1996, Covalent attachment of synthetic DNA to self-assembled monolayer films. *Nucleic Acids Res.*, 24, 3031-3039.

22. Chrisey, L. A., O'Ferrall, E., Spargo, B. J., Dulcey, C. S., Calvert, and J. M., 1996, *Nucleic Acids Res.*, 24, 3040-3047.

23. Oh, S. J., Cho, S. J., Kim, C. O., and Park, J. W., 2002, Characteristics of DNA Microarrays Fabricated on Various Aminosilane Layers, *Langmuir*, 18, 1764-1769.

24. Westphal, P., and Bornmann, A., 2002, Biomolecular detection by surface plasmon enhanced ellipsometry. *Sens. Actuators B*, 84, 278-282.

25. Bassil, N., Maillart, E., Canva, M., Levya, Y., Millot, MC., Pissard, S., Narwa, R., and Goossens, M., 2003, One hundred spots parallel monitoring of DNA interactions by SPR imaging of polymer-functionalized surfaces applied to the detection of cystic fibrosis mutations. *Sens. Actuators B*, 94, 313-323.

26. Zhu-, X.M., Lin, P.H., Ao, P., and Sorensen, LB., 2002, Surface treatments for surface plasmon resonance biosensors. *Sens. Actuators B*, 84, 106-112.

27. Rich, R., and Myszka, D.G., 2000, Survey of the 1999 surface plasmon resonance biosensor literature. *J. Mol. Recognition*, 13, 388-407.

28. Green, R.J., Fraizer, R.A., Shakesheff, K.M., Davies, M.C., Roberts, C.J., and Tendler, S.J.B., 2000, Surface plasmon resonance analysis of dynamic biological interactions with biomaterials. *Biomaterials*, 21, 1823-1835.

29. Southern, E., Mir, K., and Shchepinov, M., 1999, Molecular interactions on microarrays. *Nature Genetics (Suppl.)*, 21, 5–9.

30. Lamture,J.B., Beattie,K.L., Burke,B.E., Eggers,M.D., Ehrlich,D.J., Fowler,R., Hollis,M.A., Kosicki,B.B., Reich,R.K., and Smith,S.R., 1994, Direct detection of nucleic acid hybridization on the surface of a charge coupled device. *Nucleic Acids Res.*, 22, 2121–2125.

31. Chen, D., Yan, Z., Cole, D.L. and Srivatsa, G.S., 1999, Analysis of internal (n–1)mer deletion sequences in synthetic oligodeoxyribonucleotides by hybridization to an immobilized probe array. *Nucleic Acids Res.*, 27, 389–395.

32. Guo,Z., Guilfoyle,R.A., Thiel,A.J., Wang,R. and Smith,L.M., 1994, Direct fluorescence analysis of genetic polymorphisms by hybridization with oligonucleotide arrays on glass supports. *Nucleic Acids Res.*, 22, 5456–5465.

33. Beier, M., and Hoheisel, J.D., 1999, Versatile derivatisation of solid support media for covalent bonding on DNA-microchips. *Nucleic Acids Res.*, 27, 1970–1977.
34. Cohen, G., Deutsch, J., Fineberg, J., and Levine, A., 1997, Covalent attachment of DNA oligonucleotides to glass. *Nucl. Acids. Res.*, 25, 911-912.
35. Kumar, A., Larsson, O., Parodi, D., and Linag, Z., 2001, Silanized nucleic acids: a general platform for DNA immobilization. *Nucl. Acids. Res.*, 28, e71.
36. Tomalia, D.A., Naylor, A.M. and Goddard, W.A., 1990, Starburst dendrimers: molecular-level control of size, shape, surface chemistry, topology and flexibility from atoms to macroscopic matter. *Angew. Chem. Int. Ed. Engl.*, 29, 138–175.
37. Benters, R., Niemeyer, C.M., Drutschmann, D., Blohm, D., and Wöhrle, D., 2002, DNA microarrays with PAMAM dendritic linker systems. *Nucleic Acids Res.*, 30, e10.
38. Lee, P. H., Sawan, S. P., Modrusan, Z., Arnold, L. J., Jr., and Reynolds, M. A., 2002, An Efficient Binding Chemistry for Glass Polynucleotide Microarrays. *Bioconjugate Chem.*, 13, 97-103.
39. Wiese, R., 2001, Simultaneous multianalyte ELISA performed on a microarray platform. *Clin. Chem.*, 47, 1451–1457.
40. Haab, B.B., 2001, Protein microarrays for highly parallel detection and quantification of specific proteins and antibodies in complex solutions. Genome Biology, 2, 0004.1–0004.13.
41. MacBeath, G. and Schreiber, S.L., 2000, Printing proteins as microarrays for high-throughput function determination. *Science*, 289, 1760–1762.
42. Arenkov, P., 2000, Protein microchips: use of immunoassay and enzymatic reactions. *Anal. Biochem.*, 278, 123–131.
43. Afanassiev, V., 2000, Preparation of DNA and protein micro arrays on glass slides with an agarose film. *Nucleic Acids Res.*, 28, e66.
44. Rowe, C.A., 1998, An array immunosensor for simultaneous detection of clinical analytes. *Anal. Chem*, 71, 433–439.
45. Strother, T., Hamers, R.J., and Smith, L.M., 2000, Covalent attachment of oligodeoxyribonucleotides to amine-modified Si (001) surfaces. *Nucleic Acids Res.*, 28, 3535-3541.
46. Waddell, T.G., Leyden, D.E. and DeBello, M.T., 1981, The nature of organosilane to silica-surface bonding. *J. Am. Chem. Soc.*, 103, 5303–5307.
47. Linford, M.R., Fenter, P., Eisenberger, P.M. and Chidsey, C.E.D., 1995, Alkyl Monolayers on Silicon Prepared from 1-Alkenes and Hydrogen-Terminated Silicon. *J. Am. Chem. Soc.*, 117, 3145–3155.
48. Mullet, W.M., Lai, E.P.C., and Yeung, J.M., 2000, Surface plasmon resonance based immunoassays. *Methods*, 22, 77-91.
49. Nuzzo R. G., and Allara, D. L., 1983, Adsorption of bifunctional organic disulfides on gold surfaces. *J. Am. Chem. Soc.*, 105, 4481-4483.
50. Nuzzo, R. G., Dubois, L. H., and Allara, D. L., 1990, Fundamental studies of microscopic wetting on organic surfaces. 1. Formation and structural characterization of a self-consistent series of polyfunctional organic monolayers. *J. Am. Chem. Soc.*, 112, 558-569.
51. Porter, M. D., Bright, T. B., Allara, D. L., and Chidsey, C. D., 1987, Spontaneously organized molecular assemblies. 4. Structural characterization of n-alkyl thiol

monolayers on gold by optical ellipsometry, infrared spectroscopy, and electrochemistry, *J. Am. Chem. Soc.*, 109, 3559-3568.

52. Strong, L., and Whitesides, G. M., 1988, Structures of self-assembled monolayer films of organosulfur compounds adsorbed on gold single crystals: electron diffraction studies. *Langmuir*, 4, 546-558.

53. Arnold, R., Terfort, A., and Woll, C., 2001, Determination of Molecular Orientation in Self-Assembled Monolayers Using IR Absorption Intensities: The Importance of Grinding Effects. *Langmuir*, 17, 4980-4989.

54. Houssiau, L., Graupe, M., Colorado, R., Kim, H. I., Lee, T. R., Perry, S. S., and Rabalais, J. W., 1998, Characterization of the surface structure of CH_3 and CF_3 terminated *n*-alkanethiol monolayers self assembled on Au{111}. *J. Chem. Phys.*, 109, 9134-9147.

55. Buscher, C. T., McBranch, D., and Li, D. Q., 1996, Understanding the Relationship between Surface Coverage and Molecular Orientation in Polar Self-Assembled Monolayers. *J. Am. Chem. Soc.*, 118, 2950-2953.

56. Riepl, M., Enander, K., Liedberg, B., Schaferling, M., Kruschina, M., and Ortigao, F., 2002, Functionalized Surfaces of Mixed Alkanethiols on Gold as a Platform for Oligonucleotide Microarrays. *Langmuir*, 18, 7016-7023.

57. Duschl, C., Sevin-Landais, A., and Vogel, H. 1996, Surface engineering: optimization of antigen presentation in self- assembled monolayers. *Biophys. J.*, 70, 1985–1995.

58. Peterlinz, K. A., and Georgiadis, R., 1996, In Situ Kinetics of Self-Assembly by Surface Plasmon Resonance Spectroscopy, *Langmuir*, 12, 4731–4740.

59. Schlenoff, J. B., Li, M., and Ly, H., 1995, Stability and Self-Exchange in Alkanethiol Monolayers. *J. Am. Chem. Soc.*, 117, 12528–12536.

60. Bain, C. D., Troughton, E. B., Tao, Y. T., Evall, J., Whitesides, G. M., and Nuzzo, R. G., 1989, Formation of monolayer films by the spontaneous assembly of organic thiols from solution onto gold. *J. Am. Chem. Soc.*, 111, 321–335.

61. Sasaki, S., Nagata, R., Hock, B., and Karube, I., 1998, Novel surface plasmon resonance sensor chip functionalized with organic silica compounds for antibody attachment. *Anal.Chim. Acta*, 368, 71–76.

62. Lu, Y. J., Zhang, F., and Sui, S. F., 2002, Specific Binding of Integrin $alpha_{IIb}beta_3$ to RGD Peptide Immobilized on a Nitrilotriacetic Acid Chip: a Surface Plasmon Resonance Study. *Biochemistry (Mosc)*, 67, 933–939.

63. Zhu, H., Bilgin, M., Bangham, R., and Hall, D., 2001, Global Analysis of Protein Activities Using Proteome Chips. *Science*, 293, 2101–2105.

64. Seong, Y.S., and Choi, C.Y., 2003, Current status of protein chip development in terms of fabrication and application. *Proteomics*, 3, 2176–2189.

65. Brockman, J.M., Nelson, B.P., and Corn, R.M., 2001, Surface plasmon reosonace imaging measurements of ultrathin organic films. *Annu. Rev. Phys. Chem.*, 51, 41-63.

66. Kremsky, J.N., Wooters, J.L., Dougherty, J.P., Meyers, R.E., Collins, M. and Brown, E.L., 1987, Immobilization of DNA via oligonucleotides containing an aldehyde or carboxylic acid group at the 5' terminus. *Nucleic Acids Res.*, 15, 2891–2909.

67. VanNess, J., Kalbfleisch, S., Petrie, C.R., Reed, M.W., Tabone, J.C. and Vermeulen, N.M.J., 1991, A versatile solid support system for oligodeoxynucleotide probe-based hybridization assays. *Nucleic Acids Res.*, 19, 3345–3350.

68. Hauser, N.C., Vingron, M., Scheideler, M., Krems, B., Hellmuth, K., Entian, K.-D. and Hoheisel, J.D., 1998, Yeast, 14, 1209-1221.
69. Predki, P.F., (Article in press), Functional protein microarrays: ripe for discovery. *Curr. Opin. Chem. Biol.*

Microarchitectural Characterization of the Aortic Heart Valve

SARAH BRODY and ABHAY PANDIT
National Centre for Biomedical Engineering Science and Department of Mechanical and Biomedical Engineering, National University of Ireland, Galway, Ireland

1.   INTRODUCTION

Cardiovascular disease has become a global epidemic. It has been estimated by the World Health Organization that 16.6 million people die from cardiovascular disease annually[1]. Aortic heart valve disease, which can lead to heart failure, constitutes a high percentage of these deaths and every year approximately 60,000 aortic heart valve replacements are implanted in the US and 170,000 worldwide[2].

The aortic heart valve is located between the left ventricle and the aorta. The valve is comprised of three cusps, which open and close allowing blood to flow from the heart. Structurally the valve is composed of three layers: the fibrosa, the spongiosa and the ventricularis. The fibrosa unfolds as the valve closes and it is located on the superior surface of the closed valve. The ventricularis is located on the inferior surface of the closed valve. The spongiosa is a loose gelatinous layer between the fibrosa and ventricularis[3] (Fig.1). All layers are mobile and can compress and shear during the valves opening and closing cycle[4,5]. The aortic valve is susceptible to failure for a number of reasons; the most frequently being aortic valve regurgitation (AVR) and aortic valve stenosis (AVS).

AVR involves the backflow of blood into the left ventricle during the diastole phase of the heartbeat and occurs because the heart valve cannot close properly. AVS causes constriction of the valve orifice, leading to increased resistance to blood flow from the left ventricle into the aorta.

Biomaterials: *From Molecules to Engineered Tissues,* edited by
N. Hasırcı and V. Hasırcı, Kluwer Academic/Plenum Publishers, 2004

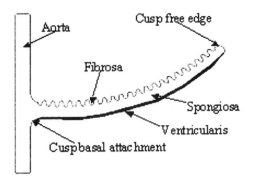

Figure 1. A cross-sectional view of the aortic heart valve cusp

Common causes of AVR and AVS are rheumatic fever, infective endocarditis, myxomatous valve disease[i], high blood pressure, Marfan syndrome[ii] and calcification. Congenitally acquired AVR and AVS occur for a number of reasons: the aortic valve is bicuspid rather than tricuspid, the valve develops in a funnel shape or the valve fails to grow at the same rate as the heart. Both conditions lead to increased demand on the heart, the left ventricle increases in size in an effort to pump more blood through the valve, and this in turn increases demand on the coronary arteries. Both conditions can lead to heart failure and are diagnosed using echocardiography. Common treatments for AVR and AVS include regular patient monitoring for mild cases, medication, surgical valve repair and in the case of AVS, balloon valvuloplasty[iii]. In more severe cases, the valve is replaced by either a mechanical or a bioprosthetic implant.

The world market for bioprosthetic implants is expected to reach US$1 billion by the end of this decade[6]; however, a fully effective aortic heart valve replacement has not been developed. Mechanical implants have good durability characteristics, but it is necessary for patients to remain on anticoagulation drugs indefinitely. Although bioprosthetic valves overcome the problems of coagulation, they can fail due to calcification and subsequent mechanical damage. Tearing, which occurs in areas experiencing high mechanical force, such as, the coaptation edges, is another form of bioprosthetic valve failure[6]. Bioprosthetic heart valves have poor durability with an average lifespan of only 10-15 years[7]. Current trends in research are focusing on the development of a tissue-engineered heart valve; which

---

[i] A disorder that weakens the valves tough connective tissue permitting the tissue to stretch abnormally.

[ii] A genetic disease of the connective tissue that affects many organs and can cause leakage of the aortic valve and dissection and widening of the aorta.

[iii] A procedure in which a balloon tipped catheter is used to expand the valve.

would overcome the problems of durability, calcification, mechanical damage and immunogenic reactions associated with the mechanical and bioprosthetic implants.

2. TISSUE ENGINEERING HEART VALVES

Although the concept of tissue engineering an aortic heart valve is futuristic in nature and currently far from clinical use, research in this area is currently underway. The design of a tissue-engineered heart valve (TEHV) can be approached from two basic concepts, the first involves the growth of autologous cells on either a biological or polymeric biodegradable scaffold followed by implantation, while the second involves the implantation of a suitable scaffold and subsequent growth of cells on the scaffold. It is hoped that both approaches would result in a valve replacement with similar functional, biochemical and mechanical properties as a healthy valve, that would grow with the patient and that would limit immunogenicity, coagulation and durability issues associated with current valve design. Development of such a TEHV involves a multi-disciplinary approach, drawing on the fields of cellular biology, biochemistry, physiology, biomaterials, biomechanics and medicine.

In the areas of cellular biology and biochemistry, recent studies have focused on the specific cell types found within the valve. Techniques for obtaining biopsies, isolating and culturing myofibroblasts that are structurally and functionally similar to those in natural aortic cusps, have been investigated[8]. Characterization of the smooth muscle cell population of the aortic valve through immunochemical analysis on cellular proteins has been investigated[9,10]. Studies have examined molecular and cellular biology of the interstitial cells and the role of integrins in cell-scaffold interactions, in particular the role that integrins play in tissue morphogenesis and adhesive interactions[11].

From a biomaterial perspective, scaffold development focuses on two specific methodologies. Scaffold development for the *in vivo* colonization method involves the generation of materials that support cellular growth after implantation. Biomaterials, such as small intestinal submucosa, have been studied for this purpose[12-14]. Various methods of decellularizing scaffolds have been investigated, as have changes to the mechanical properties of the scaffolds following the decellularization process[7,15,16]. The *in vitro* method, as the name suggests, focuses on scaffolds that support cellular growth outside the body. Recent studies on the suitability of polymeric scaffolds have indicated that polyhydroxyalkanoate, a thermoplastic polymer and hydrogels based on poly (vinyl alcohol) (PVA) may be advantageous due to their elastic material properties, biodegradability and biocompatibility[17,18].

Microarchitectural characterization of the aortic valve structure provides a critical parameter for consideration in TEHV scaffold design. Characterizing the microarchitecture of the collagen and elastin fibers within the valve and replicating them in the scaffold design will create a scaffold, which bears a microarchitectural resemblance to that of the natural valve.

Collagen is an elongated structural protein, found in the form of long fibrils, which aggregate to form fibers throughout the body. The fibers are crimped and are an important source of strength for tissues during tensile loading. Various different types of collagen exist and the particular type found in each organ is dependent on the organ in question[19]. Heart valves predominantly contain type IV collagen. Due to the large length-diameter ratio associated with tubular collagen fibers, they do not play a role in resisting compression loads. Elastin, another structural protein, plays an important role in ensuring tissue elasticity. It is believed that elastin is responsible for the elastic recoil of the heart valve[20]. However, the valve contains a relatively low amount of elastin compared to that of other cardiovascular tissues such as the aorta. The aortic valve contains 50% collagen and 13% elastin by dry weight[21].

Characterization of the microstructure and the configuration of structural components such as collagen and elastin are vital as they are directly related to biomechanical properties. For example, the collagen fiber orientation in pericardium is directly related to the functional demands of the pericardial sac. Pericardium experiences a broad range of loading conditions depending on the heart's activity and body position, and not surprisingly, tests have shown that collagen appears to have the ability to alter its configuration depending on loading direction[22]. Similarly, the surface of the aortic valve has tendonous chords which are oriented circumferentially. Under small loads the collagen fibers in these chords respond by slightly uncrimping in an accordion-like manner increasing the space between the fibers dramatically. The circumferential nature of these chords restricts circumferential deformation of the valve but allows radial deformation, which in turn allows sufficient coaptation[23].

Using various experimental techniques, the configuration of collagen and other structural proteins and their response to forces may be evaluated. In particular, collagen and elastin have been viewed by a number of techniques including electron microscopy[24,25], X-ray diffraction[26], immuno-histochemistry[27], transmission electron microscopy[28] and small-angle light scattering (SALS)[29,30]. The birefringent property of collagen allows it to be viewed using polarized light microscopy (PLM)[31].

2.1   Microstructure

PLM is a an ideal method for viewing collagen fibers and in particular collagen fiber orientation in translucent samples of hard and soft tissues since it does not necessitate the treatment of samples with stains, fluorescent labels and dyes and it does not alter tissue samples. Elastin is also a birefringent material, but its birefringence is somewhat lower than that of collagen fibers.

Birefringence is an optical property of a material. When light is passed through a polarizing filter and then through a birefringent material, its orientation is altered. The polarized light separates into two separate rays, an ordinary ray and an extraordinary ray (Fig. 2). The distance between the rays, or the degree by which one ray slows down relative to the other, generates a phase difference and is measured by a second polarizing filter (the analyzer) through which the light passes having passed through the specimen. The analyzer is oriented at 45° to the first polarizer. If there is no birefringent material on the microscope stage, no form of light can pass through both filters and no image will be created. When a birefringent material is present it alters the direction of the polarized light allowing it to be detected by the analyzer and an image of the birefringent material is created.

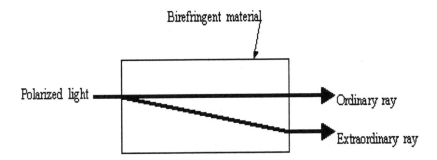

Figure 2. Polarized light passes through a birefringent material and separates into two rays, an ordinary and an extraordinary ray

However, the brightness of the image generated is dependent on the orientation of the birefringent material relative to the transmission axes of the filters[31]. To overcome this problem circular rather than linear analysis is used, whereby the stage of the microscope is rotated and various images of the birefringent material are taken. Therefore, circularly polarized light eliminates some of the orientation related effects normally associated with linearly polarized light. The use of circularly polarized light is particularly important in the case of collagen because it has a crimped shape[32].

PLM has been used widely in the area of biomedical research. On a cellular level, PLM has been used to visualize the dynamics and architecture of the mitotic spindle during cell division[33]. Extracellular matrix (ECM) has a low level of birefringence and therefore its use in the field of PLM has been limited, however, highly sensitive measuring techniques have been developed which allow the quantification of optical anisotropy in the ECM of unstained articular cartilage[34]. Polarized light has also been used, as an alternative to histological evaluation, to view the margins of cancer[35]. The degree of backscatter from the polarized light can distinguish cancerous from non-cancerous cells[36]. It has also been found that correlations exist between cardiac heart function and the ability of myocardial fibers to respond to ATP and calcium, the latter of which can be measured using quantitative birefringence measurements[37].

However, the use of PLM in the area of cardiovascular microstructural characterization has been limited. Much of the microstructural characterization has been done using SALS. Using SALS, it has been shown that the preferred fiber direction in the valve is the circumferential direction, with only a small percentage of the fibers oriented in the radial direction[38]. This knowledge of the collagen fiber distribution is an important consideration in TEHV design as it can be used in conjunction with mechanical tests to design mechanical valves to a higher standard and to compare bioprosthetic heart valves to healthy heart valves.

2.2   Biomechanics

The biomechanics of planar tissues is complex. Tissue biomechanics involves both the straightening of crimped fibers and the rotation of fibers towards the stretch axis[39]. An understanding of the biomechanics of natural heart valves is essential to the design of optimum mechanical, bioprosthetic and tissue engineered replacement valves. Although uniaxial tests are often used to determine tissue mechanical properties[40], results generated are of limited value. Tissues *in vivo* are rarely subjected to forces in only one direction and they are all anisotropic composites. Therefore, uniaxial tests do not give an accurate representation of loading conditions *in vivo*. Uniaxial loading does not attempt to recreate the full relationship between all stress and strain components. Biaxial loading involves the application of force along two axes. Triaxial loading (testing tissues in three directions) is impractical due to the presence of edge effects; however, in planar tissues, plain stress can be assumed, eliminating the need to test tissues in a third direction.

Various techniques for biaxial testing have been employed in recent studies. One of the most basic method is a ball burst test which is used to infer the in-plane biaxial strength of a material by measuring the resistance of the material to force applied via a metal ball[12]. However, results are limited, as intrinsic material properties cannot be calculated. The contact area that the ball has with the material increases as the force applied increases, so the relationship between stress and strain cannot be calculated because the area over which the force is applied is constantly changing. Another method of biaxial testing, involves load transmission to the tissue via twelve axes, each 30° apart[41].

The most commonly used method for biaxially testing tissues involves clamping a square-shaped specimen on four sides and applying a load along two axes. In such a system, designed by Fung[42], all samples are clamped along four edges and mounted in a water bath with saline solution (pH 7.4) at a specified temperature. Four force-distributing platforms operate from outside the water-bath on a sliding mechanism and apply a constant force. A video dimension analyzer consisting of a television camera, a video processor and a television monitor measures strain in the system. The force distributing platforms can operate independently or can be coordinated and strain is measured in two directions[42].

A number of factors must be considered when analyzing tissues biaxially. Carew *et al.* [43] investigated effects of specimen size and aspect ratio[i]. Results showed that specimen stiffness increased and extensibility decreased as the aspect ratio increased. Conversely, it was shown that as specimen size was reduced stiffness decreased and extensibility increased. Specimen shape and size is therefore an important factor to consider when comparing biaxial test results[43]. Other testing parameters that can vary between studies include methods of preconditioning, tissue clamping mechanisms, tissue markers and location of the tissue markers and the testing environment including testing temperature and the solutions used to test the tissues.

Extensive studies have been carried out on the response of cardiovascular tissues to biaxial force. Results on pericardium tests have indicated that stress-strain curves for pericardium are highly non-linear[44-48]. Pericardium exhibits high extensibility at low strains and rapid stiffening at high strains[46]. The biaxial response of other soft tissues, such as skin, aorta and small intestinal submucosa, has also been investigated[12,14,41,44].

Aortic valve mechanical behavior is anisotropic. Studies on the strain-pressure relationship in the aortic valve indicate that valve extensibility differs in the radial and circumferential directions, and that uniaxial tests overestimate strain in the radial and circumferential directions by a factor of

---

[i] The ratio between the width and length of a specimen.

two[49]. Studies have also shown that the ratio of radial to circumferential stretch in the aortic cusps is 6.0 ± 1.1[50].

The primary drawback of biaxially testing heart valves is the difficulty associated with generating stress-strain curves. Inconsistencies in tissue thickness make accurate stress calculation difficult. The specimen size and aspect ratio can influence the results[43], as can the gauge length or method used to estimate the appropriate gauge length[51]. Problems associated with estimating the gauge length arise because the excised heart valve assumes a more folded or wrinkled state and therefore the length of the tissue *in vitro* is not the same as the length at which the tissue begins to resist loading *in vivo*. In biaxial tests where markers are glued onto the tissue and their displacement is tracked, difficulties arise in ensuring that stress and strain states remain constant in the area of the tissue under examination, i.e. the area with the markers.

Biaxial tests have been used to compare the allograft aortic cusps to native aortic valve cusps to determine the effects of gluteraldehyde crosslinking on bioprosthetic heart valves. Explanted allograft cusps have less extensibility than fresh porcine aortic cusps[52]. Billar *et al.* have shown that chemically crosslinking cusps reduces their compliance and induces a higher level of anisotropy, compared to those of the fresh cusps[38,53]. The 'locking' effect of the crosslinking is thought to be responsible for the decrease in compliance and is related to the pressure at which fixation occurs[6]. Biaxial fatigue tests on glutaraldehyde crosslinked bioprosthetic heart valves have suggested that under long term cyclic loading the collagen in the valves undergoes structural changes[54].

2.3   Combinational Approach Using Microstructure and Biomechanics to Characterize Scaffold Properties

The geometrical configuration of collagen fibers and their interaction with the non-collageneous tissue components form the basis of the mechanical properties of biologic tissues[55]. Using a combination of PLM to determine the geometrical configuration of structural proteins and biaxial testing to study material response to applied loads, the mechanical properties of aortic heart valves may be elucidated.

Tower *et al.* developed a method for generating fiber alignment maps of tissues during uniaxial testing using PLM[56,57]. The collagen fiber direction was either parallel to or perpendicular to the direction of force application. Tissue equivalents were tested and the effects of crosslinking and preconditioning on fiber alignment were examined. For the tissue equivalents most of the fiber realignment occurred in the low force region, in the preconditioning tissues fiber realignment occurred during the stress free return to zero displacement and the crosslinked tissues exhibited a significant reduction in fiber motility. This method, which correlates collagen fiber structural changes with tissue biomechanics, can only be used in slow speed testing due to the time required to generate an image and results are limited due to the use of uniaxial rather than biaxial testing[58].

Other studies have involved a combination of biaxial tests and collagen imaging techniques, such as SALS. Billiar *et al.*[59] used a biaxial machine in conjunction with SALS to develop a system that would quantify fiber kinematics during stretch. A custom built biaxial device was designed and mounted vertically on the SALS device. Deformation was applied by turning four orthogonally positioned lead screws. Graphite markers were attached to the tissue using a cyanoacrylate adhesive and were tracked using a camera. Samples tested were bovine pericardium and porcine aortic valves and these were attached to the biaxial rig with sutures. It was found that for the loads tested the tissues have noticeably different fiber geometry and mobility. The porcine aortic valve has a highly oriented collagen fiber distribution, which does not alter significantly with stretch, but the bovine pericardium has a single broad fiber distribution when unloaded and when loaded is markedly changed[59]. Sacks *et al.*[60] have also conducted similar tests on the mitral valve by combining SALS with biaxial tests. Results indicated that collagen fibers in the mitral valve are designed to allow valve closure followed by a dramatic increase in stiffness to prevent leakage. The dramatic increase in stiffness can be attributed to the complete straightening of the collagen fibers in the leaflet. Similar studies by Kunzelman *et al.* used a combination of uniaxial tests, SALS, PLM and histological examinations to show that collagen fiber density is greater and stiffness is higher in chordae than in cusps of the mitral valve[40].

3.   RATIONALE

Using the phenomenon of birefringence PLM may be used to map out the microarchitecture of collagen fibers in the entire aortic valve cusp. Biaxial tests are an effective means of observing the biomechanical response of aortic heart valve cusps and can be used to validate the identified collagen fiber structural orientation. Microarchitectural characterization of the collagen fiber configuration as a basis for deriving the mechanical properties of the tissue, through biomechanical testing, will deduce optimum scaffold parameters. Studying the aortic valve microarchitecture will expand the scientific knowledge in the area of cardiovascular research and incorporating the microstructural parameters deduced into the design process has the potential to improve existing TEHV scaffold design.

4.   AIMS

The overall objective of the study was to identify optimum parameters for TEHV scaffold design. Specific aims under this objective were:

- To characterize the collagen fiber microarchitecture in the aortic valve using the PLM technique.
- To identify the structural differences in collagen fiber configuration in the three cusps of the aortic valve.
- To design and construct a biomechanical testing machine, which can be used to validate the microarchitecture.

5.   MATERIALS AND METHODS

5.1   PLM Sample Preparation and Analysis

Six porcine aortic hearts were acquired from a local abattoir. The hearts were dissected and the aortic valves were harvested. This procedure yielded six left non-coronary cusps, six left coronary cusps and six right coronary cusps. The cusps were rinsed with distilled water and crosslinked at room temperature by 0.5% glutaraldehyde in a 0.067 M phosphate buffer solution (pH 7.4) at atmospheric pressure for 24 h [48,61]. The cusps were then immersed in 0.067 M phosphate buffer solution for storage purposes prior to microscopic analysis.

Figure 3. A heart valve cusp mounted on a glass slide prior to microscopic examination

To prevent dehydration, samples were mounted on glass slides and covered with a thin layer of glycerol (Fig. 3). Due to the large area of the cusp, an average of thirty-three views of each valve cusp were taken. Each

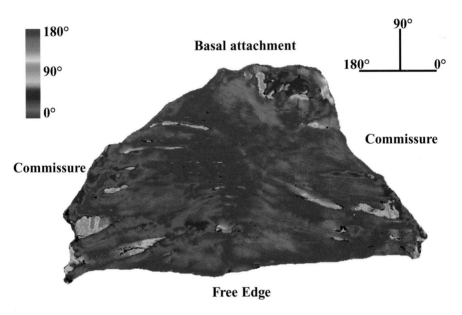

Figure 4. A compound image of an aortic heart valve cusp, depicting collagen fiber orientation. Collagen fiber angles are represented by colors, which correspond to the legend in the top left hand corner.

view consisted of three images and each image represented a quantity measured by the microscope: an orientation image, a birefringence image and a transmittance image.

- Orientation image: the angle of the birefringent material in the image.
- Birefringence image: a quantity of optical anisotropy that is related to the retardation of light passing through the specimen, i.e. a measure of birefringence.
- Transmittance image: light intensity through the system in the absence of any birefringence.

All images were saved and converted into bitmaps. The bitmaps were transferred to MS Paint™ (Microsoft Inc., Washington, USA) and assembled to create a composite image of the heart valve cusp. The composite image gave a pictorial representation of the angle of the collagen in the heart valve cusps. The angle was represented by colors which corresponded to the colors in the legend (Fig. 4).

Image analysis and statistical analysis was then carried out on the cusps to determine the collagen fiber configuration in the tissue. The image analysis software, MCID® (Imaging Research Inc., Ontario, Canada), calculated the number of pixels of a particular color in each image. The percentage area of each cusp covered with fibers in a particular fiber angle direction was then calculated and the results were statistically analyzed.

A preferred collagen fiber angle was also calculated. This is the average angle of the fibers covering the total area of all of the cusps. Biaxial loading was applied in the direction of the preferred fiber angle.

5.2   Biaxial System Design

A biaxial testing system was designed in collaboration with Zwick-Roell® (Zwick GmbH & Co. KG, Ulm, Germany). The system was designed as a fixture which could be used on the Zwick Z010® model (Fig. 5). Main features of the system included four 10 N load cells. The load cells were attached, via a specially built frame, to the crosshead of the machine. This mechanism ensured that all load cells moved at the same rate during testing. All tests were carried out in a water-bath, which was heated to 37°C, to simulate *in vivo* conditions. The water-bath was attached to the base of the biaxial system machine by four bolts. Four stainless steel cables were used to connect the load cells to the sample in the water-bath via pulleys located on the test platform (base) of the water-bath.

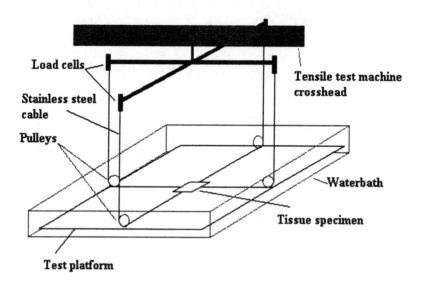

Figure 5. Schematic of the biaxial testing fixture

5.3 Biaxial Testing

The size of load cell (10 N) was chosen in order to minimize the margin of error and because the force required to test the samples to failure was not expected to exceed 10 N. An important aspect of any tensile or biaxial test is the clamping mechanism, especially in tests such as this where small strain rates are expected. Therefore, sandpaper was used in the clamping mechanism to prevent slippage (Fig. 6).

Figure 6. (a) An image of the biaxial system, (b) the pulleys, cables and clamping mechanism exerting force on the tissue specimen.

6. RESULTS

Characterization of collagen fiber configuration in the porcine aortic heart valve was deduced using polarized light. For statistical analysis purposes, the data generated was divided into specific 20° ranges, e.g. 10°-30°, 30°-50°. It was calculated that 76.5% of the area of the cusps contained fibers oriented in either the 10°-30° or 170°-10° ranges. Only 23.5% of the area of the cusp contained fibers oriented between 30° and 170° (Fig. 7). Therefore, the majority of the area is covered with fibers oriented in the circumferential direction.

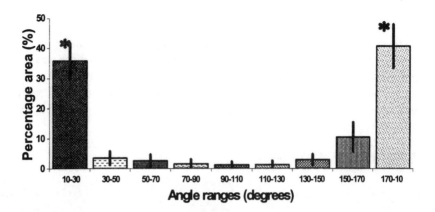

Figure 7. Average percentage area of the cusp covered by collagen fibers in each angle range. A statistical difference was found between the results for the 170°-10° and 10°-30° ranges compared to those of the other ranges (indicated by *) (n=18, p<0.05).

Collagen fiber orientation was also compared in the three cusps of the aortic valve and it was calculated that there is no statistical difference between the cusps; in the percentage areas covered by collagen fibers oriented in specific angle ranges (Fig. 8).

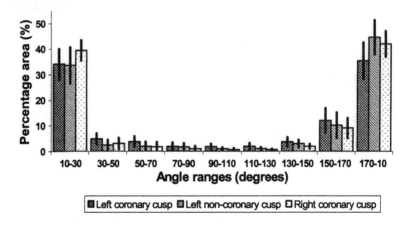

Figure 8. Average percentage area of the left coronary cusp, left non-coronary cusp and right coronary cusp covered collagen fibers in each angle range (n=18, p<0.05).

This is a semi-quantitative study since the actual number of fibers oriented in each particular direction is not known. A preferred collagen fiber angle of 10° was calculated using a weighted average method. From these results, it was hypothesized that the direction of maximum tensile strength is in the circumferential rather than the radial direction. In order to prove this hypothesis biaxial tests were conducted on fresh samples.

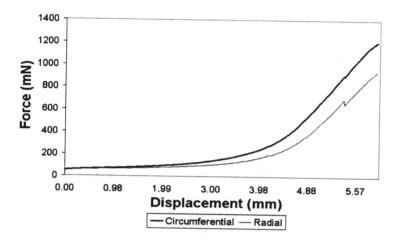

Figure 9. Force-displacement curve generated through biaxial testing. As hypothesized following the PLM study, aortic heart valve cusps have a greater Young's modulus in the circumferential direction than in the radial direction

For each sample tested, the biaxial machine generated four force-displacement curves, one curve per load cell. The average force-displacement curve in the radial and circumferential directions was then calculated (Fig. 9). Results of the biaxial tests supported those of the PLM analysis; stiffness is greater in the circumferential direction than in the radial direction.

7.   DISCUSSION AND CONCLUSION

The microarchitecture deduced through PLM indicates that the collagen fibers in the aortic heart valve are primarily oriented in the circumferential direction. It is possible that the downward pressure of the blood acting on the valve influences this orientation. A similar microarchitecture has previously been deduced using the SALS technique on bioprosthetic aortic heart valves[30]. The results of the biaxial tests are also representative of the biomechanics of the valve during opening and closing. Stiffness is lower in the radial direction than in the circumferential direction because during coaptation the valve extends further in the radial direction than in the circumferential direction. These results are similar to those deduced by Lo and Vesely[49]. The pressure versus strain relationship in fresh porcine aortic valve cusps was investigated and anisotropic distensibility was found. Maximal distensibility was greater in the radial direction than in the circumferential direction[49]. Christie and Barratt-Boyes[50], through tests on strips of fresh porcine aortic valve tissue, found similar results on aortic valve cusp anisotropic extensibility.

Although porcine and human hearts are comparable in size, correlations must be drawn between collagen orientation in human and porcine aortic valve cusps. This study was based on the assumption that the aortic valve cusp has uniform thickness. Sections of the valve were not taken for PLM analysis as it was felt that the difficulties associated with taking sections parallel to the ventricular surface would compromise any results generated.

This study was semi-quantitative. The percentage area of the aortic heart valve covered with collagen fibers in each angle range was calculated, however, the actual quantity of collagen fibers in each angle range was not measured. Therefore, it is desirable to quantify the level of collagen in the aortic valve, the levels of collagen in each angle range in the aortic valve and the collagen fiber distribution in the aortic valve in future studies.

The microarchitecture of collagen fibers in the aortic valve was characterized using PLM and alignment maps of the collagen fibers within the tissue were created for each cusp examined. Following both image and statistical analysis it was found that on average 76.5% of the collagen fibers lie between -10° and 30° and that the overall preferred collagen fiber angle direction is 10°. Collagen orientation in the left coronary cusp, left non-coronary cusp and right coronary cusp was also compared. The confidence interval calculated did not suggest a statistically significant difference in collagen fiber orientation in the three types of cusp. It is thought that the three cusps of the aortic valve are subjected to equal loads and have the same physiological function.

A biaxial system was designed, fabricated and used to validate the results generated through PLM analysis. As hypothesized, following calculation of the preferred fiber angle (10°), it was found that stiffness was greater in the circumferential direction than in the radial direction.

Tissue engineered heart valve design is a complex process requiring a multidisciplinary approach. The characterization of collagen fiber microarchitecture provides some of the parameters required for tissue engineering scaffold design. By incorporating these parameters in the scaffold design process, we may move one step closer to developing a TEHV that overcomes failures associated with current implant designs.

ACKNOWLEDGEMENTS

This study was supported by the Irish Council for Science, Engineering and Technology (IRCSET). Thanks also to Mr. Matthew Wallen and Mr Patrick Kelly, Department of Mechanical and Biomedical Engineering, National University of Ireland, Galway, Ireland.

REFERENCES

1. WHO, *The World Health Report 2002, Reducing risks, promoting healthy life*. 2002, World Health Organisation. p. 188 (Annex 2).
2. Schoen, F.J.,1995, Approach to the analysis of cardiac valve prostheses as surgical pathology or autopsy specimens. *Cardiovasc Pathol*, 4: 241-255.
3. Gross, L. and Kugel, M.A., 1931, Topographic anatomy and histology of the valves in the human heart. *Am J Pathol*, 7: 445-456.
4. Vesely, I. and Boughner, D., 1989, Analysis of the bending behaviour of porcine xenograft leaflets and of neutral aortic valve material: bending stiffness, neutral axis and shear measurements. *J Biomech*, 22(6-7): 655-671.
5. Song, S.H., Vesely, I., and Doughner, D.R., 1990, Effects of dynamic fixation on the shear behaviour of porcine xenograft valves. *Biomaterials*, 11: 191-196.
6. Vesely, I., 2003, The evolution of bioprosthetic heart valve design and its impact on durability. *Cardiovasc Pathol*, 12(5): 277-286.
7. Booth, C., Korossis, S.A., Wilcox, H.E., Watterson, K.G., Kearney, J.N., Fisher, J., and Ingham, E., 2002, Tissue engineering of cardiac valve prostheses I: development and histological characterization of an acellular porcine scaffold. *J Heart Valve Dis*, 11(4): 457-462.
8. Maish, M.S., Hoffman-Kim, D., Krueger, P.M., Souza, J.M., Harper, J.J., 3rd, and Hopkins, R.A., 2003, Tricuspid valve biopsy: a potential source of cardiac myofibroblast cells for tissue-engineered cardiac valves. *J Heart Valve Dis*, 12(2): 264-269.
9. Cimini, M., Rogers, K.A., and. Boughner, D.R., 2003, Smoothelin-positive cells in human and porcine semilunar valves. *Histochem Cell Biol*, 120(4): 307-317.
10. Flanagan, T.C., Mulvihill, A., Black, A., Jockenhoevel, S, and Pandit, A., 2003, Characterisation of mitral valve interstitial cells and endocardial cells in 2-D and 3-D culture. *Int Artif Organs*, 26: 584.
11. Wiester, L.M. and Giachelli, C.M., 2003, Expression and function of the integrin alpha9beta1 in bovine aortic valve interstitial cells. *J Heart Valve Dis*, 12(5): 605-616.
12. Whitson, B.A., Cheng, B.C., Kokini, K., Badylak, S.F., Patel, U., Morff, R., and O'Keefe, C.R., 1998, Multilaminate resorbable biomedical device under biaxial loading. *J Biomed Mater Res*, 43(3): 277-281.
13. Sacks, M.S. and Gloeckner, D.C., 1999, Quantification of the fiber architecture and biaxial mechanical behaviour of porcine intestinal submucosa. *J Biomed Mater Res*, 46: 1-10.
14. Gloeckner, D.C., Sacks, M.S., Billiar, K.L., and Bachrach, N., 2000, Mechanical evaluation and design of a multilayered collagenous repair biomaterial. *J Biomed Mater Res*, 52(2): 365-373.
15. Cebotari, S., Mertsching, H., Kallenbach, K., Kostin, S., Repin, O., Batrinac, A., Kleczka, C., Ciubotaru, A., and Haverich, A., 2002, Construction of autologous human heart valves based on an acellular allograft matrix. *Circulation*, 106(12 Suppl 1): 163-168.

16. Korossis, S.A., Booth, C., Wilcox, H.E., Watterson, K.G., Kearney, J.N., Fisher, J., and Ingham, E., 2002, Tissue engineering of cardiac valve prostheses II: biomechanical characterization of decellularized porcine aortic heart valves. *J Heart Valve Dis,* 11(4): 463-471.

17. Sodian, R., Sperling, J.S., Martin, D.P., Egozy, A., Stock, U., Mayer, J.E., Jr., and Vacanti, J.P., 2000, Fabrication of a trileaflet heart valve scaffold from a polyhydroxyalkanoate biopolyester for use in tissue engineering. *Tissue Eng,* 6(2): 183-188.

18. Nuttelman, C.R., Henry, S.M., and Anseth, K.S., 2002, Synthesis and characterization of photocrosslinkable, degradable poly(vinyl alcohol)-based tissue engineering scaffolds. *Biomaterials,* 23(17): 3617-3626.

19. Dorland, I. and Newman, W.A., 2003, *Dorland's Illustrated Medical Dictionary,* Saunders: Philadelphia, 388.

20. Vesely, I., 1998, The role of elastin in aortic valve mechanics. *J Biomech,* 31:115-123.

21. Bashey, R.I., Torii, S., and Angrist, A., 1967, Age-related collagen and elastin content of human heart valves. *J Gerontol,* 22(2): 203-208.

22. Chew, P.H., Yin, F.C., and Zeger, S.L., 1986, Biaxial stress-strain properties of canine pericardium. *J Mol Cell Cardiol,* 18(6): 567-578.

23. Thubrikar, M.J., 1990, *The aortic valve,* CRC Press: Boca Raton, FL., 23-24.

24. Petersen, W. and Tillmann, B., 1998, Collagenous fibril texture of the human knee joint menisci. *Anat Embryol (Berl),* 197(4): 317-324.

25. Goldman, H.M., Blayvas, A., Boyde, A., Howell, P.G., Clement, J.G., and Bromage, T.G., 2000, Correlative light and backscattered electron microscopy of bone--part II: automated image analysis. *Scanning,* 22(6): 337-344.

26. Alvisi, C., Bigi, A., Pallotti, C., Pallotti, G., Re, G., and Roveri, N., 1982, Carotid wall as an isotropic mechanical system. *Neurol Res,* 4(1-2): 47-61.

27. Chang, S.L., Howard, P.S., Koo, H.P., and Macarak, E.J., 1998, Role of type III collagen in bladder filling. *Neurourol Urodyn,* 17(2): 135-145.

28. Xia, Y. and Elder, K., 2001, Quantification of the graphical details of collagen fibrils in transmission electron micrographs. *J Microsc,* 204(Pt 1): 3-16.

29. Sacks, M.S. and Smith, D.B., 1998, Effects of accelerated testing on porcine bioprosthetic heart valve fiber architecture. *Biomaterials,* 19(11-12): 1027-1036.

30. Sacks, M.S. and Schoen, F.J., 2002, Collagen fiber disruption occurs independent of calcification in clinically explanted bioprosthetic heart valves. *J Biomed Mater Res,* 62(3): 359-371.

31. Whittaker, P., 1995, Polarised light microscopy in biomedical research. *Microscopy and analysis,* (1): 15-17.

32. Whittaker, P., Kloner, R.A., Boughner, D.R., and Pickering, J.G., 1994, Quantitative assessment of myocardial collagen with picrosirius red staining and circularly polarized light. *Basic Res Cardiol,* 89(5): 397-410.

33. Inoue, S., 1953, Polarisation optical studies of the mitotic spindle. 1. The demonstration of spindle fibres in living cells. *Chromosoma,* 5: 487-500.

34. Arokoski, J.P., Hyttinen, M.M., Lapvetelainen, T., Takacs, P., Kosztaczky, B., Modis, L., Kovanen, V., and Helminen, H., 1996, Decreased birefringence of the superficial zone collagen network in the canine knee (stifle) articular cartilage after long distance running training, detected by quantitative polarised light microscopy. *Ann Rheum Dis,* 55(4): 253-264.

35. Jacques, S.L., Roman, J.R., and Lee, K., 2000, Imaging superficial tissues with polarized light. *Lasers Surg Med,* 26(2): 119-129.

36. Hielscher, A.J., Eick, A.A., Mourant, J.R., Shen, D., Freyer, J., and Bigio, I.J., 1997, Diffuse backscattering Mueller matrices of highly scattering media. *Opt Express*, 1: 441-153.

37. Darracott-Cankovic, S., Stovin, P.G., Wheeldon, D., Wallwork, J., Wells, F., and T.A. English, T.A., 1989, Effect of donor heart damage on survival after transplantation. *Eur J Cardiothorac Surg*, 3(6): 525-532.

38. Billiar, K.L. and Sacks, M.S., 2000, Biaxial mechanical properties of the natural and glutaraldehyde treated aortic valve cusp-Part I: Experimental results. *J Biomech Eng*, 122(1): 23-30.

39. Millington, P.F., Gibson, T., E.J. H., and Barbenel, J.C., 1971, Structural and mechanical aspects of connective tissue. *Adv Biomed Eng*, 1: 189-248.

40. Kunzelman, K.S. and Cochran, R.P., 1992, Stress/strain characteristics of porcine mitral valve tissue: parallel versus perpendicular collagen orientation. *J Card Surg*, 7(1): 71-78.

41. Reihsner, R. and Menzel, E.J., 1998, Two-dimensional stress-relaxation behavior of human skin as influenced by non-enzymatic glycation and the inhibitory agent aminoguanidine. *J Biomech*, 31(11): 985-993.

42. Fung, Y.C., 1993, *Biomechanics - Mechanical properties of living tissues*. New York: Springer, 295-298.

43. Carew, E.O., Patel,J., Garg, A., Houghtaling, P., Blackstone, E., and Vesely, I., 2003, Effect of specimen size and aspect ratio on the tensile properties of porcine aortic valve tissues. *Ann Biomed Eng*, 31(5): 526-535.

44. Vito, R.P., 1980, The mechanical properties of soft tissues-I: A mechanical system for bi-axial testing. *J Biomech*, 13: 947-950.

45. Vito, R.P.,1979, The role of the pericardium in cardiac mechanics. *J Biomech*, 12(8): 587-592.

46. Choi, H.S. and Vito, R.P., 1990, Two-dimensional stress-strain relationship for canine pericardium. *J Biomech Eng*, 112(2): 153-159.

47. van Noort, R., Yates, S.P., Martin, T.R., Barker, A.T., and Black, M.M., 1982, A study of the effects of glutaraldehyde and formaldehyde on the mechanical behaviour of bovine pericardium. *Biomaterials*, 3(1): 21-26.

48. Langdon, S.E., Chernecky, R., Pereira, C.A., Abdulla, D., and Lee, J.M., 1999, Biaxial mechanical/structural effects of equibiaxial strain during crosslinking of bovine pericardial xenograft materials. *Biomaterials*, 20(2): 137-153.

49. Lo, D. and Vesely, I., 1995, Biaxial strain analysis of the porcine aortic valve. *Ann Thorac Surg*, 60(2 Suppl): S374-378.

50. Christie, G.W. and Barratt-Boyes, B.G., 1995, Mechanical properties of porcine pulmonary valve leaflets: How do they differ from aortic valve leaflets? *Ann Thorac Surg*, 60: 195-199.

51. Carew, E.O. and Vesely, I., 2003, A new method of estimating gauge length for porcine aortic valve test specimens. *J Biomech*, 36(7): 1039-1042.

52. Christie, G.W. and Barratt-Boyes, B.G., 1995, Biaxial mechanical properties of explanted aortic allograft leaflets. *Ann Thorac Surg*, 60: 160-164.

53. Billiar, K.L. and Sacks, M.S., 2000, Biaxial mechanical properties of the native and glutaraldehyde-treated aortic valve cusp: Part II-A structural constitutive model, *J Biomech Eng*, 122(4): p. 327-35.

54. Wells, S.M. and Sacks, M.S., 2002, Effects of fixation pressure on the biaxial mechanical behavior of porcine bioprosthetic heart valves with long-term cyclic loading. *Biomaterials*, 23(11): 2389-2399.

55. Ozkaya and Nordin, 1999, *Fundamentals of biomechanics (Equilibrium, motion and deformation)*. Springer-Verlag Inc: New-York, 205-206.

56. Tower, T.T. and Tranquillo, R.T., 2001, Alignment maps of tissues: 1. Microscopic elliptical polarimetry. *Biophys J*, 81: 2954-2963.
57. Tower, T.T. and Tranquillo, R.T., 2001, Alignment maps of tissues: 2. Fast harmonic analysis for imaging. *Biophys J*, 81: 2964-2971.
58. Tower, T.T., Neidert, M.R., and Tranquillo, R.T., 2002, Fiber alignment imaging during mechanical testing of soft tissues. *Ann Biomed Eng*, 30: 1221-1233.
59. Billiar, K.L. and Sacks, M.S., 1997, A method to quantify the fiber kinematics of planar tissues under biaxial stretch. *J Biomech*, 30(7): 753-756.
60. Sacks, M.S., He, Z., Baijens, L., Wanant, S., Shah, P., Sugimoto, H., and Yoganathan, A.P., 2002, Surface strains in the anterior leaflet of the functioning mitral valve. *Ann Biomed Eng*, 30: 1281-1290.
61. Duncan, A.C., Boughner, D., and Vesely, I., 1996, Dynamic glutaraldehyde fixation of a porcine aortic valve xenograft. I. Effect of fixation conditions on the final tissue viscoelastic properties. *Biomaterials*, 17(19): 1849-1856.

Biodynamic Modeling of Human Articulating Joints

ALİ ERKAN ENGİN
Department of Mechanical Engineering, University of South Alabama, Mobile, AL, USA

1. INTRODUCTION

A substantial difficulty in theoretical modeling of human joints arises from the fact that the number of unknowns are usually far greater than the number of available equilibrium or dynamic equations. Thus, the problem is an indeterminate one. To deal with this indeterminate situation, optimization techniques have been employed in the past[1,2]. Another technique dealing with the indeterminate nature of joint modeling considers the anatomical and physiological constraint conditions together with the equilibrium or dynamic equations. The different techniques used by various researchers mainly vary as to the method of applying these conditions. At one extreme, all unknowns are included in the equilibrium or dynamic equations. A number of unknown forces are then assumed to be zero to make the system determinate so that the reduced set of equations can be solved. This process is repeated for all possible combinations of the unknowns, and the values of the joint forces are obtained after discarding the inadmissible solutions[3].

The majority of joint models in the literature are either static or quasi-static. That is, the equilibrium equations together with the inertia terms are solved for a known kinematic configuration of the joint. Complexity of joint modeling becomes paramount when one considers a true dynamic analysis of an articulating joint structure possessing realistic articulating surface geometry and nonlinear soft tissue behavior. Although the literature[4] cites mathematical joint models that consider both the geometry of

Biomaterials: *From Molecules to Engineered Tissues,* edited by
N. Hasırcı and V. Hasirci, Kluwer Academic/Plenum Publishers, 2004

187

the joint surfaces and behavior of the joint ligaments, these models are quasi-static in nature, and employ the so-called inverse method in which the ligament forces caused by a specified set of translations and rotations along the specified directions are determined by comparing the geometries of the initial and displaced configurations of the joint. Furthermore, for the inverse method utilized in reference[4], it is necessary to specify the external force required for the preferred equilibrium configuration. Such an approach is applicable only in a quasi-static analysis. For a dynamic analysis, the dynamic equilibrium configuration *preferred* by the joint is the unknown and the mathematical analysis is required to provide that dynamic equilibrium configuration.

In this chapter first, a formulation of a three-dimensional mathematical dynamic model of a general two-body-segmented articulating joint is presented. The two-dimensional version of this formulation subsequently is applied to the human knee joint to investigate the relative dynamic motion between the femur and tibia as well as the ligament and contact forces developed in the joint. This mathematical joint model takes into account the geometry of the articulating surfaces and the appropriate constitutive behavior of the joint ligaments. With an improved solution method, the two-body segmented joint model is extended to a three-body segmented formulation, and an anatomically based dynamic model of the knee joint which includes patello-femoral articulation is presented to assess patello-femoral contact forces during kicking type of activity.

2.   GENERAL THREE-DIMENSIONAL DYNAMIC JOINT MODEL

The articulating joint is modeled by two rigid body segments connected by nonlinear springs simulating the ligaments. It is assumed that one body segment is rigidly fixed while the second body segment is undergoing a general three-dimensional dynamic motion relative to the fixed one. The coefficients of friction between the articulating surfaces are assumed to be negligible. This is a valid assumption due to the presence of synovial fluid between the articulating surfaces.[5] Accordingly, the friction force between the articulating surfaces will be neglected.

2.1 Representation of the Relative Positions

The position of the moving body segment 1 relative to the fixed body segment 2 is described by two coordinate systems as shown in Fig. 1. The inertial coordinate system *(x, y, z)* connected to the fixed body segment and a moving coordinate system (x', y', z') attached to the center of mass of the moving body segment.

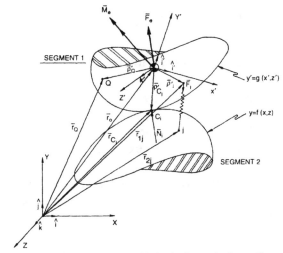

Figure 1. A two-body segmented joint is shown in three dimensions.

The (x', y', z') coordinate system is also taken to be the principal axis system of the moving body segment. The motion of the (x', y', z') system relative to the fixed (x, y, z) system may be characterized by six quantities: the translational movement of the origin of the (x', y', z') system along the x, y, and z directions, and θ, ϕ, and ψ rotations with respect to the x, y, and z axes.

2.2 Joint Surfaces and Contact Conditions

Assuming rigid body contacts between the two body segments are occurring at points C_i ($i = 1, 2$), let us represent the contact surfaces by smooth mathematical functions of the following form:

$$y = f(x,z), \quad y' = g(x',z') \tag{1}$$

As implied in Eq. (1), y and y' represent the fixed and the moving surfaces, respectively. The position vectors of the contact points C_i ($i = 1, 2$) in the base $(\hat{i}, \hat{j}, \hat{k})$, $\bar{r}_{c_i}$, and in the base $(\hat{i}', \hat{j}', \hat{k}')$, $\bar{\rho}'_{c_i}$, are related as

$$\{r_{c_i}\} = \{r_o\} + [T]^T \{\rho'_{c_i}\} \tag{2}$$

where matrix [T] is a 3 x 3 orthogonal transformation maxtrix.

This is a part of the geometric compatibility condition for the two contact surfaces. Furthermore, the unit normals to the surfaces of the moving, $\overline{n}'_{C_i}$, and fixed body segments, $\overline{n}_{C_i}$ at the points of contacts must be collinear, i.e.

$$\overline{n}_{C_i}\left\{n_{c_i}\right\}=\left[T\right]^T\left\{\hat{n}'_{c_i}\right\} \quad \text{or} \quad \left(\hat{n}'_{c_i} \times T^T\hat{n}'_{c_i}\right)=0. \qquad (3)$$

2.3 Ligament and Contact Forces

During its motion the moving body segment is subjected to the ligament forces, contact forces, and externally applied forces and moments (Fig. 1). The contact forces and the ligament forces are the unknowns of the problem and the external forces and moments will be specified.

The ligaments are modeled as nonlinear elastic springs. To be more specific, for the major ligaments of the knee joint, the following force-elongation relationship can be assumed:

$$F_j = K_j\left(L_j - \ell_j\right)^2 \text{ for } L_j > \ell_j \qquad (4a)$$

in which K_j is the spring constant, L_j and ℓ_j are, respectively, the current and initial lengths of the ligament j. The tensile force in the jth ligament is designated by F_j. It is assumed that the ligaments cannot carry any compressive force; accordingly,

$$F_j = 0 \quad \text{for } L_j < l_j \qquad (4b)$$

The stiffness values, K_j, are estimated according to the data available in the literature.[6, 7] The contact forces, N_j, acting on the moving body segment are given by

$$\overline{N}_i = \left|N_i\right|\left[\left(n_{c_i}\right)_x \hat{i}+\left(n_{c_i}\right)_y \hat{j}+\left(n_{c_i}\right)_z \hat{k}\right] \qquad (5)$$

where $|N_i|$, represent the unknown magnitudes of the contact forces and $(n_{c_i})_x$, $(n_{c_i})_y$, and $(n_{c_i})_z$, are the components of the unit normal, $\overline{n}_{c_i}$, in the x, y, z directions, respectively.

In general, the moving body segment of the joint is subjected to various external forces and moments whose resultants at the center of mass of the moving body segment are given as

$$\overline{F}_e = \left(F_e\right)_x \hat{i}+\left(F_e\right)_y \hat{j}+\left(F_e\right)_z \hat{k}, \quad \overline{M}_e = \left(M_e\right)_x \hat{i}+\left(M_e\right)_y \hat{j}+\left(M_e\right)_z \hat{k}. \quad (6)$$

2.4 Equations of Motion

The equations governing the forced motion of the moving body segment are:

$$\left(F_e\right)_x + \sum_{i=1}^{q}\left|N_i\right|\left(n_{c_i}\right)_x + \sum_{j=1}^{p}F_j\left(\lambda_j\right)_x = m\ddot{x}_o \quad \sum M_{x'x'} = I_{x'x'}\,\dot{\omega}_{x'} + \left(I_{z'z'} - I_{y'y'}\right)\omega_{y'}\,\omega_{z'}$$

$$\left(F_e\right)_y + \sum_{i=1}^{q}\left|N_i\right|\left(n_{c_i}\right)_y + \sum_{j=1}^{p}F_j\left(\lambda_j\right)_x = m\ddot{y}_o \quad \sum M_{y'y'} = I_{y'y'}\,\dot{\omega}_{y'} + \left(I_{x'x'} - I_{z'z'}\right)\omega_{z'}\,\omega_{x'} \quad (7)$$

$$\left(F_e\right)_z + \sum_{i=1}^{q}\left|N_i\right|\left(n_{c_i}\right)_z + \sum_{j=1}^{p}F_j\left(\lambda_j\right)_z = m\ddot{z}_o \quad \sum M_{z'z'} = I_{z'z'}\,\dot{\omega}_{z'} + \left(I_{y'y'} - I_{x'x'}\right)\omega_{x'}\,\omega_{y'}$$

where p and q represent the number of ligaments and the contact points, respectively. and $I_{x'x'}, I_{y'y'}, I_{z'z'}$, are the principal moments of inertia of the moving body segment about its centroidal principal axis system $(x',\ y',\ z')$, and ω_x, ω_y, and ω_z, are the components of the angular velocity vector along the principal axes.

Equations (7) form a set of six nonlinear second-order differential equations; along with the contact conditions (2) and (3) they become, assuming two contact points, i.e., (i=1,2), a set of 16 nonlinear equations with 16 unknowns. These unknowns are: three rotations (θ, ϕ, ψ), three components of position vector, $\overline{r}_o$, eight components of two contact points, and the magnitudes of two contact forces.

3. SOLUTION TECHNIQUES & APPLICATIONS

This formulation of three-dimensional dynamic modeling of an articulating joint was originally introduced by the author and one of his graduate students[8] at the Third International Conferences on Mathematical Modeling in 1981.Differential equations of motion were reduced to a set of nonlinear simultaneous algebraic equations by applying the Newmark method of differential approximation. By subsequent application of Newton-Raphson iteration process the same equations were converted to a set of simultaneous linear algebraic equations. Details of this dynamic modeling were provided in a journal article[9]. The two-dimensional version was subsequently applied[10] to the knee to investigate the dynamic motion between femur and tibia as well as the ligament and contact forces when a transient external force was applied perpendicular to the long bone axis of the tibia. Figure 2 shows the response of the anterior cruciate ligament and joint contact force for an exponentially decaying sinusoidal force pulse of various durations.

Although the solution of the two-dimensional reduced form (made of up three equations of motion, one contact condition and two geometric compatibility conditions) was successfully obtained by the iteration process[10],

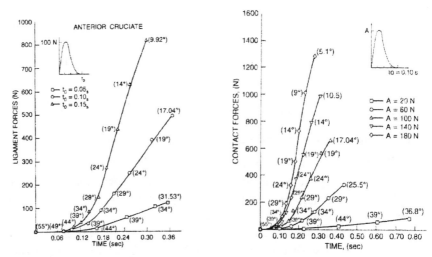

Figure 2. Anterior cruciate ligament and joint contact forces for an exponentially decaying pulse of various durations applied to the tibia.

it is found to be too cumbersome to yield a solution for the originally posed three-dimensional problem. The aforementioned difficulty of obtaining solutions of dynamic modeling of articulating joints led the author and his colleague to consider different approaches which yielded two completely different methods[11]. The first method which was called *Method of Excess Differential Equations* (EDE) involves conversion of the algebraic constraint equations to differential equations via second derivatives. The basic postulate here is that if the constraints are satisfied initially, then, satisfying the second derivatives of the constraints in future time steps would also satisfy the constraints themselves. Considering the 2-D case we will now have 6 differential equations instead of three. Simultaneous solution of these equations can be accomplished by means of any of the well established numerical integration methods (e.g., Euler's Method). The excess differential equations method has *no iteration* and the only numerical procedure is the integrations. The second proposed method which was named *Method of Minimal Differential Equations* (MDE) has a philosophy quite opposite to that of the first method. The basic idea here is the reduction of the number of differential equations in *closed form* by satisfying constraint equations as well as their derivatives and solving only the resulting nonlinear differential equations via numerical integration. These innovative solution techniques were utilized to solve the equations of a three-body segment model of the human knee joint involving tibio-femoral and patello-femoral joint articulations[12].

3.1 Three-Body Segment Dynamic Model of the Knee Joint

In an *Applied Mechanics Review* article, Hefzy and Grood[13] discussed both phenomenological and anatomically based models of the knee joint and stated, "To date, all anatomically based models consider only the tibio-femoral joint and neglect the patello-femoral joint, although it is an important part of the knee." In fact, only a few patello-femoral joint models are cited in the literature,[14,15,16] and they are restricted to the study of orientations and static forces related to the patella. Hirokawa's three-dimensional model[14] of the patello-femoral joint has some advanced features over the models of Van Eijden et al.[15] and Yamaguchi and Zajac.[16] Nevertheless, these models consider the patello-femoral articulation in isolation from the dynamics of the tibio-femoral articulation.

In this section, patello-femoral and tibia-femoral contact forces exerted during kicking type of activities are presented by means of a dynamic model of the knee joint which includes tibio-femoral and patello-femoral articulations and the major ligaments of the joint. Major features of the model include two contact surfaces for each articulation, three muscle groups (quadriceps femoris, hamstrings, and gastroc-nemius), and the primary ligaments (anterior cruciate, posterior cruciate, medial collateral, lateral collateral, and patellar ligaments). For a quantitative description of the model, as well as its mathematical formulation, three coordinate systems as shown in Fig. 3 are introduced. An inertial coordinate system (x, y) is attached to the fixed femur with the x axis directed along the anterior-posterior direction and the y axis coinciding with the femoral longitudinal axis. The moving coordinate system (u, v) is attached to the center of mass of the tibia in a similar fashion. The second moving coordinate system (p, q) is connected to the attachment of the quadriceps tendon, with its p axis directed toward the patella's apex.

Since we are dealing here with an anatomically based model of the knee joint, the femoral and tibial articulating surfaces as well as posterior aspect of the patella and intercondylar groove must be represented realistically. This is achieved by utilizing previously obtained polynomial functions.[9,17,18]. For ligaments we use the para-bolic and linear force-strain relationship developed by Wismans.[19] The motion of the tibia relative to the femur is described by three variables: the position of its mass center (x_{o2}, y_{o2}), and angular orientation θ_2. Equations of motion of the tibia can be written in terms of these three variables, along with the mass of the lower leg (m), its centroidal moment of inertia (I), the patellar ligament force (F_p), the tibio-femoral contact force (N_2), the hamstrings and gastrocnemius muscle forces (F_H,

F_G), the weight of the lower leg (W), and any externally applied force on the lower leg (F_E). The contact conditions at the tibio-femoral articulation and at the patello-femoral articulation are expressed as geometric compatibility and colinearity of the normals of the contact surfaces. The force coupling between the tibia and patella is accomplished by the patella ligament force F_P.

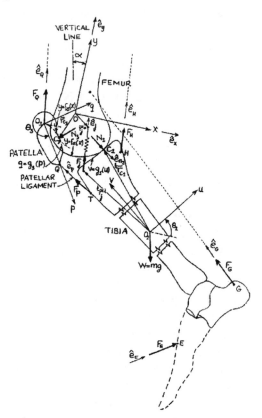

Figure. 3 Three-body segment model of the knee joint showing various coordinate systems, forces, and articulating surface functions.

The model has three nonlinear differential equations of motion and eight nonlinear algebraic equations of constraint. The major task in the solution algorithm involves solution of the three nonlinear differential equations of the tibia motion along with three coupled nonlinear algebraic equations of constraint associated with the tibio-femoral articulation. This is accomplished by following solution techniques developed by the author and his colleague[11,12] which are briefly described in Section 3.

The kicking type of lower limb activity is a rather complex activity that involves most of the muscles of the lower limb.[20] The model shown in Fig. 3 is general enough to include major muscles associated with knee motion, namely, the quadriceps femoris, hamstrings, and gastrocnemius muscle groups.[18] Any dynamic activity can be simulated with this model provided magnitudes, durations, and relative timings of these muscle groups are supplied. In this section, some results for the extension phase of the knee under the activation of the quadriceps femoris muscle group are presented. The force activation of the quadriceps muscle group during the final extension of the knee is taken in the form of an exponentially decaying sinusoidal pulse. For a quadriceps force of 0.1 sec and 2650 N amplitude, the maximum angular acceleration of the lower leg reaches to 360 rad/s^2 which is a typical value for a vigorous lower limb activity such as kicking. Fig. 4 shows for the final phase of knee extension variations of the tibio-femoral and patellofemoral contact forces with time. The aforementioned quadriceps pulse is applied when the knee flexion is at 55°. The values in parentheses indicate the flexion angles at the corresponding times; thus, behaviors of the patello-femoral and tibio-femoral contact forces are shown from the flexion angle of 55° to 5.5°. It is quite interesting to note that under such dynamic conditions, the patello-femoral contact force is higher than the tibio-femoral contact force.

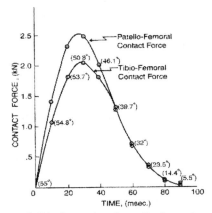

Figure 4. Variations of tibio-femoral and patello-femoral contact forces with time and flexion angles

4. CONCLUDING REMARKS

It is a well-established fact that in a class of activities such as stair climbing, rising from a seated position, or similar activities, large patello-femoral contact forces naturally accompany large knee-flexion angles. For these large knee-flexion angles, a rough estimate of the patello-femoral contact force can be easily obtained by considering a simple static equilibrium of the patella with patella tendon force and quadriceps femoris force. According to the static analysis, at full extension of the knee this force is practically zero. Results presented here indicate that the patella can be subjected to very large patello-femoral contact forces during a strenuous lower limb activity such as kicking even under conditions of small knee-flexion angles. Finally, under such dynamic conditions the patello-femoral contact force can be higher than the tibio-femoral contact force.

REFERENCES

1. Seirek, A. and Arvikar, R.J., A mathematical model for evaluation of forces in lower extremities of the musculo-skeletal system, *J. Biomechanics,* 6, 313, 1973.
2. Seirek, A. and Arvikar, R.J., The prediction of muscular load sharing and joint forces in the lower extremities during walking, *J. Biomechanics,* 8, 89, 1975.
3. Chao, E.Y., Opgrande, J.D., and Axmear, F.E., Three-dimensional force analysis of finger joints in selected isometric hand functions, J. *Biomechanics, 9, 387, 1976.*
4. Wismans, J., Veldpaus, F., Janssen, J., Huson, A., and Struben, P., A three-dimensional mathematical model of the knee joint, *J. Biomechanics,* 13m 677m 1980.
5. Engin, A.E., Mechanics of the knee joint: guidelines for osteotomy in osteoarthritis, in *Orthopaedic Mechanics: Procedures and Devices,* Ghista, D.N. and Roaf, R., Eds., Academic Press, London, 1978, 55.
6. Kennedy, J.C., Hawkins, R.J., Willis, R.B., and Danylchuk, K.D., Tension studies of human knee joint ligaments, *J. Bone Jt. Surg.,* 58A, 350, 1976.
7. Trent, P.S., Walker, P.S., and Wolf, B., Ligament length patterns, strength and rotational axes of the knee joint, *Clin. Orthoped. Rel. Res.,* 117, 263, 1976.
8. Engin, A.E. and Moeinzadeh, M.H., Dynamic modeling of human articulating joints, *Proc. Third Internal Conf Math. Modeling,* 1981, 58.
9. Engin, A.E. and Moeinzadeh, M.H., Dynamic modeling of human articulating joints, *Mathematical Modelling,* 4, 117, 1983.
10. Engin, A.E. and Moeinzadeh, M.H., Two-dimensional dynamic modeling of human joints, *Dev. Theor. Appl. Mech.,* 11, 287, 1982.
11. Engin, A.E. and Tumer, S.T., An innovative approach to the solution of highly nonlinear dynamics problems associated with joint biomechanics, *ASME Biomechanics Symp.,* 120, 225, 1991.
12. Engin, A.E. and Tumer., S.T., Improved dynamic model of the human knee joint and its response to impact loading on the lower leg, *J. Biomech. Eng.,* 115, 137, 1993.
13. Hefzy, M.D. and Grood, E.S., Review of knee models, *Appl. Mech. Rev.* 41, 1, 1988.
14. Hirokawa, S., Three-dimensional mathematical model analysis of the patellofemoral joint, *J. Biomechanics,* 24, 659, 1991.
15. Van Eijden, T.M., Kouwenhoven, E., Verburg, J., and Weijs, W.A., A mathematical model *of* the patellofemoral joint, *J. Biomechanics,* 19, 219, 1986.
16. Yamaguchi, G.T. and Zajack F.E., A planar model of the knee joint to characterize the knee extensor mechanism, *J. Biomechanics,* 22, 1, 1989.
17. Engin, A.E. and Moeinzadeh, M.H., Modeling of human joint structures, Report AFAMRL-TR81-117, Wright- Patterson Air Force Base, OH, 1982.
18. Tumer, S.T. and Engin, A.E., Three-body segment dynamic model of the human knee, *J. Biomechanical Eng.,* 115, 350, 1993.
19. Wismans, J., A., A three-dimensional mathematical model of the human knee joint, doctoral dissertation, Eindhoven University of Technology, The Netherlands, 1980.
20. Lindbeck, L., Impulse and moment of impulse in the leg joints by impact from kicking, *J. Biomechanical Eng.,* 105, 108, 1983.

Novel Bioabsorbable Antibiotic Releasing Bone Fracture Fixation Implants

MINNA VEIRANTO*, ESA SUOKAS[#], NUREDDIN ASHAMMAKHI*, and PERTTI TÖRMÄLÄ*

*Institute of Biomaterials, Tampere University of Technology, Tampere, FINLAND;
#Linvatec Biomaterials Ltd., Tampere, FINLAND

1. INTRODUCTION

Bone infection is a serious complication that may follow any surgery or trauma. The risk of infection is especially high in immunocompromised patients such as diabetics, aged and those on immunosuppressive drugs[1-3]. Around 50% to 70% of patients with open (compound) fractures have bacterial contamination of their wounds and potential risk of a bone infection[1,4]. In 2001 the estimated number of fracture fixation device-associated infections was 100 000 – 200 000[5].

The primary treatment for bone infection is systemically administered antibiotics that penetrate bone and joint cavities. Chronic bone infection in most cases can be treated successfully only when all foreign bodies are removed and surgical debridement is used in combination with systemically administered antibiotic therapy[1, 6-8]. To avoid bone infection after surgical procedures or trauma, prophylactic administration of antibiotics is important[1, 9].

The systemic antibiotic therapy is generally long lasting requiring multiple antibiotics. Depending on the mechanism of infection and age of the patient one or two different antibiotics are commonly used in systemic therapy. This systemic antibiotic therapy normally lasts four to six weeks in acute cases, and in chronic cases it may last for more than three months[1, 17].

Adequate local drug levels may be difficult to achieve in infected or traumatised bone using systemically administered antibiotics because the blood supply to the infected or traumatised area may be compromised[10-13]. High doses of systemic antibiotics are thus needed to achieve sufficient local concentrations, which may cause unwanted side effects such as an organ toxicity[10, 14]. To achieve high concentrations of antibiotics locally and non-toxic levels systemically some surgeons prepare antibiotic-containing bone cements by adding antibiotic powder to the bone cement with subsequent hand mixing. However, bone cements prepared by hand mixing do not have consistent and controllable quality[15, 16].

Methods for local delivery of antibiotics have been widely studied and some of them have been clinically used in the last two decades. For example, polymethylmethacrylate (PMMA) impregnated with gentamicin has been commercially available in Europe since 1977 in the form of cement and beads. The concentration of gentamicin around those implants remains high in the area of two to three centimetres around the beads or cement. The concentration of gentamicin in the serum and organs remains very low - almost undetectable[14]. The major disadvantage of PMMA based systems is that PMMA is a non-resorbable polymer and the system must be removed in a second surgical procedure several weeks after implantation[14, 17].

Bioabsorbable local antibiotic-releasing systems do not require surgical removal. While bioabsorbable antibiotic-releasing systems based on synthetic bioabsorbable polymers have been widely studied in the last two decades, according to our knowledge, they are not yet commercially available. For example polylactides (PLA)[18, 19], copolymers of lactide and glycolide (PLGA)[7], polyanhydrides (PAN)[10,16] and polycaprolactone (PCL)[20] have been used to prepare implants that contain different antibiotics. Bioabsorbable polymers which have been studied for those systems are mainly low or medium molecular weight polymers with quite poor mechanical properties and such materials are not suitable for the manufacture of fracture fixation implants.

Besides general measures and systemic antibiotic therapy, local bioabsorbable antibiotic-releasing fixation implants may be advantageous for successful treatment of infected and potentially infection sensitive fractures. Recently, our group developed bioabsorbable antibiotic-releasing implants, which have their primary function as fixation devices with additional secondary function of drug delivery. After implantation, such implants provide stable fixation of fractured bone and the release of therapeutic antibiotic levels at the implantation site. These implants are bioabsorbable and thus do not need a second operation for their removal. In this paper we describe materials and manufacture, mechanical and drug release properties of these recently-developed antibiotic releasing bone fracture fixation implants, and their effect on bacterial attachment and biofilm formation *in vitro*.

2.   MATERIALS

Two different high molecular weight bioabsorbable polymers were studied as matrix polymers for antibiotic releasing fixation implants.

The first was the commercial Resomer® LR708 (Boehringer Ingelheim, Germany). It is an amorphous synthetic bioabsorbable copolymer of L-lactide and DL-lactide with monomer ratio 70L, 30DL. The measured inherent viscosity of P(L/DL)LA 70:30 was 6.3 dl g^{-1} and weight average molecular weight (M_w) was 910 000 g mol^{-1}.

Another studied matrix polymer was PuraSorb®PLG (Purac Biochem bv., Gorinchem, Netherlands), which is a semicrystalline bioabsorbable copolymer of L-lactide and glycolide with monomer ratio 80L, 20G. The measured inherent viscosity of PLGA 80:20 was 6.4 dl g^{-1}.

The antibiotic was ciprofloxacin, which is a commonly used synthetic antibiotic. It is a light yellow crystalline powder with molecular weight of 331.4 g mol^{-1}. It belongs in the family of fluoroquinolone antibiotics and is active against common bone infection causing bacteria such as *Staphylococcus aureus, Staphylococcus epidermis* and *Pseudomonas aeruginosa*. Ciprofloxacin penetrates bone and joint cavities well. The minimum inhibitory concentration (MIC) of ciprofloxacin against most strains is less than 2.0 µg ml^{-1} [21, 22]. The potentially detrimental level that can adversely affect growth of osteoblast-like cells is more than 20 µg ml^{-1} [23, 24].

3.   MANUFACTURE OF THE IMPLANTS

Matrix polymer and antibiotic powder were extruded with a small laboratory scale mixer into billets, which were die-drawn into self-reinforced (SR) rods. During self-reinforcement the isotropic structure of the extruded bioabsorbable polymer changes into a highly anisotropic composite structure where the uniaxially oriented amorphous and crystalline phases have the same chemical composition. Implants based on self-reinforced high molecular weight bioabsorbable PLAs and PLGAs have been widely studied and used for bone fracture fixation in many different applications. Initial mechanical properties of SR implants coincide with those of bone and during the healing they degrade gradually and slowly transfer load to the healing bone[25-27].

Studied bioabsorbable antibiotic-releasing bone fracture fixation implants (Fig. 1), rods, and screws with the same geometry as BioSorb™PDX 1.5 Screws (SR-PLGA) and BioSorb™FX 2.0 Screws (SR-P(L/DL)LA) (Linvatec Biomaterials Ltd., Tampere, Finland), were machined from SR-rods. The finished implants were washed, dried under vacuum, packed into aluminium foil pouches and sterilised with γ-irradiation with a nominal dose of 25 kGy (Willy Rüsch AG, Kernen-Rommelshausen, Germany). After manufacture and sterilization ciprofloxacin was bacteriologically proved to be bioactive (unpublished data).

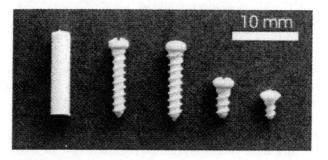

Figure 1. Examples of recently developed novel bioabsorbable ciprofloxacin releasing bone fracture fixation implants.

4.   MECHANICAL PROPERTIES *IN VITRO*

Mechanical properties were measured with an Instron 4411 universal testing machine (Instron Ltd., High Wycombe, England) at room temperature. Five to six parallel samples were tested at each sampling time. Shear strength of the screws was measured with a tool, which was constructed by modifying the device presented in the standard method ASTM B 769-94[28]. The crosshead speed was 10 mm min[-1]. Bending strength of the samples was determined with a three-point bending test according to the modified procedure of the standard methods ASTM D 790-84 and DIN 53452[29,30]. The crosshead speed was 2 mm min[-1] and the gauge length was 22 mm. Radius of the loading nose and the supports was 5.0 mm and 1.5 mm, respectively. Torsion strength of the samples was measured using a modified procedure of the standard method ASTM F 117-79[31]. The crosshead speed was 50 mm min[-1].

To determine the bending and shear strength of developed implants *in vitro*, samples were placed in a phosphate buffer solution (PBS) KH_2PO_4 (0.05 M) and NaOH (0.04 M) at a pH of 7.4 and kept stationary in closed brown flasks at 37 °C for 12 to 52 weeks depending on studied implant. Buffer solution was changed every one to two weeks and the pH was measured using a Mettler Toledo MP225 pH meter (Mettler Toledo AG, Greifensee, CH). At the sampling times, specimens were removed from the solution, rinsed with distilled water and tested immediately. Initial samples were tested dry. Besides antibiotic-releasing implants, plain SR-P(L/DL)LA and SR-PLGA based implants were tested as reference controls.

Measured initial mechanical properties of the studied screws were as good as those of commercially-available SR bioabsorbable fixation screws and much higher than, e.g. the mechanical properties of injection molded P(L/DL)LA 70:30 fixation pins, which have reported initial bending strength of 126 MPa.[32] Torsion strength of the studied antibiotic-releasing screws was in the case of SR-P(L/DL)LA screws lower than values of the reference screws and in the case of SR-PLGA as high as reference screws. The measured initial shear, bending and torsion strengths of the studied screws are shown in Table 1.

Table 1. Initial mechanical properties of studied γ-sterilised plain (*P*) and ciprofloxacin-releasing (*C*) self-reinforced poly-L/DL-lactide [SR-P(L/DL)LA] 70:30 and self-reinforced polylactide-co-glycolide (SR-PLGA) 80:20 screws.

Sample	Shear strength [MPa]	Bending strength [MPa]	Torsion strength [MPa]
P SR-P(L/DL)LA	185 ± 10	171 ± 5	58 ± 5
C SR-P(L/DL)LA	152 ± 15	151 ± 10	46 ± 2
P SR-PLGA	184 ± 20	-	65 ± 10
C SR-PLGA	172 ± 20	186 ± 20	65 ± 10

The most important requirement for polymeric bone fracture fixation implants is that they retain their strengths until bone healing is achieved. Normally, this requires a period of four to eight weeks[33]. The studied ciprofloxacin-releasing implants retain their mechanical properties at the level that ensures their fixation properties for at least 12 weeks (ciprofloxacin releasing SR-P(L/DL)LA screws)[34] and 9 weeks (ciprofloxacin releasing SR-PLGA rods, Ø 3.0 mm) *in vitro*. Shear strength retention of the ciprofloxacin-releasing SR-P(L/DL)LA screw and SR-PLGA rod are shown in Figure 2 as an example.

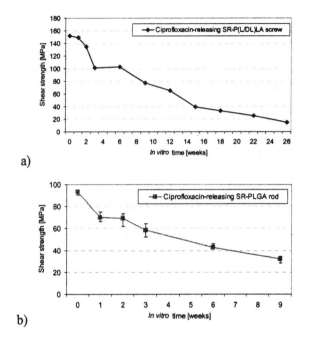

a)

b)

Figure 2. Shear strength retention of studied bioabsorbable γ-sterilized ciprofloxacin-releasing SR-P(L/DL)LA screw (a) and SR-PLGA rod Ø 3.0 mm (b).

5.   PULL-OUT STRENGTH PROPERTIES OF ANTIBIOTIC-RELEASING SCREWS

Besides shear, bending and torsion strength measurements a pullout test to measure the holding power of an implant is a standard method for evaluating mechanical properties of the implants. Tiainen et al.[35] and Leinonen et al.[36] recently compared the pullout forces of developed ciprofloxacin releasing and plain screws and studied insertion properties of the developed screws. In both studies, 50 plain and 50 antibiotic-releasing screws were inserted into human cadaver bones. The pullout strength properties of the antibiotic-releasing SR-PLGA and SR-P(L/DL)LA screws were studied in human cadaver cranial bone and fibulae, respectively. The screws were inserted so that the application simulated the clinical indication of a bioabsorbable screw and plate. A universal tensile testing machine was used to measure the screw pullout force from the bone.

Both studies demonstrated that the studied ciprofloxacin-releasing screws had a lower pullout strength than corresponding plain conventional SR-P(L/DL)LA and SR-PLGA screws. The pull-out forces of the ciprofloxacin-releasing SR-P(L/DL)LA and SR-PLGA screws were 142.9 ± 25.9 N and 66.8 ± 4.9 N (significant difference, P < 0.001), respectively. The most common cause of failure was thread breakage for the ciprofloxacin-releasing SR-P(L/DL)LA screw and screw-shaft breakage for the ciprofloxacin-releasing SR-PLGA screw. The pullout forces of plain SR-P(L/DL)LA and the SR-PLGA screws were 162.7 ± 37.8 N and 96.3 ± 9.3 N, respectively, and the most common cause of failure was thread breakage.

The studied screws were easy to insert. Nevertheless, prior training in the insertion techniques of bioabsorbable screws was recommended. According to Tiainen et al. the studied ciprofloxacin-releasing SR-PLGA screws can be applied in unloaded or slightly load-bearing indications or may be used along with conventional plain screws for bone fixation and antibiotic therapy, e.g. in craniomaxillofacial surgery. Leinonen et al. reported ciprofloxacin-releasing SR-P(L/DL)LA screws to be suitable for osteofixation and infection prophylaxis, especially in trauma cases.

6.   *IN VITRO* CIPROFLOXACIN RELEASE PROPERTIES

To determine released ciprofloxacin concentrations *in vitro,* studied screws (500 mg) were placed into 50 ml of phosphate buffer solution (PBS) KH_2PO_4 (0.05 M) and NaOH (0.04 M) at a pH of 7.4. Five parallel samples in brown drug bottles were kept in an incubator shaker at 37°C. At the specific sampling times, released ciprofloxacin concentrations were measured using a UV-spectrometer (UV-2501PC, Shimadzu, Japan and UNICAM UV 540, Thermo Spectronic, Cambridge, UK). Analysis was carried out by measuring the absorbance values at UV spectra maximum of ciprofloxacin in PBS at λ = 270.5 nm. The concentrations of released ciprofloxacin were calculated according to the Beer-Lambert law.

All the loaded ciprofloxacin was released from the studied screws after 44 weeks (SR-P(L/DL)LA)[34] and 23 weeks (SR-PLGA)[37] *in vitro* (Fig. 3). During this time, the concentration of released ciprofloxacin per day remained in the range of 0.06 – 8.7 µg/ml for SR-P(L/DL)LA and 0.6 - 11.6 µg/ml for SR-PLGA after the start-up burst peak. The maximum concentration value of the released ciprofloxacin was recorded in the 15th week for P(L/DL)LA and 8th week for PLGA *in vitro*. Ciprofloxacin was proven to be bacteriologically bioactive throughout the *in vitro* drug release test (unpublished data).

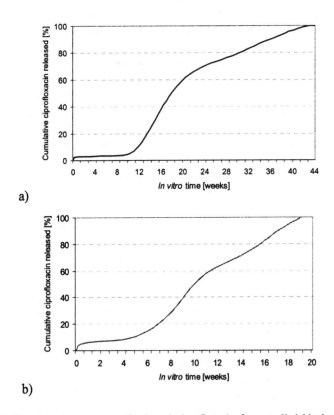

a)

b)

Figure 3. Cumulative percentage of released ciprofloxacin from studied bioabsorbable γ-sterilized ciprofloxacin-releasing SR-P(L/DL)LA (a) and SR-PLGA (b) screws

7.   BACTERIAL GROWTH INHIBITION TEST ON AGAR PLATES AND BIOFILM FORMATION *IN VITRO*

Biofilm formation is a serious problem that follows bacterial colonization of biomaterials. Niemelä et al.[38] have recently evaluated whether developed bioabsorbable ciprofloxacin-releasing PLGA 80:20 have any advantage over plain PLGA and commonly used titanium in preventing *Staphylococcus epidermidis* attachment and biofilm formation *in vitro*.

Niemelä et al.[38] studied cylindrical specimens of titanium, PLGA and ciprofloxacin-releasing PLGA. The cylinders were examined in triplicate for their effect on biofilm-forming *S. epidermidis* ATCC 35989 attachment and biofilm formation after incubation with bacterial suspension for 1, 3, 7, 14 and 21 days using scanning electron microscopy. Bacterial growth inhibition properties of the cylinders were also tested on blood agar plates containing *S. epidermidis* ATCC 35989 suspensions. The plates were incubated overnight at 35°C and the diameters of the inhibition areas were measured.

Ciprofloxacin-releasing PLGA was superior to plain PLGA and to titanium in preventing bacterial attachment and biofilm formation. During the entire study period no biofilm was observed in 74% - 93% situations for titanium, 57%-78% for PLGA and 98%-100% for ciprofloxacin-releasing PLGA. Ciprofloxacin-releasing PLGA inhibited bacterial growth on blood agar plates, showing an average diameter of growth inhibition area of 34 mm, while no inhibition was seen with plain PLGA and titanium specimens.

8. CONCLUSION

The studied novel ciprofloxacin-releasing bone fracture fixation implants based on self-reinforced bioabsorbable polymers have sufficient strength retention and antibiotic release properties *in vitro*. Ciprofloxacin-releasing PLGA appeared to reduce attachment and biofilm formation by *Staphylococcus epidermidis* significantly *in vitro* when compared with plain PLGA or titanium. These novel controlled drug delivery implants may have a promising future in clinical indications, e.g. for prevention of infections in trauma surgery or in the treatment of bone infections.

In vivo studies of these novel antibiotic-releasing bioabsorbable materials and bone fracture fixation implants are currently in progress and results will be published in the near future.

ACKNOWLEDGEMENTS

We greatly appreciate research grants from the Technology Development Center of Finland (TEKES).

The authors express their appreciation to Ms. Kaija Honkavaara, Mr. Kimmo Ilen, Ms. Kaisa Knuutila, and Ms. Raija Reinikainen for helping with *in vitro* mechanical and drug release measurements, and Ms. Mirja Törmälä, and Ms. Hilkka Losoi for administrative assistance.

REFERENCES

1.   Mader, J.T., and Calhoun, J.H., 2003, Staging and Staging Application in Osteomyelitis, In *Musculosceletal Infections* (J.H. Calhoun and J.T. Mader, eds.), Marcel Dekker, Inc., New York, pp.63-77.
2.   Wang, J., Calhoun, J.H., Mader, J.T., and Lazzarini, L., 2003, Musculoskeletal Infection in Systemically and Locally Immunocompromised Hosts. In *Musculosceletal Infections* (J.H. Calhoun and J.T. Mader, eds.), Marcel Dekker, Inc., New York, pp.79-114.
3.   Lazzarini, L., Mader, J.T., and Calhoun, H., 2003, Diabetic Foot Infection. In *Musculoskeletal Infections* (J.H. Calhoun and J.T. Mader, eds.), Marcel Dekker, Inc., New York, pp.325-339.
4.   Patzakis, J., Zalavras, C.G., and Holtom, P.D., 2003, Open Fractures, Current Concepts of Management. In *Musculosceletal Infections* (J.H. Calhoun and J.T. Mader, eds.), Marcel Dekker, Inc., New York, pp.131-147.
5.   Darouiche, R.O., 2001, Review. Device-associated infections: a macroproblem that starts with microadherence, *Clin. Infect. Dis.* 33(9):1567-1572.
6.   Mader, J.T., Landon, G.C., and Calhoun, J. 1993, Antimicrobial Treatment of Osteomyelitis, *Clinical Orthopaedics and Related Research* 295: 87-95.
7.   Garvin, K.L., Miyano, J.A., Robinson, D., Giger D., Novak, J., and Radio, S., 1994, Polylactide/Polyglycolide Antibiotic Implants in the Treatment of Osteomyelitis, A Canine Model, *The Journal of Bone and Joint Surgery* 76-A, 10:1500-1506.
8.   Stengel, D., Bauwens, K., Sehouli, J., Ekkernkamp, A., and Porzsolt, F., 2001, Review. Systemic rewiev and meta-analysis of antibiotic therapy for bone and joint infections, *Lancet Infectious Diseases* 1: 175-188.
9.   Wiesel, B.B., and Esterhai, J.L., Jr., 2003, Prophylaxic of Musculoskeletal Infection, In *Musculoskeletal Infections* (J.H. Calhoun and J.T. Mader, eds.), Marcel Dekker, Inc., New York, pp.115-129.
10.   Laurencin, C.T., Gerhart, T., Witschger, P., Satcher, R., Domb, A., Rosenberg, A.E., Hanff, P., Edsberg, L., Hayes, W., and Langer, R., 1993, Bioerodible Polyanhydrides for Antibiotic Drug Delivery: In Vivo Osteomyelitis Treatment in a Rat Model System, *Journal of Orthopaedic Research* 11:256-262.
11.   Wu, X.S., 1996, Controlled Drug Release Systems (Part I), Technomic Publishing Co., Inc., pp. 1-149.
12.   Graham, N.B., and Wood, D.A., 1984, Polymeric inserts and implants for the controlled release of drugs, In *Macromolecular Biomaterials* (G.W. Hastings, and P. Ducheyne, eds.), CRC Press, Inc., Boca Raton, pp.181-208.
13.   Li, V.H.K., Robinson, J.R., and Lee, V.H.L., 1987, Influence of Drug Properties and Routes of Drug Administration on the Design of Sustained and Controlled Release Systems. In *Controlled Drug Delivery, Fundamentals and Applications*, (J.R. Robinson and V.H.L Lee, eds.), Marcel Dekker, Inc., New York, pp.3-94.
14.   Kanellakopoulou, K., and Giamarellos-Bourboulis, E.J., 2000, Carrier Systems for the Local Delivery of Antibiotics in Bone Infections, *Drugs* 59(6):1223-1232
15.   Nijhof, M.W., Dhert, J.A., Fleer, A., Vogely, H.C., and Verbout, A.J., 2000, Prophylaxis of implant-related staphylococcal infections using tobramycin-containing bone cement, *Journal of Biomedical Materials Research* 52(4):754-761.
16.   Stephens, D., Li, L., Robinson, D., Chen, S., Chang, H-C., Liu, R.M., Tian, Y., Ginsburg, E.J., Gao, X., and Stultz, T., 2000, Investigation of the in vitro release of gentamicin from a polyanhydride matrix, *Journal of Controlled Release* 63:305-317.
17.   Lew, D.P., and Waldvogel, F.A., 1995, Quinolones and Osteomyelitis: State-of-the-Art, *Drugs* 49(2):100-111.

18. Andreopoulos, A.G., Korakis, T., Dounis, E., Anastasiadis, A., Tzivelekis, P., and Kanellakopoulou, K., 1996, In vitro Release of New Quinolones from Biodegradable Systems: A Comparative Study, *Journal of Biomaterials Applications* 10:338-347.
19. Dounis, E., Korakis, T., Anastasiadis, A., Kanellakopoulou, K., Andreopoulos, A., and Giamarellou, H., 1996, Sustained release of fleroxacin in vitro from lactic acid polymer, *Hospital for Joint Diseases* 55(1):16-19.
20. Burd, T.A., Anglen, J.O., Lowry, K.J., Hendricks, K.J., and Day, D., 2001, In Vitro Elution of Tobramycin From Bioabsorbable Polycarbolactone Beads, *Journal of Orthopaedic Trauma* 15(6):424-428.
21. Lambert, H.P., and O'Grady, F.W., 1992, Quinolones, In *Antibiotic and Chemotherapy*, pp.245-262.
22. Christian, J.S., 1996, The Quinolone Antibiotics, *Prim Care Update Ob/Gyns* 3:87-92.
23. Miclau, T., Edin, M.L., Lester, G.E., Lindsey, R.W., and Dahners, L.E., 1998, Effect of Ciprofloxacin on the Proliferation of Osteoblast-like MG-63 Human Osteosarcoma Cells In Vitro, *Journal of Orthopaedic Research* 16:509-512.
24. Holtom, P.D., Pavkovic, S.A., Bravos, P.D., Patzakis, M.J., Shepherd, L.E., and Frenkel, B., 2000, Inhibitory Effects of the Quinolone Antibiotics Trovafloxacin, Ciprofloxacin, and Levofloxacin on Osteoblastic Cells In Vitro, *Journal of Orthopaedic Research* 18:721-727.
25. Törmälä, P., and Rokkanen, P., 2001, Review. Bioabsorbable Implants in the fixation of fractures, *Annales Chirurgiae et Gynaecologiae* 90:81-85.
26. Törmälä, P., Pohjonen, T., and Rokkanen, P., 1998, Bioabsorbable polymers: materials technology and surgical applications, *Proc Instn Mech Engrs* 212(H):101-111.
27. Suuronen, R., Kallela, I., and Lindqvist, C., 2000, Invited Review. Bioabsorbable Plates and Screws: Current State of the Art in Facial Fracture Repair, *Journal of Cranio-Maxillofacial Trauma* 6(1):19-27.
28. ASTM B 769-94, 2000, Standard Test Method for Shear Testing of Aluminum Alloys
29. ASTM D 790-00, 2000, Standard Test Method for Flexural Properties of Unreinforced and Reinforced Plastics and Electrical Insulating Materials
30. DIN 53452, 1977, Testing of plastics; bending test
31. ASTM F 117-79, 1985, Standard Test Method for Driving Torque of Self-Tapping Medical Bone Screws
32. Claes, L., Rehm, K., and Hutmacher, D., 1992, The Development of a New Degradable Pin for the Refixation of Bone Fragments, *Fourth World Biomaterials Congress*, April 24-28, Berlin, 205.
33. An, Y.H., Woolf, S.K., and Friedman, R.J., 2000, Pre-clinical in vivo evaluation of orthopaedic bioabsorbable devices, *Biomaterials* 21:2635-2652
34. Veiranto, M., Suokas, E., and Törmälä, P., 2002, In vitro mechanical and drug release properties of bioabsorbable ciprofloxacin containing and neat self-reinforced P(L/DL)LA 70/30 fixation screws, *Journal of Materials Science: Materials in Medicine* 13:1259-1263.
35. Tiainen, J., Veiranto, M., Suokas, E., Törmälä, P., Waris, T., Ninkovic, M., and Ashammakhi, N., 2002, Bioabsorbable ciprofloxacin-containing and plain self-reinforced polylactide-polyglycolide 80/20 screws: pullout strength properties in human cadaver parietal bones, *J Craniofac Surg* 13(3):427-433.
36. Leinonen, S., Suokas, E., Veiranto, M., Törmälä, P., Waris, T., and Ashammakhi, N., 2002, Holding Power of Bioabsorbable Ciprofloxacin-Containing Self-Reinforced Poly-L/DL-lactide 70/30 Bioactive Glass 13 Miniscrews in Human Cadaver Bone, *J Craniofac Surg* 13(2):212-218.

37. Veiranto, M., Suokas, E., and Törmälä, P., 2002, Controlled Release of Ciprofloxacin from Fixation Screws Based on Poly(lactide-co-glycolide), *Society for Biomaterials 28th Annual Meeting*, April 24-27, Tampa, Florida, USA, 637.
38. Niemelä, S-M., Ikäheimo, I., Koskela, M., Syrjälä, H., Veiranto, M., Suokas, E., Waris, T., Törmälä, P., and Ashammakhi, N., 2003, Multifunctional Bioabsorbable Ciprofloxacin-Releasing Devices Inhibit Bacterial Attachment and Biofilm Formation. *Biofilms in Medicine*, October 17-19, Elsinor, Denmark.

Advances in Analgesic Drug Design and Delivery:
A Current Survey

IWONA MASZCZYNSKA BONNEY[*], DILEK SENDIL KESKIN[#],
ANDRZEJ W. LIPKOWSKI[&], VASIF HASIRCI[#] and DANIEL B. CARR[*]
[*]Department of Anesthesia, Tufts-New England Medical Center, Boston, MA, USA;
[#]Department of Engineering Sciences, Middle East Technical University, Ankara, TURKEY;
[&]Neuropeptide Laboratory, Medical Research Centre, Polish Academy of Sciences, Warsaw,
Poland; [#]Biotechnology Research Unit, Departments of Biological Sciences and
Biotechnology, Middle East Technical University, Ankara, TURKEY

1.   INTRODUCTION

Increasingly, pain is being recognized as a major medical problem. Pain is defined by the International Association for the Study of Pain (IASP) as "an unpleasant sensory and emotional experience associated with actual or potential tissue damage, or described in terms of such damage" [1]. Pain is always subjective. Each individual learns the application of the word through experiences related to injury in early life[1]. Pain is widely underestimated and undertreated. In Western countries, 10% of population have acute or chronic pain, and up to 30% of those with disabilities suffer from chronic pain[2]. The most common types of chronic noncancer pain are arthritis, low back pain, other musculoskeletal pain or stiffness, and fibromyalgia. Chronic pain, defined by the IASP as "pain that persists beyond the normal time of healing..."[3], is a significant problem associated with many long-term diseases, including musculoskeletal disorders, various neurological conditions such as diabetic, HIV-related, or postherpetic neuropathies, and cancer. Most patients find oral medications effective. However, some patients with chronic pain have inadequate pain relief, intolerable side

Biomaterials: From Molecules to Engineered Tissues, edited by
N. Hasırcı and V. Hasırcı, Kluwer Academic/Plenum Publishers, 2004

effects, progressive lack of efficacy (tolerance) or allergy associated with oral medications[4].

The ability of current medical science to treat pain effectively is limited by a lack of understanding of the mechanisms of pain signaling, and the high prevalence of side effects after systemic administration of available analgesics[5, 6]. In order to improve pain management, in recent years there has been an increasing interest to develop new analgesic molecules based on evolving research on mechanisms of pain processing, new drug-delivery systems, as well as improvements on old ones.

Pain treatment requires active ingredients to be delivered to the required site of action in a controlled manner to produce maximum relief with minimum side effects. Drug delivery technologies can offer a solution.

Drug delivery technologies offer greater efficiency and efficacy while minimizing side effects. Delivery of medications directly into the spinal column has progressed from catheter administration to implantable, programmable infusion devices. These devices are now used to administer opioids, local anesthetics and other medications. When given spinally, these drugs can be delivered in very low doses, thereby reducing the risk of side effects. Many drugs can be delivered directly through the skin, mouth, nose and lungs. Immediate effectiveness is a very important characteristic and newer delivery systems allow for the rapid onset of analgesia without the need for repeated injections. Site-specific delivery enables a therapeutic concentration of a drug to be administered to the desired target without exposing the entire body to a similar concentration, that often limits the long-term (or even acute) use of many drugs.

2. POLYMERS FOR PAIN TREATMENT

The most active area of current research using biodegradable polymers is in the controlled delivery of bioactive agents. The most promising alternative to electromechanical pumps for infusing drugs for prolonged periods at constant or tailored rates is the use of implantable polymeric systems. Various approaches to the preparation of slow release formulations of local anesthetics and opioids are available, to permit patient mobility and avoid reliance on pumps and catheters of any type. Among the most promising of these are biodegradable polymer- and lipid-based systems[7]. The thermal and mechanical properties of polymers, may be manipulated to increase their possible application sites and methods. There are two primary systems for delivering active agents using biodegradable polymers: microspheres and devices. The devices can be rods, films or other shapes based on the specific product needs. Biodegradability is an important advantage of polymer-based drug delivery systems in that the release system does not require removal upon depletion[8]. Biodegradation durations vary depending on implant surface area, porosity, the properties of the polymer (e.g., molecular weight, composition, and crystallinity) and environmental conditions such as pH, temperature, humidity, etc. The end products of some polymers such as polyhydroxyalkanoate (PHA) and poly(lactic acid-co-glycolic acid) (PLGA), are carbon dioxide and water, making these materials suitable for human application.

Microcapsules and microspheres are the most intensively studied drug delivery forms among all the types of systems designed with the aim of controlling pain[9]. One possible reason for this involves their applicability to various targets in the body, such as peripheral nerves or the brain, with ease due to their small size (and, therefore, injectability). This property allows for their minimally invasive application, and their ability to be applied to sites with irregular contours. A second reason is the feasibility of control over drug release properties by adjusting the polymer or drug characteristics. Finally, their compatibility with both water-soluble and lipid-soluble drugs enhances their preferability over the other systems.

A biodegradable, biocompatible controlled release system for the delivery of local anesthetics to obtain prolonged, reversible nerve blockade when implanted or injected was developed by Sackler et al.[10]. The system consisted of microspheres of polylactic acid (PLA) and poly(lactic acid-co-glycolic acid) (PLGA) and was designed to release bupivacaine as well as other local anesthetics. In a similar study to develop a controlled release system, DL-polylactic acid (PDLLA), and PLGA microspheres were loaded with bupivacaine and some other local anesthetics[11]. Encapsulation efficiency was found to be highly dependent on the lipophilicity of the drugs. The influence of the molecular weight of PLGA on the release rate and on the release mechanism was found to depend on the drug and its physical state within the polymeric matrix. Diffusion-controlled release was obtained in various formulations and was manifested as linearity of release as a function of the square root of time. PLA microspheres prepared via solvent evaporation as a means to achieve sustained release of butamben, tetracaine and dibucaine have been examined *in vitro*[12]. The influence of preparation conditions on loading and the rate of release from these microspheres have been studied. Drug contents of about 25% and yields of about 75% were obtained with diffusion-based release patterns. Release could be prolonged to about 600 h. DL-polylactic acid (PDLLA) microspheres containing lidocaine and dibucaine were prepared and their drug release patterns as well as their anesthetic effects were examined[13]. The release patterns of dibucaine from microspheres varied significantly with dibucaine content, and the release profile was greatly influenced by the disintegration of the microsphere. The local anesthetic effect of dibucaine in microspheres lasted much longer (300 h) than that of the dibucaine hydrochloride solutions. However, a common problem with drug release, the difference between the *in vivo* profile of the local anesthetic effect and the *in vitro* release profile, was evident in this study.

PLGA (65:35) microspheres loaded with 75% (w/w) bupivacaine alone or with 0.05% w/w dexamethasone were prepared by solvent evaporation[14]. Injected into rats, they produced sciatic nerve block ranging from 10 h to 5.5 days shown with thermal sensory testing as well as motor testing. The presence of dexamethasone increased the block duration about 13-fold, attesting to the possibility of increasing the duration of activity by the use of a combination of drugs, and the potential value of inhibiting local inflammation. Motor block was also tested in rabbits that received bupivacaine-loaded poly(d,l-lactic acid) microspheres via a chronically implanted epidural catheter[15]. Significant delay in obtaining maximum effect, and significant prolongation of motor block (24%) were observed when a dose of 5 mg of bupivacaine-loaded microspheres were injected.

Soo et al.[16] studied PLGA microspheres less than 32 μm in diameter containing fentanyl in situ, and obtained a high encapsulation efficiency (61.5 - 99.8%) for this analgesic.

Among other polymers tested are polycarbonate microspheres that were loaded with benzocaine, lidocaine, and dibucaine prepared by a solvent evaporation technique[17]. *In vitro* releases of up to 400 h with relatively constant release rates were obtained. This is a sufficiently long duration for most acute pain management applications, such as postoperative care. As expected from the surface-to-volume ratio, decreasing the size increased the rate of release from these microspheres.

Grossman and colleagues have developed a subcutaneous implant for the controlled release of hydromorphone[18]. This non-abusable, non-inflammatory, biocompatible and non-biodegradable implant delivered hydromorphone with near zero order kinetics, in other words, at a constant rate. The cylindrical implant consisted of a poly(ethylene vinyl acetate) core and a polymethyl methacrylate coat with an opening along the axis.

Sendil et al.[19] and Hasirci et al.[20], have tested a biodegradable, fiber-type, controlled release system made of PLGA and loaded with various analgesics (codeine, morphine, hydromorphone) and an anesthetic (bupivacaine). These fibers (diameter 1.3 mm) were used either to deliver a single drug or combinations of an analgesic and an anesthetic in the treatment of chronic pain in rats created via sciatic nerve ligation. The use of two types of drugs simultaneously showed a better alleviation of pain than the use of a single drug. Alleviation of this chronic and severe neuropathic pain could be maintained for about 3-4 days when "dual drug" (analgesic and anesthetic) fibers were used. This duration was twice that of single-drug fibers. Also, the use of dual drug rods with half the dose of each single drug enhanced the degree of analgesia measured on the first day of experiment. This study also showed the importance of drug release rate on efficacy of the release systems as the results were completely consistent with the differences of *in vitro* drug release rates from the rods. *In vivo* histological studies showed the biocompatibility and biodegradability of this well-known polymer and further confirmed its suitability for controlled release of analgesics or anesthetics [19-21].

Smaller diameter versions of these fibers with similar properties were also prepared for the intrathecal application of the analgesic (hydromorphone), anesthetic (bupivacaine) and a peptide type pain reliever (biphalin). *In vivo* studies showed potent, prolonged analgesia in comparison to controls. Analgesic synergy was observed with hydromorphone and bupivacaine. The antinociceptive effects lasted for up to 4 days for the combination of half doses of hydromorphone + bupivacaine or full doses of bupivacaine alone, and up to 5 days for the full dose of hydromorphone. With further refinements of drug release rate, these fibers may offer a clinically relevant alternative for intrathecal analgesia[22].

3. MULTITARGET APPROACH TO PAIN TREATMENT

The traditional approach to searching for new drugs is to evaluate compounds that will be easy to administer orally, intravenously or intramuscularly; will readily penetrate biological barriers such as the gut-blood and the blood-brain barriers; and that reach target receptors in the central nervous system (CNS). Progress in interventional techniques during the past twenty years allowed the application of drugs directly into the site of desired effect. In 1978, morphine was first injected intrathecally in humans. Since that time, several techniques of direct application of the analgesics to the spinal cord and brain have been developed[23]. These techniques are characterized by fewer side effects at equianalgesic medication doses than occur with traditional systemic drug administration. The introduction of patient controlled epidural and intrathecal analgesia (PCA) and implantable, programmable pumps for central drug delivery has provided an additional impetus for the popularization of site-specific drug delivery. Site-specific techniques for central drug administration allow the use of low doses of substances whose spectrum of receptor affinities is very broad. Therefore in clinical practice, the combination of multireceptor-targeted drugs delivered using modern techniques of site-specific application is a most promising evolution of pain management[24]. Although this approach also permits a combination of drugs to be used, the distinct pharmacokinetic and pharmacodynamic profiles of different agents limit the use of many potential mixtures. An attractive solution to this problem is to develop compounds designed with a broad spectrum of receptor affinities[25]. Because a single such molecule comprises covalently fixed pharmacophores, the balance of activities of each pharmacophore is the most critical factor determining the analgesic properties of the entire molecule. Because of structural cross reactivities, the receptor affinities of a new molecule that hybridizes various pharmacophores is not the simple combination of each component. The analysis of necessary elements of each pharmacophore and simulation of interference with the structural elements of other hybridized pharmacophores is an important step in the design of new multitarget drugs.

3.1 Biphalin

Twenty years ago, a dimeric opioid peptide peptidomimetic with two tetrapeptide pharmacophores connected through a hydrazide bridge was synthesized[26]. This compound, termed biphalin, expresses a broad spectrum of opioid receptor affinities, i.e. high, equal affinity for mu and delta opioid receptors and lower but significant affinity for kappa receptors[27]. Structural analysis showed that both pharmacophores are flexible and can easily adopt conformations to bind to all opioid receptor types[28]. Its broad spectrum of affinities was for many years off-putting to pharmacologists focused upon developing receptor-selective opioid ligands.

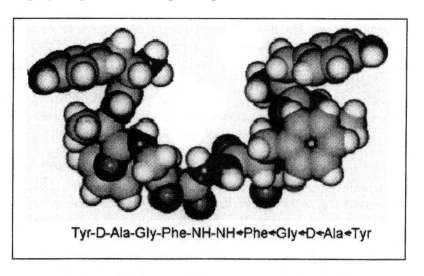

Tyr-D-Ala-Gly-Phe-NH-NH-Phe-Gly-D-Ala-Tyr

Figure 1. Amino acid sequence and solid state structure of biphalin

Recent evolution in understanding the pain system has revealed that nociceptive modulation is a complex process in which multiple opioid receptors participate[29]. This insight has reactivated pharmacological studies of biphalin, the results of which indicate a unique pharmacological profile. Although it is a uniquely potent analgesic when applied centrally[30], it has very low dependence liability[31]. This constellation of properties, together with its low toxicity in preclinical models, argues for its clinical development as an analgesic drug.

3.2   Opioid-substance P Hybrids

Extensive, prolonged study of peptidomimetic SP antagonists led to the disappointing conclusion that such compounds themselves have to date proven ineffective as analgesics. Nevertheless, although counter-intuitive, it has been shown that small quantities of an SP antagonist significantly potentiate the antinociceptive effect of opioid peptides[32]. This observation initiated the search for chimeric compounds that may simultaneously interact with SP receptors as an antagonist and opioid receptors as an agonist.

The first compound of this series hybridized an N-terminal fragment of a casomorphin-related molecule with a C-terminal fragment of an SP antagonist[33]. This chimeric compound indeed had high potency as an analgesic but its low solubility limited further studies. A recently synthesized new compound, AA501, with such dual properties hybridizes "head-to-head" an opioid tetrapeptide with carbobenzyloxy-tryptophan through a hydrazide bridge. In this molecule, opioid agonist and substance P antagonist pharmacophores partially overlap.

AA501 expresses affinities for mu opioid and NK1 receptors, and acts in vivo as an analgesic in models of inflammatory and neuropathic pain[34]. These promising results prompted further studies of SP-opioid hybrids. Although SP in general transmits nociceptive signals from the periphery to the brain, and the endogenous opioids modulate that signal, prior studies indicate that autofeedback occurs within SP-containing pathways[35]. Fragments of substance P[36] may induce antinociception in certain models[37] and potentiate morphine analgesia[38]. Therefore, the potential interaction of an SP agonist pharmacophore hybridized with opioid agonist pharmacophore seemed worthy of study. As expected, the final pharmacological properties of such SP agonist/opioid agonist hybrids depend upon the relative activities of these two physiologically antagonistic pharmacophores. A compound in which the tachykinin pharmacophore dominates induces hyperalgesia[39,40], whereas another compound in which the opioid pharmacophore dominates is antinociceptive[41]. In addition to these expected effects, one compound of this series, termed ESP7, showed the intriguing property of achieving analgesia in animals previously made tolerant to the analgesic effects of morphine. The latter unique property suggests that such hybrid compounds offer a previously unexplored avenue to develop drugs for opioid-tolerant patients, or may find a role in the treatment of opioid dependence.

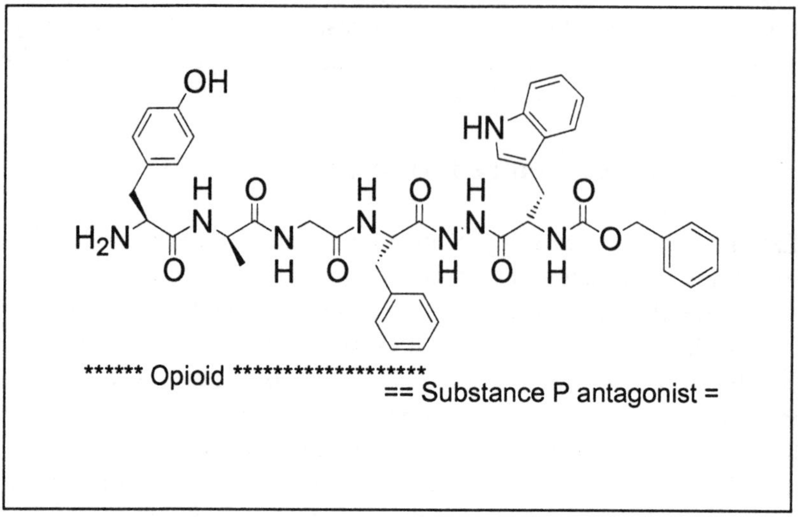

Figure 2. Chemical structure of AA501

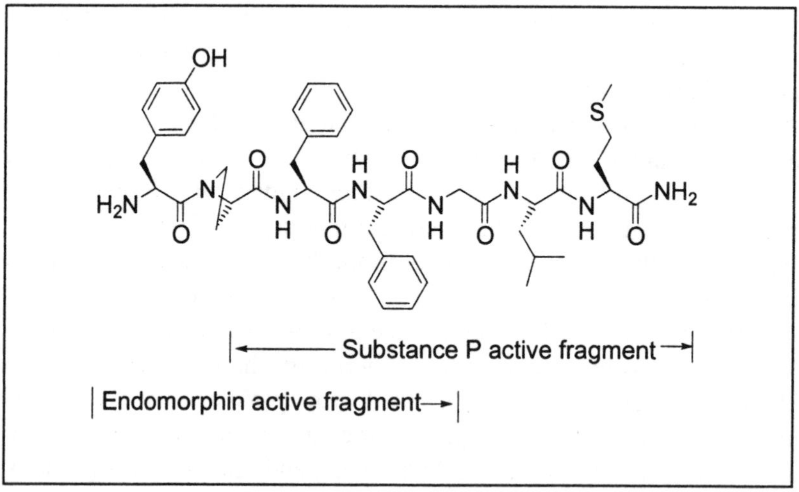

Figure 3. Chemical structure of ESP7

4. CONCLUSION

Pain, in addition to being a symptom of underlying disease, may itself become chronic, and in so doing may assume features of a disease in its own right. Pain management will have an increasingly important place in the future healthcare system because its resultant disability and economic loss, in addition to personal suffering, require that this therapeutic area receive as much attention as other diseases. Significant progress has occurred in the search for new analgesic molecules and their site-specific delivery, creating a hope for improving current inadequate pain management.

ACKNOWLEDGEMENTS

Support for this work was provided by Richard Saltonstall Charitable Foundation and Evenor Armington Fund.

REFERENCES

1. Merskey, H., and Bogduk, N., 1994, Classification of Chronic Pain, Second Edition, IASP Task Force on Taxonomy (Merskey, H., and Bogduk, N., eds.), IASP Press, Seattle, pp. 209-214.
2. Harstall, C., and Ospina, M., 2003, How prevalent is chronic pain? *Pain: Clinical Updates* Vol. XI, No.2.
3. Verhaak, P.F.M., Kerssens, J.J., Dekker, J., Sorbi, M.J., Bensing, J.M., 1998, Prevalence of chronic benign pain disorder among adults: a review of the literature. *Pain* 77: 231-239.
4. Kloke, M., Rapp, M., Bosse, B., Kloke, O., 2000, Toxicity and/or insufficient analgesia by opioid therapy: risk factors and the impact of changing the opioid. A retrospective analysis of 273 patients observed at a single center. *Supportive Care in Cancer* 8: 479-486.
5. Goudas, L.C., Carr, D.B., Bloch, R., Balk, E., Ioannidis, J.P.A., Terrin, N., Gialeli-Goudas, M., Chew, P., Lau, J., October 2001, Management of cancer pain. Volume 1 and 2. Evidence Report/Technology Assessment No. 35, Agency for Healthcare Research and Quality (AHRQ) Publication No. 02-E002.
6. Carr, D.B., Goudas, L.C., Lawrence, D., Pirl, W., Lau, J., DeVine, D., Kupelnick, B., Miller, K., July 2002, Management of cancer symptoms: pain, depression, and fatigue. Evidence Report/Technology Assessment No. 61, Agency for Healthcare Research and Quality (AHRQ) Publication No. 02-E032.
7. Renck, H., and Wallin, R., 1996, Slow release formulations of local anesthetics and opioids. *Current Opinion in Anesthesiology* 9: 399-403.
8. Hasirci, V., Sendil, D., Goudas, L.C., Carr, D.B., Wise, D.L., 2000, Controlled release pain management systems. In: *Handbook of pharmaceutical controlled release technology* (Wise D.L., ed.), Marcel Dekker, Inc., New York, Basel, pp. 787-806.
9. Carr, D.B., and Sendil, D., 2002, Current approaches to analgesic drug delivery for chronic pain. *Technology & Health Care* 10: 227-235.

10. Sackler, R., Goldenheim, P., Chasin, M., 1998, Prolonged local anesthesia with colchicine, US Patent 5,747,060.

11. Le Corre, P., Rytting, J.H., Gajan, V., Chevanne, F., Le Verge, R., 1997, In vitro controlled release kinetics of local anesthetics from poly(D,L-lactide) and poly(lactide-co-glycolide) microspheres. *J Microencapsul* 14: 243-255.

12. Wakiyama, N., Juni, K., Nakano, M., 1981, Preparation and evaluation in vitro of poly lactic acid microspheres containing local anesthetics. *Chem Pharm Bull* 29: 3363-3368.

13. Wakiyama, N., Juni, K., Nakano, M., 1982, Influence of physicochemical properties of polylactic acid on the characteristics and in vitro release patterns of polylactic acid microspheres containing local anesthetics. *Chem Pharm Bull* 30: 2621-2628.

14. Curley, J., Castillo, J., Hotz, J., Uezono, M., Hernandez, S., Lim, J-O., Tigner, J., Chasin, M., 1996, Prolonged regional nerve blockade. Injectable biodegradable bupivacaine/ polyester microspheres. *Anesthesiology* 84: 1401-1410.

15. Malinovsky, J.M., Bernard, J.M., Le Corre, P., Dumand, J.B., Lepage, J.Y., Le Verge, R., Souron, R., 1995, Motor and blood pressure effects of epidural sustained-release bupivacaine from polymer microspheres: a dose-response study in rabbits. *Anesth Analg* 81: 519-524.

16. Soo Choi, H., Seo, S.A., Khang, G., Rhee, J.M., Lee, H.B., 2002, Preparation and characterization of fentanyl-loaded PLGA microspheres: in vitro release profiles. *International Journal of Pharmaceutics* 234: 195-203.

17. Kojima, T., Nakano, M., Juni, K., Inoue, S., Yoshida, Y., 1984, Preparation and evaluation in vitro of polycarbonate microspheres containing local anesthetics. *Chem Pharm Bull* 32: 2795-2802.

18. Grossman, S.A., Leong, K.W., Lesser, G.J., Lo, H., 1997, Subcutaneous Implant, US Patent 5,633,000.

19. Sendil, D., Wise, D.L., Hasirci, V., 2002, Assessment of biodegradable controlled release rod systems for pain relief applications. *J. Biomat. Sci. Polym. Ed.* 13: 1-15.

20. Hasirci, V., Bonney, I., Goudas, L.C., Shuster, L., Carr, D.B., Wise, D.L., 2003, Antihyperalgesic effect of simultaneously released hydromorphone and bupivacaine from polymer fibers in the rat chronic constriction injury model. *Life Sci* 73: 3323-3337.

21. Sendil Keskin, D., Altunay, H., Wise, D.L., Hasirci, V., 2003, In vivo pain relief effectiveness of analgesic-anesthetic carrying biodegradable controlled release rod systems. *J. Biomaterials Sciences, Polymer Edition*, in press.

22. Sendil, D., Maszczynska Bonney, I., Carr, D.B., Lipkowski, A.W., Wise, D.L., Hasirci, V., 2003, Antinociceptive effects of hydromorphone, bupivacaine and biphalin released from PLGA polymer after intrathecal implantation in rats. *Biomaterials* 24: 1969-1976.

23. Carr, D.B., and Cousins, M.J., 1998, Spinal route of analgesia. Opioids and future options. In *Neural Blockade in Clinical Anesthesia and Management of Pain* (Cousins, M.J., Bridenbaugh, P.O., eds.) 3rd Ed., Lippincott-Raven Publishing, pp. 915-983.

24. Walker, S., Goudas, L.C., Cousins, M.J., Carr, D.B., 2002, Combination spinal analgesic chemotherapy: a systematic review. *Anesthesia Analgesia* 95: 674-715.

25. Kosson, D., Lachwa, M., Maszczynska, I., Misicka, A., Carr, D.B., Lipkowski, A.W., 1999, Potential application of peptides as the new generation analgesics. *Anestezjologia Intensywna Terapia* 31 (Suppl. II): 55-56.

26. Lipkowski, A.W.,. Konecka, A.M,. Sroczynska, I., 1982, Double-enkephalins-synthesis, activity on guinea-pig ileum, and analgesic effect. *Peptides* 3: 697-700.

27. Lipkowski, A.W.,. Konecka, A.M,. Sroczynska, I,. Przewlocki, R,. Stala, L, Tam, S.W., 1987, Bivalent opioid peptide analogues with reduced distances between pharmacophores. *Life Sci.*, 40, 2283-2288.

28. Flippen-Anderson, J.L.,. Deschamps, J.R,. George, C, Hruby, V.J., Misicka, A., Lipkowski, A.W., 2002, Crystal structure of biphalin sulfate: a multireceptor opioid peptide. *J. Peptide Res.* 59, 123-133.
29. Lipkowski, A.W., Carr, D.B., Bonney, I., Misicka, A., Payza K., 2003, Biphalin: A Multireceptor Opioid Ligand Binding and Activity of Opioid Ligands at the Cloned Human Delta, Mu, and Kappa Receptors. In *The Delta Receptor.* (Chang, K-J., Porreca, F., and Woods, J.H., eds.), Marcel Dekker, Inc., New York, in press.
30. Horan, P.J., Mattia, A., Bilsky, E., Weber, S., Davis, T.P., Yamamura, H.I.,. Malatynska, E, Appleyard, S.M.,. Slaninova, J, Misicka, A., Lipkowski, A.W., Hruby, V.J., Porreca F., 1993, Antinociceptive profile of biphalin, a dimeric enkephalin analog. *J. Exp. Pharm. Ther.* 265: 1446-1454.
31. Yamazaki, M., Suzuki, T., Narita, M., Lipkowski, A.W., 2001, The opioid peptide analogue biphalin induces less physical dependence than morphine. *Life Sci.* 69, 1023-1028.
32. Misterek, K., Maszczynska, I., Dorociak, A., Gumulka, S.W., Carr, D.B., Szyfelbein, S.K., Lipkowski, A.W., 1994, Spinal co-administration of peptide substance P antagonist increases antinociceptive effect of the opioid peptide biphalin. *Life Sci.* 54, 936-944.
33. Lipkowski A.W., Carr D.B., Misicka A., Misterek K., 1994, Biological activities of peptide containing both casomorphine-like and substance P antagonist structural characteristics. In *β-Casomorphins and related peptides: Recent developments* (Brantl, V., Teschemacher, H., eds.), VCH, Weinheim, pp. 113-118.
34. Maszczynska Bonney, I., Foran, S.E.,. Marchand, J.E, Lipkowski, A.W., Carr, D.B., 2004, Spinal antinociceptive effects of AA501, a novel chimeric peptide with opioid agonist and neurokinin antagonist moieties. *Eur. J. Pharmacol*, in press.
35. Lipkowski, A.W., Osipiak, B., Czlonkowski, A., Gumulka, W.S., 1982, An approach to the elucidation of self-regulatory mechanism of substance P action. I. Synthesis and biological properties of pentapeptides related both to the substance P C-terminal fragment and enkephalins. *Polish J. Pharmacol. Pharm.* 34, 63-68.
36. Klusha, V.E., Abissova, N.A., Muceniece, R.K., Svirskis, S.V., Bienert, M., Lipkowski, A.W., 1981, Comparative study of substance P and its fragments: analgesic properties, effect on behavior and monoaminergic processes. *Bull. Exp. Biol. Med. Eng. Tr.,* 92, 1665-1667.
37. Maszczynska, I., Lipkowski, A.W., Carr, D.B., Kream, R.M., 1998, Dual functional interactions of substance P and opioids in nociceptive transmission: review and reconciliation. *Analgesia* 3: 259-268.
38. Kream, R.M., Kato, T., Shimonaka, H., Marchand, J.E., Wurm, W.H., 1993, Substance P markedly potentiates the antinociceptive effects of morphine sulfate administered at the spinal level. *Proc. Natl. Acad. Sci. USA* 90, 3564-3568.
39. Lipkowski, A.W., Osipiak, B., Gumulka, W.S., 1983, An approach to the self regulatory mechanism of substance P actions: II. Biological activity of new synthetic peptide analogs related both to enkephalin and substance P. *Life Sci.* 33 (Suppl. I). 141-144.
40. Lei, S.Z., Lipkowski, A.W., Wilcox, G.L., 1991, Opioid and neurokinin activities of substance P fragments and their analogs. *Eur. J. Pharmacol.* 193, 209-215.
41. Foran, S.E., Carr, D.B., Lipkowski, A.W., Maszczynska, I., Marchand, J.E., Misicka, A., Beinborn, M., Kopin, A., Kream, R.M., 2000, A substance P-opioid chimeric peptide as a unique nontolerance-forming analgesic. *Proc. Natl. Acad. Sci. USA*, 97, 7621-7626.

Drug Delivery to Brain by Microparticulate Systems

H. SÜHEYLA KAŞ *

Department of Pharmaceutical Technology, University of Hacettepe, Sıhhiye, Ankara, Turkey

The site specific delivery of chemotherapeutic agents allows maximum concentration of an agent at a desired body site. This area specific drug delivery decreases the unwanted systemic distribution and decreases toxicity of the administered drugs.

Blood-Brain Barrier (BBB) is considered to be an obstacle in delivering large number of drugs to brain. The endothelial cells forming the tubular capillaries in the brain are cemented together by intercellular tight junctions. In this way, the BBB has an important role in providing a stable extracellular environment in the central nervous system[1]. Lack of fenestrations, very few pinocytotic vesicles, and more mitochondria are other differences of the brain capillaries which play important role in transport of drugs to brain (Fig.1).

The purpose of this paper is to summarise the methods for BBB permeability modifications and to focus on various examples in delivering drugs, especially neuroncology and neuroactive drugs, to brain by microparticulate systems.

1. BBB PERMEABILITY MODIFICATIONS

Certain drugs due to their physicochemical features cannot readily penetrate the brain. Lipid solubility, amount of unionized drug at physiological pH, low degree of protein binding and low molecular weight (<700 Da) are the main characteristics of a drug which promote its entry into the brain. BBB permeability can be modified by the methods shown in Table 1.

---

*retired - 2001

Biomaterials: *From Molecules to Engineered Tissues,* edited by
N. Hasırcı and V. Hasırcı, Kluwer Academic/Plenum Publishers, 2004

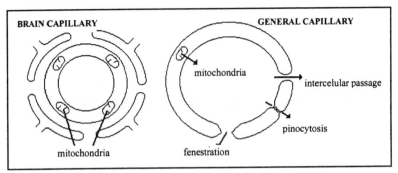

Figure 1. Differences between the general and brain capillaries.

Neurosurgical strategies are invasive procedures which require access to brain ventricules in the carotid artery. Osmotic shocks, humoral means and chemical disruption transiently open BBB and give way to the entrance of drugs to the brain[2,3]. Conversion to lipid soluble prodrugs, and encapsulation into liposomes are two other pharmacological strategies for increasing BBB transport of water soluble drugs. Drugs are also retained in the brain through cationic–anionic interactions with the BBB. This method for enhancing the transcellular delivery is cationization. In this strategy, the isoelectric point of the protein is raised and the cationic moieties bind to anionic groups of the brain capillary endothelial cells. Peptides and oligopeptides being water soluble cannot pass BBB in general. However, chimeric peptides, after conjugation with vectors (modified proteins or monoclonal antibodies) can easily be transported to brain through receptor mediated transcytosis.

Table 1. BBB permeability modifications

A.	Neurosurgical			B.	Pharmacological
	1.	Intraventricular infusion			1. Lipidization
	2.	BBB disruption			2. Liposomes
		a.	Osmotic shock	C.	Physiological
		b.	Humoral		1. Cationization
		c.	Chemical		2. Chimeric peptides

This table is from reference [2].

Limited infusion distance by use of catheters, high rates of infection, unpredictable release rates and obstruction due to clogging are the factors which contribute to the failure of direct infusion to brain. Transcapillary transfer from blood to brain, ventriculocisternal perfusion, catheter perfusion, microdialysis, intracranial implantation and stereotaxic implantation are the transport pathways to brain.

2. MICROPARTICULATE SYSTEMS

Since the administration of therapeutic drugs complicated due to the presence of BBB, interest has been focused in the use of microparticulate systems as localized delivery devices. High levels of drug concentration in the targeted area, reduced systemic toxicity, controlled release and protection of labile drugs from degradation are the advantages of these delivery systems. Synthetic (e.g. polyesters, polyanhydrides) and natural (e.g. human serum albumin-HSA, bovine serum albumin-BSA, alginate, chitosan) polymers that are completely biodegradable and biocompatible are used to transport drugs to brain. In the following sections, examples of especially, neuroncology and neuroactive drugs, encapsulated into microspheres, microcapsules, nanospheres, nanoparticles, liposomes, mini-pellets and slabs applied to brain will be reviewed.

2.1 Anticancerogen Drugs

5-Fluorouracil (5-FU), carmustine (BCNU), nimustin hydrochloride (ACNU), mitomycin, doxorubicin (DXR), oxantrazole, carboplatin, etoposide, methotrexate, bromocryptine mesylate (BM) are some examples of neuroncology drugs transported to brain by microparticulate carriers. Kubo et al[4] encapsulated 5-FU, adriamycin, mitomycin and ACNU by polymethacrylic methyl acid (PMMA) and administered these implants to the malignant glioma by stereotaxic method (Fig. 2). By this method, pellets were inserted exactly into residual tumors and drugs release for about 30 – 40 days.

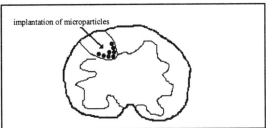

Figure 2. Schematic demonstration of implantation of microparticles to brain.

Menei et al[5] implanted stereotaxically 5-FU incorporated poly (lactide-co-glycolide) = PLGA microspheres to rat brains and showed a marked anticancer activity. 5-FU loaded and chitosan coated D,L-PLA and PLGA microspheres as biodegradable drug carriers for cerebral tumors were prepared by Chandy et al[6] . Their work demonstrated that the prepared carrier system is a good candidate for targeted delivery of antitumor drugs to cerebral tumors.

BCNU is an antineoplastic agent used for brain tumors. Menei et al[7] and Painbeni et al[8] prepared BCNU loaded PLGA microspheres as intracerebral implants to obtain sustained high local concentrations in the brain without systemic toxicity Valtonen et al[9] investigated intracranially implantable GLIADEL (approved by FDA) for controlled delivery of BCNU in 32 patients. Prolonged survival has been demonstrated in a Phase III study of patients with recurrent glioma. In the brain tumor laboratories, major efforts have been directed towards the use of magnetically responsive microspheres or slabs for brain tumors. Using radioactively labeled microspheres or slabs, it has been proven that the spheres can be concentrated in the brain 100 fold compared to spheres studied without a magnetic field[10]. Hassan and Gallo[11] have prepared magnetic chitosan microspheres containing Oxantrazole. Application of magnetic field increased the accumulation of the active substance in the brain due to cationization–cationic and anionic interaction.

Kreuter et al[12] prepared DXR nanoparticles by polybutylcyanoacrylate (PBCA) and coated them with polysorbate 80. These nanoparticles were applied to rat brain and the concentration of DXR is measured at time intervals. The authors have shown that the polysorbate 80 coated nanoparticles were taken up by endocytosis, increased plasma concentration, prolonged circulation time and reached detectable concentrations in the brain when compared to uncoated nanoparticles. Brigger et al[13] also showed that poly(ethylene glycol) coated hexadecylcyanoacrylate nanospheres displayed a combined effect for brain tumor targeting. Zhou et al[14] investigated the antivasculative effects of DXR containing liposomes in an intracranial rat brain tumor model. Koukourakis[15] applied ^{99m}Tc labeled stealth liposomal DXR in glioblastomas and metastatic brain tumors.

Chen and Lu[16] and Whateley et al[17] prepared and characterized an injectable and biodegradable brain implant of carboplatin, a potent anticancer agent. However, systemic administration of the drug for the treatment of malignant glioma requires high dose due to the limited brain entry of the drug. Such a high dose causes side effects. Incorporation of carboplatin into PLGA and their direct intracerebral implantation to the brain tumor region minimized the side effects, sustained release, and improved efficacy. Whateley et al[17] also prepared PLGA implants of etoposite.

Arica et al[18,19] have prepared biodegradable bromocryptin mesylate microspheres using PLGA, poly(L-lactide) = L – PLA, poly(D,L-lactide) = D,L-PLA for the purpose of prolonging its therapeutic activity after stereotaxic implantation to rat brain and hypophysis. (Fig.3). Histological examinations showed no signs of toxicity. These studies proved that these formulations would be potentially useful in the therapy of brain and hypophysis tumor.

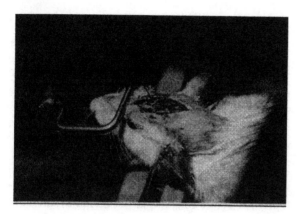

Figure 3. Stereotaxic implantation of microparticles

Delivery systems of neuroncology drugs are shown in Table 2.

Table 2. Delivery systems of neuroncology drugs

CARRIER SYSTEM	ACTIVE DRUG	POLYMER
Pellet	5-FU	polydimethylsiloxane
Pellet	5-FU	PMMA
Microsphere	5-FU	chitosan / D,L-PLA
Microsphere	5-FU	chitosan / PLGA
Cylinder	carmustine	EVAc
Disc	carmustine	polyanhydride copolym
Microsphere	carmustine	PLGA
Microsphere	BM	PLGA / D, L-PLA / L-PLA
Nanoparticle	DXR	PBCA
Liposome / Stealth – Liposome	DXR	
Pellet	mitomycin	PMMA
Pellet	adriamycin	PMMA
Pellet	nimustin	PMMA
Pellet	methotreksate	PMMA
Disc	taxol	polyanhydride copolym
Disc	cyclophosphamide	polyanhydride copolym
Microsphere	carboplatin	PLGA
Microsphere	BCNU	PLGA
Microspheres	etoposide	PLGA
Microsphere	oxantrazole	chitosan – magnetic

This table is modified from references 7 and 20.

2.2   Neuroactive Drugs

Levodopa (L-Dopa), carbidopa, dopamine (DA), PC 12 cells, bethanechol and nerve growth factor (NGF) are some examples of neuroactive   drugs transported to brain by microparticulate carriers. Filippovic-Grcic et al.[21] encapsulated L-Dopa in alginate–chitosan microspheres in order to prolong the release. They observed 10^3 fold difference between the drug released from microspheres and that released from the unbound form. Maysinger et al.[22] implanted L-Dopa–HSA microspheres into the rats brain and followed drug release in vivo for up to 60 hours by brain microdialysis method. Arica et al[23,24] prepared PLGA, L-PLA and D,L-PLA microspheres loaded with antiParkinsonian drugs L-Dopa and Carbidopa and implanted these microspheres stereotaxically to Parkinson established male albino rats. Animal studies were carried out in rats to confirm the sustained release properties and to observe the rotational behavior. The authors observed reduction in the rotational behavior in vivo. Yurasov et al[25] investigated the effect of long term administration of L-Dopa loaded liposomes on the turnover of dopamine in the striatum of mice with experimental Parkinson's syndrome. Dopamine delivery from brain implants may be of benefit to Parkinsonian patients. Dopamine when taken orally does not cross the BBB. However, when slabs (EVAc) containing up to 50 % DA were implanted adjacent to the striatum in rats, striatal extracellular fluid concentration of dopamine were found to be elevated over 200 times than the control for two month period[26]. These implanted slabs, significantly decreased (nearly 50 %) apomorphine induced rotational behavior. Zhigaltzov et al[27] prepared dopamine entrapped liposomes for dopamine delivery into the brain of Parkinsonian mice. PC 12, is the cell line frequently used to synthesize, store and release large amounts of dopamine. Water soluble polyelectrolytes were utilized for membrane encapsulation of dopamine secreting PC 12 cells[28]. These cells have been transplanted into striatum of animals where Parkinsonism is induced experimentally. From 1 - 4 weeks post implantation, a significant reduction in rotational behavior under apomorphine challenge was observed with PC 12 cell loaded microcapsules when compared to empty microcapsules. Valbacka et al[29] implanted the microencapsulated dopamine producing cells in rat brains. Bethanechol is an acetylcholinesterase resistant cholinomimetic. Microspheres of bethanechol when implanted into the hippocampus of rats which had previously undergone a cholinergic denervation, were able to reverse lesion induced memory deficits[30]. Nerve Growth Factor, a biologically active peptide, has been incorporated into EVAc, alginate-polylysine and PLGA polymers. Koseki et al[31] have developed collagen based minipellets of NGF. These minipellets were intracerebrally implanted into cat brain by stereotaxic implantation. Distribution and content of NGF within the brain parenchyma was studied and controlled release was observed for a period of two weeks. Delivery systems of neuroactive drugs are shown in Table 3.

Table 3. Delivery systems of neuroactive drugs

CARRIER SYSTEM	ACTIVE DRUG	POLYMER
Pellet	dopamine	polydimeyhylsiloxane
Disc	dopamine	EVAc
Rod	dopamine	EVAc
Liposome	dopamine	
Microcapsule	PC – 12	polyelectrolytes
Microsphere	L - Dopa	alginate / chitosan, HSA
Microsphere	L - Dopa	PLGA / D, L-PLA / L-PLA
Liposome	L – Dopa	
Microsphere	carbidopa	PLGA / D, L-PLA / L-PLA
Rod	NGF	EVAc
Disc	NGF	EVAc
Cylinder	NGF	collagen
Microsphere	NGF	alginate / PLGA
Microsphere	bethanecol	polyanhydride copolym

This table is modified from references 7 and 20.

2.3   Other Drugs

Drugs, other then neuroncology and neuractive properties, applied to brain will be discussed under this section. These drugs are dalargin, tubocurarine, dexamethasone, dexamethasone sodium phosphate, and lorepamide. Dalargin is a hexapeptide. Kreuter et al[32] investigated dalargin targeting to brain by polysorbate 80 coated nanoparticles. Dalargin transport across BBB was accomplished and analgesic activity was measured by the tail flick test. Alyautdin et al[33] studied the possibility of using polysorbate–80 coated PBCA nanoparticles to deliver low molecular polar hydrophilic drugs to the central nervous system. A quaternary ammonium salt, tubocurarine, which cannot penetrate the normal intact BBB, was selected as the active drug. Loading of tubocurarine onto the polysorbate–80 coated nanoparticles enabled the transport of the compound through the BBB. Alyautdin et al[34] also prepared polysorbate–80 coated PBCA nanoparticles of lorepamide and studied delivery across the BBB. Dexamethasone implants of PLGA are prepared for treating cerebral oedema associated with brain tumors[35]. The release of dexamethasone from implants were measured for up to 192 days. Dexamethasone sodium phosphate (DSP) is a corticosteroid widely used in the treatment of brain oedema. The aim of the studies of Eroglu et al[36,37] was to minimize its adverse effects, extend its duration of release, and to demonstrate its effectiveness in brain oedema by preparing biodegradable, biocompatible BSA, PLGA and L-PLA microspheres of DSP. The authors generated brain oedema by cold lesion method and investigated the effectiveness of the microspheres in brain oedema treatment by the wet-dry weight, lipid peroxidation ratios, and histological evaluations[37] (Table 4).

Table 4. Water content and lipid peroxidation ratios of the brain tissue

TREATMENT	WATER CONTENT (%)	LIPID PEROXIDATION RATIO (nmol/g wet tissue)
i,p, DSP solution	1.41	69.66 ± 22.12
microsphere implantation	0.29	46.67 ± 10.34

The degree of oedema was significantly different from the control group for the wet-dry weight method and lipid peroxidation ratio ($p < 0.05$). Similarly, histological evaluation of the tissues showed that degree of oedema was significantly decreased with respect to the control group. All these results showed that implantation of microspheres was significantly more effective with respect to the systemic administration of DSP in the treatment of brain oedema associated with brain tumors. Delivery systems of other drugs are shown in Table 5.

Table 5. Delivery systems of other drugs

CARRIER SYSTEM	ACTIVE DRUG	POLYMER
nanoparticle	dalargin	dextran
nanoparticle	tubocurarine	PBCA
nanoparticle	lorepamide	PBCA
microsphere	dexamethasone	PLGA
microsphere	DSP	BSA / PLGA / L-PLA

3.   CONCLUSION

The methods for modifying drug transport across BBB have several limitations. Many investigators working on transport to brain propose that consideration must be given to the development of polymer delivery systems rather than local approaches such as surgery or radiation therapy[38]. Polymeric microparticulate systems are being investigated in order to minimize these limitations, to decrease adverse effects, to increase patient compliance, and to deliver and distribute drugs to brain space. The obtained results are very promising and the research is continuing in this field.

ACKNOWLEDGEMENT

The author is grateful to Dr. Betul Arıca and Hakan Eroglu (M.Sc) for the experimental work during the study of their thesis. The author also thanks Ms. Ozge Kas for her valuable assistance in the preparation of figures.

REFERENCES

1. Brightman, M.W., 1997, Morphology of blood-brain interface. *Exp. Eye Res.* 25:1-25.
2. Partridge, W.M., 1996, Brain drug delivery and blood-brain barrier transport. *Drug Delivery* 3: 99-115.
3. Partridge, W.M., 1999, Vector mediated drug delivery to the brain. *Adv. Del. Reviews* 36: 299-321
4. Kubo, O., Tajika, Y., Muragaki, Y., 1994, Local chemotherapy with slow releasing anticancer drug-polymers for malignant brain tumors. *J. Cont. Rel.* 32: 1-8.
5. Menei, P, Boisdron-Celle, M, Croue, A., 1996, Effect of stereotaxic implantation of biodegradable 5-fluorouracil-loaded microspheres in healthy and C6 glioma-bearing rats. *Neurusurgery* 39: 117-124.
6. Chandy, T., Das, G.S., Rao, G.H.R., 2000, 5-Fluorouracil loaded polylactic acid microspheres as biodegradable drug carrier for cerebral tumors, *J. Microencapsulation* 17: 625-638.
7. Menei, P., Venier-Julienne, M.C., Benoit, J.P., 1997, Drug delivery into the brain using implantable polymeric systems. *S.T.P. Pharma Sci.* 7: 53-61.
8. Painbeni, T., Venier- Julienne, M.C., Benoit, J.P., 1998, Internal morphology of poly (D,L-lactide-co-glycolide) BCNU-loaded microspheres: Influence on drug stabiliy. *Eur. J. Pharm. Biopharm.* 45: 31-39.
9. Valtonen, S., Timonen, U., Tolvanen, P., 1997, Interstitial chemotherapy with carmustin loaded polymers for high grade gliomas: A randomized double-blind study. *Neurosurgery* 41: 44-48.
10. Langer, R., 1990, New methods of drug delivery. *Science* 249: 1527-1533.
11. Hassan, E.E. and Gallo, J.M., 1993, Targeting anticancer to the brain. 1. Enhanced brain delivery of oxantrazole following administration in magnetic cationic microspheres. *J. Drug Targeting* 1: 7-14.
12. Kreuter, J., 1998, Recent advances in nanoparticles and nanospheres. In *Biomedical Science and Technology* (A.A. Hincal and H.S. Kas, eds.) Plenum Press, NY, pp. 31-39.
13. Brigger, L., Morizet, J., Aubert, G., Chacun, H., Terrier-Lacombe, M.J., Couvreur, P., Vassal, G., 2002, Poly (ethylene glycol) coated hexadecylcyanoacrylate nanospheres display a combined effect for brain tumor targeting. *J. Pharm. Exp. Therapeutics* 303: 928-936.
14. Zhou, R., Mazurchuk, R., Straubinger, R.M., 2002, Antivasculature effects of doxorubicin containing liposomes in an intracranial rat brain tumor model. *Cancer Res.* 62: 2561-2566.
15. Koukourakis, M.I., 2002, ^{99m}Tc labeled stealth liposomal doxorubicin (Caelyx[R]) in glioblastomas and metastatic brain tumors. *British J. Cancer* 86: 660-661.
16. Chen, W. and Lu, D.R., 1999, Carboplatin loaded PLGA microspheres for intracerebral injection: formulation and characterisation. *J. Microencapsulation* 16: 551-563.
17. Whateley, T.L., Fallon, P.A., Robertson, L., 1995, In vivo study of PLGA implants for sustained delivery of carboplatin and etoposide to the brain. *Proceed. Intern. Symp. Contol. Rel. Bioact. Mater.* 22: 582-583.
18. Arica, B., Kas, H.S., Orman, M.N., Hincal, A.A., 2002, Biodegradable bromocriptine mesylate microspheres prepared by a solvent evaporation technique I. Evaluation of formulation variables on microsphere characteristics for brain delivery. *J. Microencapsulation* 19: 473-484.
19. Arica, B., Kas, H.S., Sargon, M.F., Acikgoz, B., Hincal, A.A., 1999, Biodegradable bromocriptine mesylate microspheres prepared by a solvent evaporation technique II. Suitability for brain and hypophysis delivery. *S.T.P. Pharma Sci.* 9: 447-455.
20. Arica, B., and Kas, H.S., 1997, Beyine ilac tasinmasi. Hacettepe Tip Dergisi 28: 4-14.

21. Filipovic-Grcic, J., Maysinger,D., Zorc, B., Jalsenjak, I., 1995, Macromolecular prodrugs IV. Alginate-chitosan microspheres of PHEA-L-Dopa adduct. *Int. J. Pharm.* 116: 39-44.
22. Maysinger,D., Jalsenjak, V., Stolnik, S., Jalsenjak, I., 1992, Release of L-Dopa from HSA microspheres into the cat brain: in vitro and in vivo characterization by microdialysis. In *Microencapsulation of Drugs* (T.L.Whateley, ed), Harwood Acad. Publ., Switzerland, pp. 269-275.
23. Arica, B., Kas, H.S., Hincal, A.A., 1997, Evaluation and formulation of biodegradable levodopa microspheres using 3^2 factorial design. *4th Int. Biomed Sci. Technol. Symp.*, pp. 9-11.
24. Arica, B., Kas, H.S., Hincal, A.A., 1997, Factors influencing the formulation parameters of biodegradable carbidopa microspheres. *11th Int. Symp. Microencapsulation* p.55.
25. Yurasov, V.V., Kucheryanu, V.G., Kudrin, V.S., Zhigal-Tsev, I.V., Nikushkin, E.V., Sandalov, Y.G., Kapkin, A.P., Shvets, V.S., 1997, Effect of long term parenteral administration of empty and L-Dopa loaded liposomes on the turnover of dopamine and its metabolites in the striatum of mice with experimental Parkinson's syndrome. *Bull. Exp. Biol. Med.*, 123: 126-129.
26. During, M,J., Freese, A., Sabel, B.A., 1989, Controlled release of dopamine from a polymeric brain implant: in vivo characterization. *Ann. Neuro.* 25: 351-356.
27. Zhigaltsov, I.V., Kaplan, A.P., Kucheryanu, V.G., Kryzhanovsky, G.N., Kolomeichuk, S.N., Shvets, V.L., Yurasov, V.V., 2001, Liposomes containing dopamine entrapped in response to transmembrane ammonium sulphate gradient as carrier system for dopamine delivery into the brain of Parkinsonian mice. *J. Liposome Res.* 11: 55-71.
28. Winn, S.R., Tresco, P.A., Zielinski, B., 1991, Behavioral recovery following intrastriatal implantation of microencapsulated PC 12 cells. *Exp. Neurology* 113: 322-329.
29. Vallbacka, J.J., Nobrega, J.N., Sefton, M.V., 2001, Tissue engineering as a platform for controlled release of therapeutic agents: implantation of microencapsulated dopamine producing cells in the brain of rats. *J. Control.Rel.* 72: 93-100.
30. Howard, M.A., Gross, A., Glady, M.S., 1989, Intracerebral drug delivery in rats with lesion-induced memory deficits. *J. Neurosurgery* 71: 105-112.
31. Koseki, N., Takemoto, O., Sasaki, Y.,1996, A new application pf peptide drug delivery system to the brain. *Proceed. Intern. Symp. Contol. Rel. Bioact. Mater.* 23: 605-606.
32. Kreuter, J., Alyautdin, R. N., Kharkevich, D. A., 1995, Passage of peptides through the blood-brain barrier with colloidal polymer nanoparticles. *Brain Res.* 674: 171-174.
33. Alyautdin, R. N., Tezikov, E.B., Ramge, P., Kharkevich, D. A., Begley, D. J., Kreuter, J., 1998, Significant entry of tubocurarine into the brain of rats by adsorption to polysorbate 80-coated polybutylcyanoacrylate nanoparticles: an in situ brain perfusion study. *J. Microncapsulation* 15: 67-74.
34. Alyautdin, R. N., Petrov, V.E., Langer, R., Berthold, A., Kharkevich, D. A., Kreuter, J., 1997, Delivery of lorapamide across the blood-brain barrier with polysorbate 80-coated PBCA nanoparticles. *Pharm. Res.* 14: 324-328.
35. Fallon P.A. and Whateley, T.L., 1994, The development and evaluation of poly(lactic-glycolic acid) implants for sustained delivery of dexamethasone to the brain. *Proceed. Intern. Symp. Contol. Rel. Bioact. Mater.* 21: 364-365.
36. Eroglu,H., Kas, H.S., Oner, L., Turkoglu, O.F., Akalan, N., Sargon, M.F., Ozer, N., Hincal, A.A., 2000. In vitro / in vivo characterization of bovine serum albumin microspheres containing dexamethasone sodium phosphate, *STP Pharm Sci.* 10:303-308.
37. Eroglu, H., Kas, H.S., Oner, L., Turkoglu, O.F., Akalan, N., Sargon, M.F., Ozer, N., 2001, The in vitro and in vivo characterization of PLGA: L-PLA microspheres containing dexamethasone sodium phosphate. *J. Microencapsulation* 18: 603-612.
38. Kas, H.S. and Arica,B., 1997, Delivery to brain by microparticles. *Proc. Symp. Particulate Systems from Formulation to Production*, pp. 123-124.

Bioadhesive Drug Carriers for Postoperative Chemotherapy in Bladder Cancer

EYLEM ÖZTÜRK*, MUZAFFER EROĞLU#, NALAN ÖZDEMİR*, and
EMİR B. DENKBAŞ*
*Hacettepe University, Chemistry Dept., Biochem. Div., Beytepe, Ankara, Turkey; #SSK
İhtisas Hospital, Ankara, Turkey

Transurethral resection (TUR) is the primary mode of therapy for both diagnosis and treatment of bladder cancer. Due to the recurrency of tumoral tissues after TUR further treatment is necessary which is usually in the form of intravesical chemotherapy or immunotherapy. But these therapies have some disadvantages such as disturbancy to patients, adjustment of the suitable dosage, loss of active agents without using. In this study, an alternative approach was proposed and pharmaco-therapeutic agent delivery systems which will supply the suitable dosage of the agent for a certain time period were designed to solve those problems. For this aim, Mitomycin-C loaded alginate and chitosan carriers were prepared to use as an alternative system in the post-operative chemotherapy in bladder cancer. The carriers were prepared in the form of cylindirical geometries to facilitate the insertion of the carrier in *in vivo* studies. The effects of some parameters (i.e., polymer MW, cross-linker concentration, Mitomycin-C/polymer ratio etc.) over the morphology, swelling behavior, bioadhesion and *in-vitro* drug release rate of the carriers were evaluated. The obtained results for chitosan and alginate carriers were concluded comparatively.

Biomaterials: *From Molecules to Engineered Tissues,* edited by
N. Hasırcı and V. Hasırcı, Kluwer Academic/Plenum Publishers, 2004

1.   INTRODUCTION

Bioadhesion in simple terms can be defined as the attachment of synthetic or biological macromolecules to a biological tissue[1]. An adhesive bond may form with either the epithelial cell layer, the continuous mucus layer or a combination of the two. The term "mucoadhesion" is used specifically when the bond involves mucous coating and an adhesive polymeric device. The mechanism of bioadhesion has been reviewed extensively[2,3].

Amongst the early pioneering work on bioadhesive systems is that of Nagai and coworkers, who showed that the local treatment of the aphthea in the oral mucosa was improved by using an adhesive tablet[4]. In addition, they observed increased systemic bioavailability of insulin when given intranasally to beagle dogs as a powder dosage form[5]. For application to the oral mucosa, novel mucoadhesive ointments were also introduced; these were in fact polymer gels based on polyacrylic acid[6] and polymethyl methacrylate[7]. Indeed, the new bioadhesion concept rapidly lead to the idea that bioadhesion could be used advantageously to improve absorption through several administration routes. Over the years, bioadhesive systems have been used for nasal, ocular, buccal, vaginal, rectal and oral drug delivery. Most of the early work on bioadhesive polymers was performed with "off-the shelf polymers", such as polyacrylic acids in the dry state, often in the form of powders[8,9], tablets[10], coated spheres[11] or dried films[12]. From these studies, rankings of polymers were made and general conclusions were drawn about the physicochemical characteristics of good bioadhesives, with respect to, e.g., molecular weight, cross-linking density and charged groups.   However in the 1990s, as the interest in polymer gels as pharmaceutical dosage forms increased, it was realized that different mechanisms involved in mucoadhesion would be important compared with those of dry dosage forms. It was pointed out that adhesion observed with dry dosage forms may, to large extent, arise from water transfer and dehydration of the mucus layer[13,14].

Originally, the advantages of mucoadhesive drug delivery systems were considered to lie in their potential to prolong the residence time at the site of absorption, and to provide an intensified contact with the underlying mucosal epithelial barrier. Later, it was discovered that some mucoadhesive polymers, such as polyacrylic acids and chitosan, possess multifunctional properties, and can, for example, modulate the permeability of the epithelial tissues by partially opening the tight junctions[15,16]. The polyacrylic acids have also shown to inhibit proteolytical enzymes[17], probably by depleting the enzymes of Ca^{+2} and Zn^{+2} ions[18,19].

Bladder cancer is the fourth most common malignant disease world wide, accounting for 4 % of all cancer cases[20]. Carcinoma of the bladder occurs within the domain of human neoplasms by many different types of etiological factors such as some aromatics and smoking cigarette[21,22]. Systemic treatment options for bladder cancer include surgery, chemotherapy, radiation, and immunotherapy. Treatment modalities are often combined (e.g. surgery with chemotherapy and/or radiation). In early bladder cancer, transurethral resection is a common mode while partial or radical cystectomy is performed for muscle invasive and locally advanced bladder cancer. Transurethral resection (TUR) of the bladder is a surgical procedure that is used both to diagnose bladder cancer and to remove cancerous tissue from the bladder. It is the most common treatment method, but, most of the time the TUR is not an absolute solution for bladder cancer because of the recurrency of the formation of tumoral tissues in 50-70 % of the cases[23]. Intravesical pharmacotherapeutics infusion is the most common protective therapy for the recurrency problem after TUR which is used periodically (each week for 6 to 36 weeks)[24-26]. Unfortunately, there are some difficulties such as the establishment of the suitable dosage, loss of bioactive agent (because, bladder is filled up and discharges periodically and the pharmacotherapeutic agent leaves from bladder) and the disturbance that is given to the patient in each agent application and the loss of pharmacotherapeutic agent without being used the therapeutic procedure needs much more improvement for practical applicability.

The main aim of this study was to prepare reservoir type pharmacotherapeutic agent carriers as an alternative way for pharmacotherapy after TUR. Mucoadhesive polymeric structures (i.e. chitosan and alginate) were used in the preparation of pharmacotherapeutic agent reservoir. Mitomycin-C which is one of the most popular pharmacotherapeutic agent used in bladder cancer chemotherapy[27-30] was selected as a model pharmacotherapeutic agent. Carriers were prepared in the form of cylindrical rod to facilitate its insertion intravesically with a catheter. Chitosan carriers were prepared by cross-linking during the solvent evaporation technique. In the cross-linking a bifunctional chemical (i.e. glutaraldehyde) was used. The alginate carriers were prepared by precipitation in $CaCl_2$ solution followed by cross-linking with a cross-linker, ethylene glycol diglycidyl ether, EGDGE. The bioadhesion tests were performed as ex vivo studies. The pharmacotherapeutic agent release experiments were performed and swelling behaviour was determined by gravimetric method.

2.   MATERIALS AND METHODS

2.1   Materials

Chitosan polymers with different molecular weights (i.e. 150 kDa: LMW; 450 kDa: MMW; and 650 kDa: HMW), glutaraldehyde, calcium chloride, sodium alginate and ethylene glycol diglycidyl ether, EGDGE were supplied by Fluka, Switzerland. Mitomycin-C was purchased from Kyowa Hakko KCGYO Co., Japan. Acetic acid was supplied by Carlo Erba, Italy..

2.2   Preparation of Chitosan and Alginate Carriers

Chitosan rods were prepared by the technique of cross-linking during solvent evaporation. Certain amount of chitosan (i.e. 250 mg) was dissolved in aqueous acetic acid solution (i.e. 5% v/v). Obtained gelous chitosan was filled into an injector and injected vertically into the plastic pipette (with the internal diameter of 5 mm) filled with cross-linker (i.e. glutaraldehyde) solution. After filling the chitosan solution, plastic pipettes were dried in vertical position at room temperature for 48 h. The rods were neutralized with sodium hydroxide for 1 h. For Mitomycin-C loaded chitosan rod fabrication, the same procedure was applied just by the addition of certain amount of Mitomycin-C (i.e. 0.25 and 0.50 g/g chitosan) into the aqueous acetic acid–chitosan solution at the beginning of the procedure.

Alginate carriers were prepared by precipitation of the polymer in $CaCl_2$ solution. 250 mg of sodium alginate was dissolved in aqueous solution (5 % v/v). Obtained gelous sodium alginate was introduced into an injector and injected into the beaker filled with $CaCl_2$ solution by needleless injector. Precipitated alginate rods were cured in the same medium for 30 minutes and dried at room temperature up to constant weight. After that rods were additionally cross-linked with different amount of EGDGE. In the concentration range of 1-20 % v/v to achieve different crosslinking densities of the alginate rods cross-linked with $CaCl_2$ previously. The same procedure mentioned above was applied in the addition of certain amount of Mitomycin-C aqueous solution (0.2, 1 and 2 mg/g alginate) into the initial aqueous solution.

2.3   Characterization of Carriers

The prepared chitosan and alginate carriers were characterized by using different techniques based on morphology, swellability, in vitro drug release and bioadhesion properties.

2.4   Morphological Evaluation

Carriers were evaluated with scanning electron microscope (SEM) for morphological characterization.

2.5   Swelling Measurements

Swelling behavior of the carriers were determined by gravimetric method. A sample of carrier (50 mg) was placed in phosphate buffers with different pH values (6.0, 7.0 and 8.0) for a particular period of time. The swollen carrier was collected and the wet weight of the swollen carrier was determined periodically. The percent swelling was calculated from the following equation, where; S_c is the percent swelling of the carrier, w_t is the weight of the carrier at time t, and w_o is the initial weight of the carrier.

$$Sc = [(w_t - w_o) \times 100]/ w_o$$

2.6   Bioadhesion Tests

Bioadhesion tests were carried out using an apparatus constructed according to a method improved by Smart[31]. The apparatus included a platform, a 5x5x10 cm glass cuvette filled with urine obtained from a fresh rabbit bladder and a microbalance. The sizes of the carriers tested were 2 mm in diameter and 1 cm in height and the sizes of the bladder were 2 cm in width and 5 cm in length. A gel sample was attached to a rope suspended from a top pan microbalance (Mettler AB 204, Switzerland) and the suspended carrier was immersed into the bladder sample. The platform was lowered at a rate of 2 mm/min until the carrier detached from the carrier and the weight changings in microbalance were recorded. Obtained data were converted to mN/cm^2 of carrier.

2.7   In-vitro Mitomycin-C Release Studies

Carriers were placed in phosphate buffer solutions and the release amounts were determined with an UV-Visible spectrophotometer (Shimadzu, Japan) at 217 nm and concentrations were calculated by using the absorbance-concentration calibration curve performed previously[32]. The release medium was kept at 37°C and shaken in constant temperature bath.

3.   RESULTS AND DISCUSSIONS

3.1   Morphological Evaluation

The cross-section micrographs of the dried and freeze dried form of swollen alginate carriers are shown in Figure 1A-B. The structure of the carrier seems to be non porous in dried form, but it has some holes occured by water uptake in freeze dried of swollen rod. For wet carriers, the size was increased according to the amount of cross-linking density (or the amount of $CaCl_2$ and/or EGDGE).

The surface of the chitosan carrier seems to be very smooth and there is no pore on both surface and cross-section[32]. In the case of wet chitosan carriers some holes and texturity occurred as seen in Fig.1c. This figure represents the cross-sectional view of the chitosan carrier when reaches the saturated swelling value. The hole size and the quantities of the holes especially depend on the amount of cross-linker.

  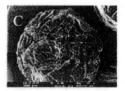

Figure 1. SEM micrographs of (A) dried (B) swollen and freeze dried alginate carrier (C) chitosan carrier (swollen and freeze dried)

3.2   Swelling Measurements

Almost all of the alginate carriers reached to the saturation swelling ratio after 2 hours and the swelling ratios changed between 30-60 % according to parameters. Swelling ratio increased by decreasing the amount of $CaCl_2$ in precipitation medium as seen in Figure 2a. This situation is due to the higher amount of cross-linking densities when $CaCl_2$ concentration increased.

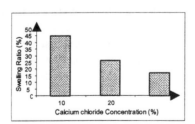 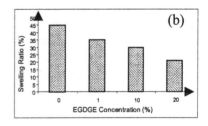

Figure 2. Effects of (a) $CaCl_2$ and (b) EGDGE concentration on swelling behavior

Amount of additional cross-linker, EGDGE, was selected as an another effective parameter over the swelling ratio. EGDGE concentration in the precipitation medium was changed in the range of 1-20 % (v/v) to investigate this effect. As can be clearly seen from Figure 2b the swelling ratio was decreased with increasing the EGDGE concentration due to the higher crosslinking densities in the case of higher amount of cross-linker.

The urine pH is changed in the range of 4.5-8[33]. Especially for the representation of the urine pH, swelling experiments were made of different pH values varied that 6-8 pH values. The swelling ratio of the sodium alginate carrier is decreased by increasing the medium pH values as seen in Figure 3a.

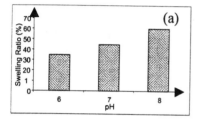 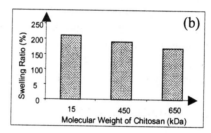

Figure 3. Effects of (a)medium pH and (b)chitosan molecular weight on swelling behavior of alginate carriers

Chitosan molecular weight, amount of crosslinker (i.e. glutaraldehyde concentration) and NaOH concenrations in neutralization media were used as the effective parameters on the swelling behaviour of chitosan carriers. The chitosan carriers reached the saturation swelling ratio after 4–6 h and changed between 190 and 210% according to the molecular weight of chitosan. This is an expected result, because it is well known that the water diffusion is more difficult in the case of higher molecular weight because of the less molecular spaces than lower molecular weights for the same volume of polymer unit.

Similar observations can be seen in Fig.4, which shows the effects of the amount of crosslinker (i.e. glutaraldehyde concentration). It means that, more glutaraldehyde causes the higher cross-linking densities subsequently causing a smaller swelling ratio as expressed in related literature[34].

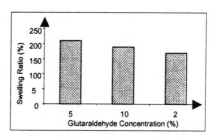

Figure 4. Effects of glutaraldehyde conc. on swelling behavior of chitosan carriers

Generally, chitosan polymers do not dissolve in aqueous solutions while freely soluble in dilute i.e. 1–10%) aqueous acetic (or formic) acid solutions[35]. When the dried chitosan carriers are placed in the aqueous media, the carriers convert into gelous form because of the residual acidic groups in the polymer. Therefore, the chitosan carriers should be neutralized with alkaline solutions. NaOH concentration will change the amount of residual acidic groups and hence it will change the swelling ratio. For example; the swelling ratio was measured as 210 % for 0.5 g NaOH/100 ml while 175 % for 4 g NaOH/100 ml. It means that the swelling ratio was decrased by increasing NaOH concentration in neutralization media.

3.3 Bioadhesion Tests

The adhesive forces are decreased by increasing the chitosan molecular weight as seen in Table 1. This can be explained by the length of polymeric chain carrying the functional groups, which is the predominant parameter for adhesive forces that increases by the increasing number of polymeric chains in the case of higher molecular weights.

Table 1. Effects of chitosan molecular weight on bioadhesion of chitosan carriers

Molecular Weight of Chitosan	Bioadhesion Force (mN/cm^2)
150.000	11
650.000	5

On the other hand, the crosslinking occurs via the functional groups of chitosan and glutaraldehyde. It means that if the amount of glutaraldehyde increases more functional groups are occupied with glutaraldehyde molecules and the decrease in the number of functional groups decreases the adhesive force as expected. This behaviour can be seen easily in Table 2.

Table 2. Effects of glutaraldehyde ratio on bioadhesion of chitosan carriers

Glutaraldehyde Ratio (ml/ml sus. medium)	Bioadhesion Force (mN/cm^2)
1/20	5.5
2/20	4.5

The preliminary results showed that the cross-linking density was the most important parameter on bioadhesion for alginate carriers. The obtained results were summarized in Table 3. The adhesive forces are decreased by increasing the calcium chloride concentration. This can be explained by increasing the cross-linking densities at higher calcium chloride concentration. This increment leads to less swelling ratio and less surface area, hence the amount of reactive groups that play an important role in bioadhesion are decreased.

Table 3. Effects of calcium chloride concentration on bioadhesion of alginate carriers

CaCl$_2$ concentration (%)	Bioadhesion Force (mN/cm^2)
10	39
20	22
40	14

3.4   In Vitro Mitomycin-C Release Studies

In the release studies, chitosan molecular weight and the initial amount of Mitomycin-C were selected as the most effective parameters on release mechanism according to the previous studies made with chitosan polymers[36]. The Mitomycin-C release was decreased by increasing the chitosan molecular weight. This behaviour depends on the higher tightness of the polymeric chains in the case of chitosan carriers prepared with higher molecular weight chitosan polymers. In this case, swelling ratio was also decreased as expressed before.

Another significant parameter was the initial amount of Mitomycin-C in chitosan carriers. The release rate is increased in the case of higher initial Mitomycin-C content. This can be explained with the generated holes by releasing the Mitomycin- C molecules. Released Mitomycin-C molecules left more space after they released in the case of higher Mitomycin-C content. This hole generation is similar with the generation of the holes during the swelling process and of course the amount of the hole numbers will affect the release rate directly.

Mitomycin-C loaded alginate carriers were prepared with different precipitation medium containing different amounts of calcium chloride. Mitomycin-C was released rapidly in the case of lower calcium chloride concentrations since the reduced cross-linking density occurred by lower calcium chloride concentrations. This release behavior is very similar with the swelling behavior for different calcium chloride concentrations discussed previously. Initial Mitomycin-C concentration was evaluated as another effective parameter on release behaviour.The release rate was increased in the case of higher initial Mitomycin-C content. This can be explained with the generated holes by releasing the Mitomycin-C molecules. Released Mitomycin-C molecules left more spaces after they released in the case of higher Mitomycin-C content. This hole generation is similar with the generation of the holes during the swelling process and of course the amount of the hole numbers will affect the release rate directly.

4.   CONCLUSIONS

Mitomycin-C loaded chitosan and alginate carriers were prepared and characterized to use in the postoperative chemotherapy in bladder cancer. The cross-linking density was obtained as the most effective parameter on the swelling ratio for chitosan and alginate. Bioadhesion data showed that alginate is more bioadhesive than chitosan, but changing the formulations changes bioadhesion values. In vitro Mitomycin-C release data showed that the cross-linking density is the most effective parameter on release rate. Cross-linking density can be changed by changing the molecular weight of polymer or cross-linker concentrations. Another effective parameter on drug release rate was determined as the Mitomycin-C content of the carrier. Release rate was increased by increasing Mitomycin-C content. All the obtained data showed that both chitosan and alginate carriers seem very promising reservoir type drug carrier system in the postoperative chemotherapy in bladder cancer after TUR. This study was summarized the comparison of the chitosan and alginate carriers for the mentioned purpose. All the other details can be found elsewhere individually for both chitosan and alginate carriers[37, 38].

REFERENCES

1. Peppas, N.A., and Buri, P.A., 1985, Surface, interfacial and molecular aspects of polymer bioadhesion on soft tissues. *J Control Release* 2:257–275.
2. Ahuja, A., Khar, R.K., Ali, J., 1997, Mucoadhesive drug delivery systems. *Drug Dev Ind Pharm* 23: 489–515.
3. Lee, J.W., Park, J.H., Robinson, J.R., 2000, Bioadhesive based dosage forms: the next generation. *J Pharm Sci* 89: 850–866.
4. Nagai, T., 1985, Adhesive topical drug delivery system. *J Control Release* 2: 121-134.

5. Nagai, T., Nishimoto, Y., Nambu, N., Suzuki, Y., Sekine, K., 1984, Powder dosage form of insulin for nasal administration. *J Control Release* 1: 15-22.

6. Ishida, M., Nambu, N., Nagai, T., 1983, Highly viscous gel ointment containing Carbopol for application to the oral mucosa. *Chem Pharm Bull* 31: 4561-4564.

7. Bremecker, K.D., Strempel, H., Klein, G., 1984, Novel concept for a mucosal adhesive ointment. *J Pharm Sci* 73: 548-552.

8. Park H., and Robinson J.R., 1985, Physico-chemical properties of water insoluble polymers important to mucin/epithelial adhesion. *J Control Release* 2:47-57

9. Chang, H.S., Park, H., Kelly, P., Robinson, J.R., 1985, Bioadhesive polymers as platforms for oral controlled drug delivery II: Synthesis and evaluation of some swelling ,water insoluble bioadhesive polymers. *J Pharm Sci* 74: 399-405.

10. Ponchel, G., Touchard, F., Duchene, D., Peppas, N.A ,1987, Bioadhesive analysis of controlled-release systems.I.Fracture and interpenetration analysis in poly(acrylic acid)-containing systems. *J Control Release* 5:129-141.

11. Teng, C.L.C., and Ho, N.F.H., 1987, Mechanistic studies in the simultaneous flow and adsorption of polymer-coated latex particles on intestinal mucus I:methods and physical model development . *J Control Release* 6: 133-149.

12. Smart, J.D., Kellaway, I.W.,Worthington, H.E.C., 1984, An in-vitro investigation of mucosa-adhesive materials for use in controlled drug delivery. *J Pharm Pharmacol* 36: 295-299.

13. Mortazavi, S.A., Smart J.D., 1993, An Investigation into the Role of Water-Movement and Mucus Gel Dehydration in Mucoadhesion. *J Control Release* 25: 197-203.

14. Smart, J.D., 1999, The role of water movement and polymer hydration in mucoadhesion, *Bioadhesive drug delivery Systems: Fundamentals,novel approaches and development* (E. Mathiowitz, D.E. Chickering III, C.M. Lehr, eds.), Marcel Dekker, New York, pp. 11-23

15. Borchard, G., Luessen, H.L., deBoer, A.G., Verhoef, J.C., Lehr, C.M., Junginger, H.E., 1996, The potential of mucoadhesive polymers in enhancing intestinal peptide drug absorption.3.Effects of chitosan-glutamate and carbomer on epithelial tight junctions in vitro. *J Control Release* 39: 131-138.

16. Schipper, N.G.M., Olsson, S., Hoogstraate, J.A., deBoer, A.G., Varum, K.M., Artursson, P., 1997, Chitosans as absorption enhancers for poorly absorbable drugs.2.Mechanism of absorption enhancement. *Pharm Res* 14: 923-929.

17. Bai, J.P.F., Chang, L.L., Guo, J.H., 1995, Effects of Polyacrylic Polymers on the Lumenal Proteolysis of Peptide Drugs in the Colon. *J Pharm Sci* 84: 1291-1294.

18. Luessen, H.L., deLeeuw, B.J., Perard, D., Lehr, C.., deBoer, A.B.G., Verhoef, J.C., Junginger, H.E., 1996, Mucoadhesive polymers in peroral peptide drug delivery.1.Influence of mucoadhesive excipients on the proteolytic activity of intestinal enzymes. *Eur J Pharm Sc* 4: 117-128.

19. Luessen, H.L., Lehr, C.M., Rentel, C.O., Noach, A.B.J., Deboer, A.G., Verhoef, J.C., Junginger, H.E., 1994, Bioadhesive Polymers for the Peroral Delivey of Peptide Drugs. *J Control Release* 29: 329-338.

20. Kamat, A.M., and Lamm, D.L., 1999, Chemoprevention of urological cancer. *J Urol* 1999, 161: 1748-60.

21. Case, R.A.M., Hosker, M.E., McDonald, D.B., Pearson, J.T., 1954. Tumours of the urinary bladder in workmen engaged the manufacture and the use of certain dyestuff intermediates in the British Chemical Industry: role of aniline, benzidine, alpha-naphthylamine. Br J Ind Med 11: 75.

22. Wynder, E.L., Goldsmith, R., 1977, The epidemology of bladder cancer: a second look. *Cancer* 40: 1246.

23. Prout, G.R..Jr., 1976, *Urol. Clin. N. Am.* 3: 149.

242   *Eylem Ozturk et al.*

24. Lamm, D.L., Thor, D.E., Stogdill, V.D., Radwin, H.M., 1982, *J Urol* 128: 931.
25. Catalona, W.J., Hudson, M.A., Gıllen, D.P., Andriole, G.L., Ratliff, T.L., 1987, *J Urol* 137: 220-224.
26. Highley, M.S., Oosterom, A.T., Maes, R.A., Brujin,E.A., 1987, *Drug Deliv Sy.* 37: 59-73.
27. Huland, H., Otto, U., Droese, M., 1984, Long term mitomycin-c instillation after transurethral resection of superficial bladder carcinoma. Influence on recurrence, progression and survival. *J Urol* 132: 27–29.
28. Herr, H.W., Laudone, V.P., Whitmore, W.F., 1987, An overwiew of intravesical therapy for superficial bladder tumors. *J Urol* 138: 1363–1368.
29. Dalton, J.T., Wientjes, M.G., Badallament, R.A., 1991, Pharmacokinetics of intravesical mitomycin C in superficial bladder cancer patients. *Cancer Res* 51: 5144.
30. Wientjes, M.G., Badalement, R.A., Wang, R.C., 1993, Penetration of Mitomycin-C in human bladder. *Cancer Res* 53: 3314–3320.
31. Smart, J.D., *1991,Int J Pharm* 73: 69.
32. Irmak, S., 2001, MSc Dissertation, Hacettepe University, Ankara, Turkey.
33. Seeley, R. R., Stephens, T. D., Tate, P. (eds.),1996, *Essentials of anatomy and physiology*, 2nd edition, WCB McGraw Hill Co., NY.
34. Denkbaş, E.B., Odabaşı, M., 2000, Chitosan micropheres and sponges: preparation and characterization. *J Appl Polym Sci* 76: 1637–1643.
35. Rinaudo, M., Domard, A., 1988, Solution.properties of chitosan, in Chitin and Chitosan. *Proceedings from the 4th International Conference on Chitin and Chitosan held in Trondheim* (G. Skjak-Braek, T. Anthonsen, P.A. Sandford, eds.), Norway, August 22–24, 1988, pp. 71.
36. Denkbaş, E.B., Seyyal, M., Pişkin, E., 1999, 5-Fluorouracil loaded chitosan microspheres designed for chemoembolization. *J Microencapsul.* 16: 741–749.
37. Eroğlu, M., Irmak, S., Acar, A., Denkbaş, E.B., 2002, Design and evaluation of a mucoadhesive therapeutic agent delivery system for postoperative chemotherapy in superficial bladder cancer, *Int J Pharm* 235: 51–59.
38. Denkbaş, E.B., Özdemir, N., Öztürk, E., Eroğlu, M., Acar A., 2004. Mitomycin-C loaded alginate carriers for bladder cancer chemotherapy, *J Bioact and Com Pol* (in press).

Three-Year Follow-up of Bioabsorbable PLLA Cages for Lumbar Interbody Fusion:
In Vitro and *In Vivo* Degradation

DEGER C. TUNC[*], MARTIJN VAN DIJK[#], THEO SMIT[#],
PAUL HIGHAM[*], ELIZABETH BURGER[#], and PAUL WUISMAN[#]
[*]*Stryker Orthopaedics, Mahwah, NJ, USA,* [#]*Vrije Universiteit Medical Center, Amsterdam, The Netherlands*

1.   INTRODUCTION

Lumbar interbody fusion cages (LIFC) were developed to reduce incidences of morbidity associated with interbody fusion using only bone graft[2,3]. Cages provide stability to the motion segment and containment for the bone graft, thereby encouraging fusion and positive remodeling. Although the use of LIFCs has been increasing due to their short term success[4-8], there are some concerns about long-term performance[9-12]. Two major variables studied for improving LIFCs are its design and material. Threaded cage designs have shortcomings with regard to orientation and slip distance in torsion[13-15]. The high stiffness of metal cages may lead to stress shielding, migration of the cage, and pseudo-arthrosis[17-22]. Long-term LIFC performance could be improved by modifying the design and material of LIFCs to better accommodate the natural motions of the spine, and bioabsorbable polymers specifically poly-(L-Lactic acid) (PLLA) may help achieve this goal. PLLA offers three advantages. First, because PLLA gradually loses strength over time, stress shielding is minimized. Second, the initial modulus of elasticity of PLLA (about 5 GPa[1]) is much closer to that of vertebral bone tissue (about 2 GPa[24]) than titanium is (100 GPa). Third, the body will absorb PLLA naturally after the intervertebral fusion is complete. Previously published results of a two-year study with PLLA cages are encouraging[25]. This report is on the performance of PLLA cages at 3-years follow-up.

Biomaterials: *From Molecules to Engineered Tissues,* edited by
N. Hasırcı and V. Hasırcı, Kluwer Academic/Plenum Publishers, 2004

2.   MATERIALS AND METHODS

The bioabsorbable polymer used to fabricate the cages was a poly-(L-Lactide), PLLA homopolymer that had an initial inherent viscosity of 7.5 dl/g, heat of fusion of 68 J/g and Tensile Strength of 57 MPal. Cages were fabricated by an Injection Molding process and were sterilized using a gas-plasma method. The LIFCs are vertical and rectangular-shaped (Figure 1).

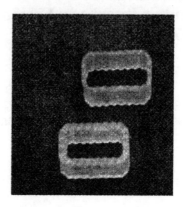

Figure 1. Two PLLA lumbar interbody fusion cages, 10x10x18mm

Two different types of cages were fabricated, stiff (axial compression stiffness of 4 kn/mm and wall thickness of 1.5 mm) and flexible (axial compression stiffness of 2 kN/mm and wall thickness of 0.75 mm). Control cages were made from titanium (axial compression stiffness of 700 kN/mm and wall thickness of 1.5 mm). The cages were studied by two methods, in vitro (84 cages) and in vivo (36 cages). forty-two stiff and 42 flexible cages were used in the in vitro study. these were evaluated after 0, 4, 8, 12, 26, 52, and 73 weeks of incubation in phosphate buffered saline (PBS) at 37 °C. Fifteen stiff and 15 flexible PLLA cages and six titanium cages were used in the in vivo study (goats).

2.1   *In Vitro* Study

PLLA cages were placed in sealed glass jars filled with phosphate buffered saline, PBS at 37 ± 1°C. After 4, 8, 12, 26, 52, and 73 weeks of incubation, six PLLA cages were removed and tested for inherent viscosity, crystallinity, weight loss, and compressive strength. Cages or cage parts retrieved from PBS were analyzed for Inherent Viscosity to evaluate the amount of degradation of the cage at the molecular level. This was done according to a previously described method[25] of determination of solution viscosity of the polymer solution in chloroform at the concentration of 0.1 gm polymer in 100 ml of chloroform at 25 +/- 0.01 °C.

Crystallinities of the cages were determined using a Perkin Elmer System-7 Differential Scanning Calorimeter with Pyris software, and weight loss of the cages as a function of incubation time was determined gravimetrically by comparing the initial weight and the dry weight of the cage after each incubation period. Before incubation and at the indicated incubation periods, cages were tested for compressive strength between two rigid aluminum plates at a compression rate of 5mm/min.using an Instron Model-4505.

2.2   *In Vivo* Study

Thirty-six mature female Dutch milk goats (2-3 years old, weighing 50 kg) were used in the study. The operative procedures and animal care were performed in compliance with the regulations of the Dutch legislation for animal research, and the protocol was approved by the Animal Ethics Committee of the Vrije Universiteit, Amsterdam, The Netherlands.

Through a left retroperitonial approach, the L3-L-4 intervertebral disc was identified and transversely penetrated by a 2-mm guide wire. An 8-mm drill bit was positioned over the guide wire, and a round channel was drilled through the intervertebral disc and adjacent vertebral endplates, leaving the anterior and posterior longitudinal ligaments intact. The intervertebral disc and approximately 2 mm of endplate and subchondral bone of both adjacent vertebral bodies within the transverse rectangular defect were then removed in a standardized way using a custom-made box gouge measuring 10x10 mm. Three different types of cages were randomly assigned and inserted into the 10x10 mm hole.

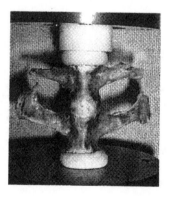

Figure 2. The Zwick 1445 material testing device was used for testing of the motion segment.

2.3   Specimen Preparation and Radiographic Analysis

Before the animals were euthanized, labeling of bone with tetracycline and calcein was performed to analyze bone remodeling and bone formation within the cages. Animals were euthanized at 3, 6, 12, 24 and 36 months after surgery. The lumbar spines were excised, trimmed of residual musculature, and radiographed. Operated motioned segments were dissected and sectioned in a standardized manner using a water-cooled band saw (EXAKT, Norderstedt, Germany), and creating parasagittal sections 3 mm and 5 mm thick. Lateral radiographs of the sectioned specimens were used to estimate interbody fusion within the cages according to a validated 3-point radiographic score[16].

Computerized comparison of defined anatomic landmarks of the motion segment on both the post operative and post-autopsy radiograph scans (accuracy 0.1 mm) was used to estimate subsidence of the operated motion segments.

Specimens for histology were sectioned into sections of 6 μm thickness and stained with Haemotoxin and Eosin. Anteroposterior and lateral contact radiographs of the lumbar spine were performed. Remnants of the PLLA cage were collected from one sagittal part without disrupting the integrity of the cage, when solid, and tested for inherent viscosity. In addition, specimens were undecalcified and embedded in poly-methylmethacrylate for histology. After sectioning, the remaining part of the PLLA cage was retrieved and if still intact, and cage height was measured (accuracy 0.01mm). This retrieved part was used in the chemical analysis of the cage and in determining the amount of the cage absorbed by the body.

3.   RESULTS

3.1   *In Vitro* Test Results

The flexible PLLA cages had a 12 % reduction in inherent viscosity after 4 weeks, 36 % after 12 weeks, and 87 % after 73 weeks of incubation in PBS (Figure 3). The stiff PLLA cages had a 13% reduction of inherent viscosity after 4 weeks, 44% after 12 weeks, and 90% after 73 weeks of incubation. The data indicated that the flexible LIFCs increased their crystallinity at a faster rate than the stiff LIFCs (Fig.4) and that no significant absorption had taken place until the 52 weeks post incubation period was reached.

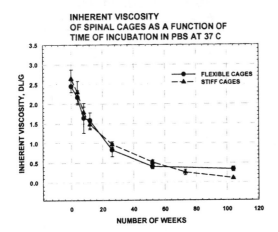

Figure 3. Inherent viscosity of spinal cages as a function of time of incubation in PBS at 37°C

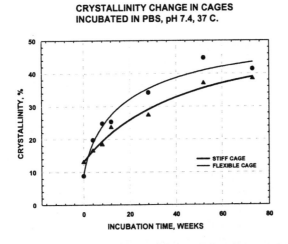

Figure 4. Crystallinity change in cages incubated in PBS pH 7.4 at 37° C.

The compressive strength of both types of cages were determined at 0 time (before incubation in the PBS) as well as at the indicated times after incubation (Figure 5). This data indicates the compressive peak load for the stiff LIFCs is about 7000 N during the first 12 weeks and about 3500 N for the flexible LIFCs.

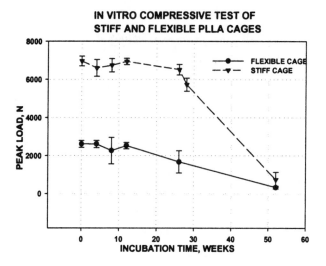

Figure 5. In vitro compressive test results of stiff vs. flexible PLLA cages.

Mechanical strengths for the PLLA cages are as follows:

Strength PLLA cages (including safety margin)		Stiffness PLLA cages	
Stiff PLLA cage	7.0 kN	Stiff PLLA cage	4.0 kN/mm
Flexible PLLA cage	3.5 kN	Flexible PLLA cage	2.0 kN/mm

The biomechanical results of *in vitro* testing suggest the following:

- Yield strength was used as the design parameter of PLLA cage, and average yield strength was 3500 N.
- Strength of the motion segment was largely not affected by implantation of PLLA cage.
- PLLA cages packed with cancellous bone are mechanically sufficient directly after implantation in axial compression. This is also the case with posterior-anterior shear and axial torsion, because the cage is embedded in the vertebral bodies.
- PLLA cages packed with bone do not appear to be mechanically inferior to titanium cages, although the material stiffness is two orders of a magnitude lower.All spinal segments showed identical shape of force-deformation curve.
- Differences in strength could be explained by variability in BMC.

- Average ultimate strength (7.5 kN) was comparable to that for a middle-aged man (6.7 kN), and yield strength was 46.4 ± 7.8 % of ultimate strength.
- Both the carbon fiber cage (stiffness, 8.3 kN/mm) and the Brantigan titanium cage (stiffness, 700 kN/mm) have higher stiffness than the PLLA cage.

3.2   *In Vivo* **Test Results**

After 12-months, an adequate amount of PLLA could be retrieved from only one specimen in the flexible PLLA group. In the stiff PLLA group, the amount of retrieved PLLA was too sparse for separate inherent viscosity determination, and therefore samples were pooled. After 24 months, no PLLA could be retrieved.

The flexible PLLA cages showed 69 % reduction in inherent viscosity at 12 weeks, 81 % at 24 weeks, and 90% after 52 weeks of implantation (Figure 6). The stiff PLLA cages showed a 64% reduction in inherent viscosity at 12 weeks, 72% at 24 weeks, and 93% after 52 weeks of implantation.

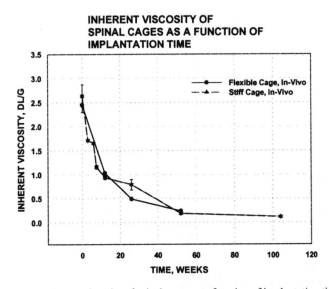

Figure 6. Inherent viscosity of spinal cages as a function of implantation time.

3.3   Radiological Evaluation

No evidence of collapse of the operated spinal segment in any PLLA group was observed after the follow-up periods. After 3 months, ingrowth of trabecular bone was observed. At six months, radiological fusion of the spinal segments was observed in 80 % (4/5) of the PLLA specimens. None of the titanium specimens showed fusion at the same period. Sequential radiographic follow-up (6 to 36 months showed successful interbody fusion in 86% (19/22) of PLLA specimens.   Specimens with Titanium cages showed successful interbody fusion in only 33% (2/6) in the same period.

Flexible

Stiff

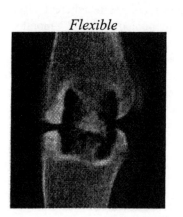
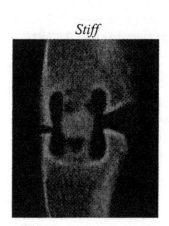

Figure 7. 3 months post-op.

Flexible

Stiff

Figure 8. At 24 months post op.

3.4   Macroscopic Findings

After 3 months, the wall thickness of the stiff and flexible PLLA cage had increased to a mean of 1.64 ± 0.09 mm and 1.01 ± 0.09 mm, respectively, and the PLLA cages were intact and had maintained their geometrical shape (height, 10 mm).

After 6 months, the wall thickness of the stiff and flexible PLLA cages measured a mean of 1.83 ± 0.23 mm, and 0.99 ± 0.0 mm, respectively, and the stiff and flexible PLLA cages were also intact and solid (height: 10 mm). The flexible PLLA cages showed formation of radial micro-cracks without displacement. The PLLA cages disclosed a whitish aspect with microcracks and the flexible cage showed a slight bending deformation.

At 12 months, the stiff and flexible PLLA cages were brittle and had disintegrated entirely into fragments with a grayish aspect; at 24 months, no PLLA could be observed macroscopically.

3.5   Microscopic Findings

Very sparse birefringent PLLA fragments could be observed in the extracellular space of the quiescent fibrous tissue enveloping the implants during the follow-up periods, but after 12 and 24 months, the PLLA cages lost their geometrical shape and disintegrated into multiple birefringent fragments of different sizes with interposition of quiescent fibrous tissue. No adverse tissue reactions could be observed.

3.6   Histology

At 3 months postoperatively (Figure 9):

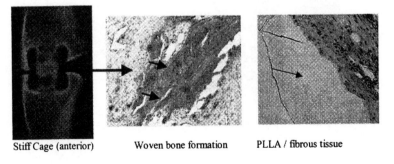

Stiff Cage (anterior)   Woven bone formation   PLLA / fibrous tissue

Figure 9.

At 6 months postoperatively (Figure 10):

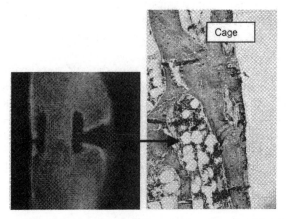

Figure 10. Quiescent fibrous tissue with bone lying sparsely against the PLLA is shown. There is no evidence of lymphocytes, plasma cells, or granulocytes.

At 12 months postoperatively (Figure 11):

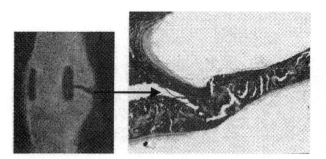

Figure 11. Increased bone remodeling is seen, and there is no evidence of any foreign body reaction.

At 24 months postoperatively (Figure 12):

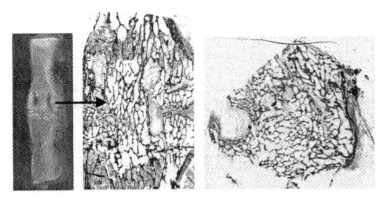

Figure 12. Most of the PLLA cage has been absorbed and partially replaced by fibrous tissue and trabecular bone.

4. DISCUSSION

The flexible and stiff PLLA cages showed a similar decline in inherent viscosity during follow-up. However, in vivo degradation was more pronounced compared to the in vitro degradation during 12 months follow-up. This may be due to the high load bearing conditions of the PLLA cages, which would diffuse body fluids into and out of the cage more rapidly and increase the rate of degradation. It might also be due to the presence of enzymes that may play a role in degradation in the in vivo case. After 3 and 6 months follow-up, the PLLA cages showed signs of degradation, but maintained their original geometrical shape and height.

After 12 and 24 months follow-up, no collapse of the spinal segment was observed, despite the fact that the PLLA cages had disintegrated into multiple fragments and no longer carried any load. After 24 months, the PLLA cage was largely absorbed. The fibrous tissue surrounding the PLLA implants showed no signs of adverse tissue reactions. After 36 months, the PLLA cage was completely absorbed and replaced by trabecular bone and quiescent fibrous tissue.

In the PLLA specimens, permanent interbody fusion with maintenance of intervertebral distance was achieved within 6 months after surgery. Titanium cages did not show any evidence of interbody fusion within this period. Radiographic evidence revealed that interbody fusion occurred in 86% (19/22) of the PLLA specimens within 6 to 36 months compared to 33% (2/6) in titanium specimens during the same time period.

5.   CONCLUSIONS

- Absorption of the PLLA cage takes place at a faster rate *in vivo*, under load, than the same cage *in vitro*, without loading.
- The stiff cage retains 94% of its initial strength and the flexible cage 64 % of its initial strength at 26 weeks *in vitro*.
- Gross evaluations of the *in vivo* implant sites indicate total absorption of the PLLA at two-years postoperatively.
- The crystallinity of the PLLA cage increases after incubation/implantation.
- The PLLA cage provides temporary stability until interbody fusion is obtained.
- Limited stiffness of PLLA cages significantly enhances the rate of interbody fusion (86%) when compared to custom-made titanium cages (33%).
- Interbody fusion was maintained during three-years follow-up.
- The biocompatibility of PLLA is excellent, with only a mild foreign body reaction.
- The PLLA cage was partially replaced by trabecular bone at 36 months postoperatively.

REFERENCES

1. Tunc DC. et al, Evaluation of Body Absorbable Bone Fixation Devices, Transactions of the 31st Annual Meeting of the Orthopaedic Society, 1985
2. McAfee PC. Interbody fusion cages in reconstructive operations on the spine. *J Bone Joint Surg [Am]* 1999; 81: 859-80.
3. Weiner BK, Fraser RD. Spine update lumbar interbody cages. *Spine* 1998;23:634-40.
4. Brantigan JW, Steffee AD. A carbon fiber implant to aid interbody lumbar fusion: Two-year clinical results in the first 26 patients. *Spine* 1993; 18: 2106-7.
6. Ray CD. Threaded titanium cages for lumbar interbody fusions. *Spine* 1997; 22: 667-79.
7. Brantigan JW, Steffee AD, Lewis ML, et al. Lumbar interbody fusion using the Brantigan I/F cage for posterior lumbar interbody fusion and the variable pedicle screw placement system: Two-year results from a Food and Drug Administration investigational device exemption clinical trial. *Spine* 2000; 25: 1437-46.
8. Kuslich SD, Danielson G, Dowdle JD, et al. Four-year follow-up results of lumbar spine arthrodesis using the Bagby and Kuslich lumbar fusion cage. *Spine* 2000; 25: 2656-62.
9. McAfee PC, Cunningham BW, Lee GA, et al. Revision strategies for salvaging or improving failed cylindrical cages. *Spine* 1999; 24: 2147-53.
10. Nillson LT, Geijer M, Neuman P, et al. The Brantigan anterior lumbar I/F cage: Two years radiological results. *Eur Spine* J 2001; 10S: 26.
11. Togawa D, Bauer TW. The Histology of Human Retrieved Intervertebral Body Fusion Cages: Good Bone Graft Incorporation and Few Particles. Presented at the Annual Meeting of the Orthopaedic Research Society, San Fransisco, California, February 25-28, 2001.
12. Tullberg T. Failure of a carbon fiber implant: A case report. *Spine* 1998; 23: 1804-6.

13. Klemme WR, Owens BD, Dhawan A, et al. Lumbar sagittal contour after posterior interbody fusion: Threaded devices alone versus vertical cages plus posterior instrumentation. *Spine* 2000; 26: 534-7.
14. Kim Y. Prediction of mechanical behaviors at interfaces between bone and two interbody cages of lumbar spine segments. *Spine* 2001; 26: 1437-42.
15. Pitzen T, Geisler FH, Matthis D, et al. Motion of threaded cages in posterior lumbar interbody fusion. *Eur Spine* J 2000; 9: 571-6.
16. van Dijk M, Smit TH, Sugihara S, et al. The effect of cage stiffness on lumbar interbody fusion: An in vivo model using poly-L-lactic acid and titanium cages. *Spine* 2002; 27: 682-8.
17. Cunningham BW, Haggerty CJ, McAfee PC. A Quantitative Densitometric Study Investigating the Stress-Shielding Effects of Interbody Spinal Fusion Devices: Emphasis on Long-Term Fusions in Thoroughbred Racehorses. Presented at the Annual Meeting of the Orthopaedic Research Society, New Orleans, Louisiana, March 9, 1998.
18. Kuslich SD, Ulstrom CL, Griffith SL, et al. The Bagby and Kuslich method of lumbar interbody fusion: History, techniques, and 2-year follow-up results of a United States prospective, multicenter trial. *Spine* 1998; 23: 1267-78.
21. Regan JJ, Aronoff RJ, Ohnmeiss DD, et al. Laparoscopic approach to L4 L5 for interbody fusion using BAK cages: Experience in the first 58 cases. *Spine* 1999; 24: 2171-4.
22. Regan JJ, Yuan H, McAfee PC. Laparoscopic fusion of the lumbar spine: Minimally invasive spine surgery: A prospective multicenter study evaluating open and laparoscopic lumbar fusion. *Spine* 1999; 24: 402-11.
24. Jee WS. Integrated bone tissue physiology: Anatomy and physiology. In Cowin SC, ed. Bone Mechanics Handbook. Boca Raton, FL: CRC Press LLC, 2001.
25. Tunc DC., etal Two-Year Follow-up of Bioabsorbable PLLA Cages for Lumbar Interbody Fusion: In Vitro and In Vivo Degradation, Transactions of the 48th Annual Meeting of the Orthopaedic Research Society, 2002.

Metals Foams for Biomedical Applications: Processing and Mechanical Properties

MUSTAFA GUDEN[*,†], EMRAH CELIK[‡], SINAN CETINER[#] and
ALPTEKIN AYDIN[#]
*Department of Mechanical Engineering,†Center for Materials Research, ‡Materials Science
and Engineering Program, Izmir Institute of Technology, Gulbahce Koyu, Urla, Izmir,
TURKEY; #Hipokrat A.Ş., 407/6 Sok., No:10, Pınarbaşı, Izmir, TURKEY

1.   INTRODUCTION

Optimized structures found in nature can be sometimes imitated in engineering structures. The recent interest in functionally graded metallic materials makes bone structures interesting because bones are naturally functionally graded[1]. The cellular structure of foam metals (Fig.1) is very similar to that of the cancellous bone; therefore, these metals can be considered as potential candidates for future implant applications if porosity level, size and shape, strength and biocompatibility aspects satisfy the design specifications of implants. Foam metals based on biocompatible metallic materials (e.g. Ti and Ti-6Al-4V) are expected to provide better interaction with bone. This is mainly due to higher degree of bone growth into porous surfaces and higher degree of body fluid transport through three-dimensional interconnected array of pores[2] (open cell foam), leading to better interlocking between implant and bone and hence reducing or avoiding the well-known *implant losening*. Furthermore, the elastic modulus of foam metals can be easily tailored with porosity level to match that of natural bone, leading to a better performance by avoiding the high degree of elastic mismatch which currently exists between conventional solid metallic implants and bone.

Biomaterials: From Molecules to Engineered Tissues, edited by
N. Hasırcı and V. Hasırcı, Kluwer Academic/Plenum Publishers, 2004   257

Foaming of metals is a complicated process in which a large number of processing and geometrical parameters have to be adjusted adequately. Currently, no complete, theoretically based understanding of all details of the foaming process has been developed. However, a set of empirical rules have been worked out that allow the production of foam metal components of a considerable quality and complexity. This has been achieved particularly for aluminium alloys[3] and manufacturing technology still needs considerable research effort for the processing of implant grade materials.

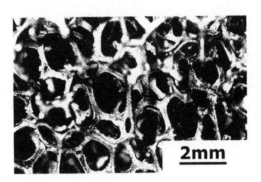

Figure 1. The cellular structure of open cell Ni foam.

2.   PROCESSING ROUTE OF IMPLANT FOAM METALS

There are two basic approaches currently available for the manufacture of foam metals: melting and powder metallurgy (PM)[4]. Designed structures are commercially manufactured via continuous or batch type casting methods, e.g. cell forming mould removal method used by DUOCEL for the production of open cell Al and Al alloy foams[4]. Self forming structures are manufactured either by gas injection through (CYMAT/HYDRO) or gas forming element addition (ALPOROS) into liquid metal[4]. Although, melting methods have been successfully applied to the manufacture of Al, Zn and Mg foams, they are not suitable for the manufacture of Ti foams due to the high melting temperature and reactivity of Ti. In the PM approach, designed structures are manufactured either by sintering of hollow spheres or by melting or partial melting of powder compacts that contain a gas evolving element (e.g.TiH_2)[3]. Since these methods unavoidably result in enclosed pores (closed cell foam), they are also not suitable for the manufacture of foamed metal implants because of the requirement of body fluid transport. Open cell implant foam metals can be however successfully manufactured by a versatile PM based process known as space holder method[5-7]. The method can be used to manufacture fully and/or partially (as coatings on solid implants for bone fixation) foamed biomedical metals. The size, level and geometry of pores can be easily altered by varying the size, amount and shape of space holder. Therefore, it is one of the appropriate methods for manufacturing designed foam metal implants.

The processing steps of space holder method are schematically presented in Fig. 2. The process starts with mixing of metal powders with a suitable space holder material, followed by a compaction step (e.g. uniaxial and isostatic pressing) that produces metal powder-space holder mixture compact. The green compact is then heat treated at a relatively low temperature to release the space holder, resulting in an unfired open cell foam metal structure. Finally, the compact is sintered at relatively high temperatures to provide structural integrity. This method allows a direct near net-shape fabrication of foamed implant components with a relatively homogeneous pore structure and a high level of porosity (60-80%).

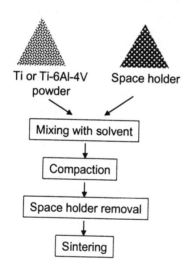

Figure 2. Processing steps of space holder method.

Ti and its alloys are known to be very reactive and can easily form interstitial solid solutions with other elements including carbon, oxygen and nitrogen. Since the presence of these elements is detrimental for the ductility, the reaction between Ti powder and the cracking products of the space holder in a temperature range of 300-600 °C must be avoided[5, 8]. It is therefore proposed that space holder should be removed at temperatures below about 200 °C[5]. Ammonium hydrogen carbonate and carbamide (urea) are the materials identified to satisfy this criterion and currently used for the processing of Ti foams[5-7].

The optimum pore size range required for the attachment and proliferation of new bone tissue and the transport of body fluids is given to be between 200 and 500 μm[9]; therefore, the particle size range of space holder must be selected and/or tailored according to the critical pore size range. In the design of foam metal final pore size range, however, pore shrinkage occurring during sintering should also be taken into consideration. For the preparation of highly porous foam parts, the particle size distribution of metal powder should be lower than the average particle size of space holder[5]. A particle size lower than 150 μm is normally sufficient for the homogeneous coating of 200-500 μm size space holder particles with Ti powder. Furthermore, the consolidation pressure of metal powder-space holder mixture must be high enough for the preparation of mechanically strong compacts that would retain their geometry throughout the foaming process. The compaction of Ti powder is usually conducted under a uniaxial pressure ranging between 100 and 200 MPa, while higher pressures, or a binder material, may be required for the compaction of the harder Ti-6Al-4V powder.

Using the space holder method, Ti and Ti-6Al-4V foam metals with 60 and 70% porosities were prepared and microscopically and mechanically characterized in our laboratory. Ti foams were prepared using angular Ti powder (<45 μm) and ammonium hydrogen carbonate (angular, 200-500 μm) as space holder. Compaction was performed by applying a uniaxial pressure of 200 MPa inside a cylindrical steel die (25 mm in diameter). The compacts were heat treated at 200 °C for 5 h to remove the space holder and then sintered at 1250 °C for 2 h. Figs. 3a-c show the microstructures of the Ti foams at various magnifications. Two different pore size ranges, macro- (200-500 μm) and micro- (1-10 μm), are clearly seen in Figs. 3b and 3c, an observation which was also made previously in Ti foams prepared by the same method[7]. Micropores are located at the cell walls (Fig. 3b), between the sintered Ti powders, and are proposed to be a result of volume shrinkage of the powder during sintering[7]. Micropores and rough cell wall surfaces were reported to be preferable in osteoinductivity[10].

Typical microstructures of prepared Ti-6Al-4V foams are shown in Figs. 4a and 4b. Ti-6Al-4V foams were prepared using the same method except the angular Ti-6Al-4V powder used (<150 μm) was compacted at a higher pressure, 400 MPa. Similar to Ti foams, micropores are also seen at the cell walls (Fig.4a). Due to the compaction pressure, the cells of the foams are observed to be preferentially aligned in the direction normal to the pressure direction, leading to an anisotropy in foam mechanical properties. The cell alignment, however, is more pronounced in Ti-6Al-4V foams because of the higher compaction pressure (Fig. 4b).

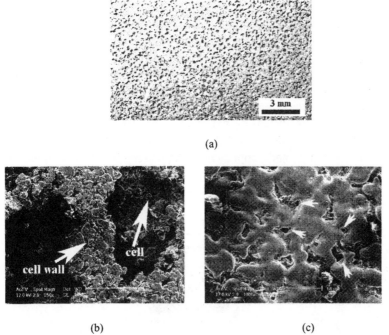

(a)

(b)   (c)

Figure 3. Scanning electron micrographs of Ti foam (70% porous) showing a) cell structure, b) cell wall and cell, and c) micropores at cell wall.

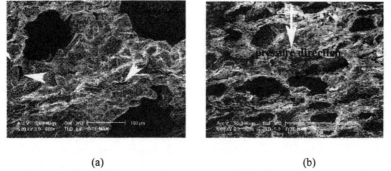

(a)   (b)

Figure 4. Scanning electron micrographs of Ti-6Al-4V foam (70% porous) of showing (a) cell wall micropores, and (b) cell alignment.

3. MECHANICAL PROPERTIES

Under compressive loads, open and closed cell metal foams show a similar, characteristic stress-strain curve composed of three distinct deformation regions[11]: linear elastic, plateau or collapse and densification as depicted in Fig. 5 for an Al closed cell foam. In the linear elastic region deformation is controlled by cell wall bending. This region is followed by a plateau or collapse region of cell wall bending and/or crushing. The onset of localization of deformation is called collapse stress (Fig. 5). Deformation is highly localized in the plateau region by the formation of a deformation band which proceeds to the undeformed regions of the sample as the strain increases. The plateau region is characterized by a plateau stress either with a constant value or increasing with increasing strain as the relative density increases. After a critical strain (ε_d) the material densifies, hence the stress increases sharply and approaches the strength of the bulk metal (densification region).

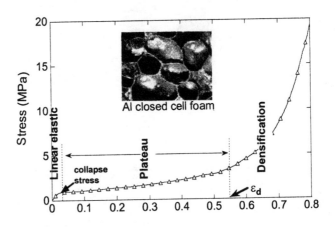

Figure 5. Compression stress-strain curve of an Al closed cell foam of 90% porosity (average cell size 3 mm)[12].

Ti and Ti-6Al-4V foams also show deformation behaviour similar to that of conventional Al foams. The compression stress-strain curves of Ti and Ti-6Al-4V foams with 60 and 70% porosities are shown in Fig. 6. In the figure, P and N refer to the compression test axis: testing parallel (P) and normal (N) to the applied pressure direction. The higher plateau stress values of 70% porous Ti foam tested in N-direction are mainly due to the cell alignment as explained in the previous section. Preliminary results have also shown that Ti-6Al-4V foam shows higher stress values than Ti foam at the same porosity level (Fig. 6). The elastic modulus of the tested foams was further found to be in the range 4-10 GPa, comparable with the elastic modulus of natural bones, 3-30 GPa[13].

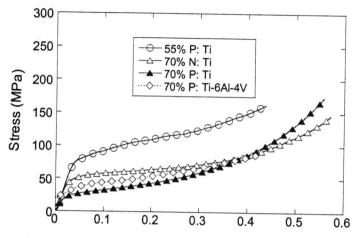

Figure 6. Compression stress-strain curves 60 and 70% porous Ti and Ti-6Al-4V foams.

The elastic modulus (E) and plateau stress (σ_{pl}) of open cell foams are usually predicted using the equations derived by Gibson and Ashby[11] and given as

$$\frac{E}{E_s} = \alpha(\frac{\rho}{\rho_s})^2 \qquad (1)$$

and

$$\frac{\sigma_{pl}}{\sigma_{ys}} = \beta(\frac{\rho}{\rho_s})^{3/2} \qquad (2)$$

where E_s, σ_{ys}, ρ_s and ρ are the elastic modulus, yield stress and density of cell wall material and density of foam, respectively. The values of constants, α and β, given in Eqns. 1 and 2 were experimentally determined to be 1 and 0.3, respectively[11]. The modulus and collapse stress of Ti foams are predicted using above equations and the following appropriate material parameters: E=105 GPa, σ_{ys}= ~700 MPa and ρ=4.5 g cm^{-3}. Fig. 7a shows the predicted modulus values of Ti foam as function of percent porosity. In Fig. 7a, the porosity range of Ti foam showing a good match with the elastic modulus of natural bones is found to be between 50 and 80% porosities. In the same range, Ti foam is predicted and experimentally shown to be stronger than the cancellous bone[6] (Fig. 7b). This is beneficial for the handling and durability of the foam implants. It should be finally noted that Eqns. 1 and 2 are applicable for open cell foams with porosities higher than 70%, therefore the predictions shown in Figs. 7a and b should be used with caution for lower values of porosities.

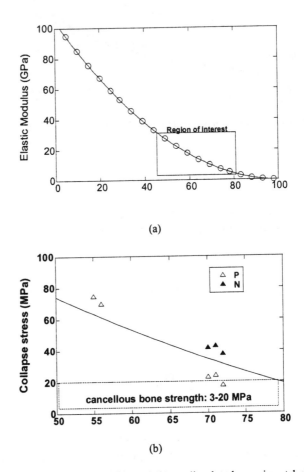

(a)

(b)

Figure 7. (a) Predicted elastic modulus and (b) predicted and experimental collapse stress values of Ti foam as function of percent porosity.

4.   CONCLUSION

Foamed metals have many potential uses in biomedical applications including fully foamed implant components and coatings on solid implants and can be manufactured by the space holder method with relatively high porosity levels. In designing implant components with biocompatible metal foams several material aspects should, however, be considered including size, shape and level of porosity and mechanical properties. The geometry of cell in the final sintered foam metal results in direction dependent mechanical properties. The mechanical aspect includes two important properties of foamed metals; elastic modulus and plateau and/or collapse stress. The former is critical for the implant loosening and the latter for handling during implantation and durability in long term service. Since the foam properties are varied with the porosity level rather than with the cell wall material, the use of scaling relations allows the manufacture of designed structures for critical applications.

ACKNOWLEDGEMENTS

The authors would like to thank the Technology Development Foundation of Turkey (TTGV) for the grant #TTGV-102/T13.

REFERENCES

1. Weiner, S., and Wagner, H. D.,1998, The material bone: structure-mechanical function relation. *Annu. Rev. Mater. Sci.* 28:271-298.
2. Pillar, R. M., 1987, Porous-surfaced metallic implants for orthopaedic applications. *J. Biomed. Mater. Res.* 21:1-3.
3. Banhart, J., 2001, Manufacture, characterisation and application of cellular metals and metal foams. *Prog. Mater. Sci.* 46:559-632.
4. Körner, C., and Singer, R. F., 2002, Processing of metal foams-challenges and opportunities. *Adv. Eng. Mater.* 2:159-165.
5. Martin, B., Stiller, C., Buchkremer, H. P., Stöver, D., and Baur, H., 2000, High purity titanium, stainless steel and superalloy parts . *Adv. Eng. Mater.* 2:196-199.
6. Wen, C. E., Mabuchi, M., Yamada, Y., Shimojima, K., Chino, Y., and Asahina, T., 2001, Processing of biocompatible porous Ti and Mg. *Script. Mat.* 45:1147-1153.
7. Wen, C. E., Yamada, Y., Shimojima, K., Chino, Y., Asahina, T., and Mabuchi, M., 2001, Processing and mechanical properties of autogenous titanium implant materials. *J. Mater. Sci.* 13:397-401.
8. Froes, F. H., 2002 Lightweight heavyweight, *Metal Powder Report* (www.metal-powder.net)
9. Clemow, A. J. T, Weinstein, A. M., Klawitter, J. J., Koeneman, J., and Anderson, J., 1981, Interface mechanics of porous titanium implants. *J. Biomea. Mate,. Res.* 15: 73-82.

10. Chang, Y., Oka, M., Kobayashi, M., Gu, H., Li, Z., Nakamura, T., and Ikada, Y., 1996, Significance of interstitial bone ingrowth under load-bearing conditions: a comparison between solid and porous implant materials, *Biomaterials* 17: 1141-1148.
11. Gibson L. J., and Ashby F., 1997, *Cellular solids: structure and properties*. Cambridge University Press.
12. Elbir, S., Yılmaz, S., Toksoy, K., Guden, M., and Hall, I. W., 2003, SiC-particulate aluminum composite foams produced by powder compacts: Foaming and compression behavior, *J. Mater. Sci.* 38:4745-4755.
13. Weiner, S., and Wagner, H. D.,1998, The material bone:structure-mechanical function relations. *Annu. Rev. Mater. Sci.* 28:271-298.

Oral Tissue Engineering of Complex Tooth Structures on Biodegradable DLPLG/β-TCP Scaffolds

ANKA LETIC-GAVRILOVIC[*], LJUBOMIR TODOROVIC[#], and KIMIO ABE[†]

[*]International Clinic for Neo-Organs, ICNO, Rome, ITALY; [#] Department of Oral Surgery, Faculty of Stomatology, University of Belgrade, YUGOSLAVIA; [†]Department of Functional Bioscience, Fukuoka Dental College, Fukuoka, JAPAN

1.   INTRODUCTION

Research for new technologies and biomaterials improving orofacial implantation and regeneration has evolved at a fast-pace[1,2,4,13-16]. Teeth are essential for survival in many vertebrates, and missing or misplaced teeth can have fatal consequences, causing some species to be unable to make use of available food supplies. Tooth loss due to periodontal disease, dental caries, trauma, or a variety of genetic disorders is one of the most severe human health problems. It is critical, therefore, that the dentition develops correctly, with the required number and type of teeth developing in specific positions in the jaws. A biological tooth substitute that could replace lost teeth would provide a vital alternative to currently available clinical treatments[5]. The purpose of this study leads to new composite constructs to be shapable, resorbable, and biocompatible multifunctional bone equivalent for applications in periodontology, oral surgery and trauma of the teeth and periodontal tissues, including bone. The goal of our compiled projects is tooth organ engineering. In order to organize such a complex project, work has been divided in three phases. Starting phase deals with experimental tools for *in vitro* and *in vivo* tooth engineering. Second phase would be selection, collection and preservation of high number of healthy teeth

Biomaterials: *From Molecules to Engineered Tissues,* edited by
N. Hasırcı and V. Hasırcı, Kluwer Academic/Plenum Publishers, 2004

extracted for orthodontic reasons, and the final phase will be to treat organizational aspects of a bank for tooth tissue. Tooth germ is an accessible source of high-quality human postnatal stem cells for research and growth factor delivery systems (GFDS). This phase requires a much better understanding of stem cells, cell-proliferation and differentiation *in vitro*. We used the approach of tooth tissue replacement utilizing culturing cells seeded onto extracellular matrices. The bioactive scaffolds (surfaces), for example, are natural or synthetic bone equivalent and/or dental implants, with bound proteins and cells on surface. The development of scaffolds and implants made of biomaterials that are not only permissive for cell growth but also equipped with a plethora of active functions is being observed: attracting or repelling cells, providing compartmentalization, having 3-D memory, allowing monitoring and modulation, regulating growth and differentiation, to name a few. Model of tooth morphogenesis is a wonderful model for studying how genes and molecules interact to create complex patterns. It is a model system for comparative studies on normal and pathological growth and differentiation of epithelial (ameloblastoma), epithelial-ectomesenchymal (ameloblastic fibroma) and ectomesenchymal (odontogenic myxoma) tissues[5,7]. Of particular interest to maxillo-facial applications are multifunctional biodegradable scaffolds. Poly(lacticacid)-based composites due to their biocompatibility and bioresorbability are used for bone plates or temporary internal fixation of damaged bone[9,22,20,24]. In the future, the availability of shapeable, biodegradable, biocompatible and bioactive constructs might be of great interest in order to avoid the inconvenient surgical insertion of large implants. Recently, the development of bio-hybrid systems (composites, copolymers, complexes, hydrogels, blends, etc.), based on natural and synthetic polymers, and their wide range of applications in biomaterial science has received tremendous attention[25]. Results from these studies will benefit scientific and clinical community by: a) organization of tooth bank, b) enabling continuous source of tooth germ pluripotent stem cells, and c) stimulating development of very specific bioactive scaffolds for tempo-spatial regulation of whole tooth differentiation which are crucial for 3-D organ engineering in general.

2.   MATERIAL AND METHODS

2.1   Composites Scaffolds

β-TCP was prepared by modified protocol of Jain[6]. Spherical granules were dried-calcined at 1,100°C/12 h. The heating and cooling rates were each 10°C/min. Evaluation of chemical composition and purity was carried out by Fourier Transform-Infrared Spectroscopy (FTIR), recorded using the KBr pellet method by Perkin Elmer 782 spectrometer. Porosity was assessed from the N_2 desorption isotherms (BET method). Three particle sizes of β-TCP (<100 μm, >100 μm and >300 μm) were used to prepare three main composite formulations, with the solvent evaporation technique. After D,L-poly(glycolic acid) DLPLG, (50:50, MW 40, 000-75, 000; Sigma-Aldrich) was dissoluted in acetone (0.5 copolymer/5 mL acetone), the solution was stirred at 100 rpm/5 h. Ceramic particles of three given sizes: a) 25-100 μm (or <100μm) b) 150-300 μm (or >100 μm) and c) >300 μm, were added to the polymer solution, until reaching the ratio (by weight) of 80/20 acetone/copolymer, was reached leading to a broad scale of the final DLPLG/βTCP ratio from 90/10-30/70. The material was mixed at 50 rpm/h. The mixture was cast in the desired shaped using chilled Teflon moulds without any pressure. For this study, discs of φ5 mm were prepared. After solvent evaporation, the samples were air dried (48 h), freeze-dried (24 h), lyophilized and sterilized with ETO. The control biomaterials (only copolymer) were processed using the same method. By adopting the same process and changing a few parameters (e.g. temperature, pressure and time), the biomaterial could be produced with various compressive strengths and porosity. Porosity of bulk composites was assessed by BET test. Structure, mechanical properties and *in vitro* degradation were measured using various techniques.

2.2   Mechanical Testing of Composites

Rectangular sheets of 2 mm thickness were made in Teflon moulds by press moulding at room temperature and dumbbell-shaped specimens for mechanical testing were cut from the sheet with a cutting die (ISO R37). The mechanical properties were determined in dry state. The elastic modulus, tensile strength and elongation at break were determined in a Houndsfield testing machine at a testing speed of 50 mm/min at room temperature. Ten specimens were used for each testing.

2.2.1   Degree of Swelling of the Composites

For swelling tests six samples/group/five experiments were cut from hot press sheets with a size of $1 \times 1 \times 0.2$ cm^3. The swelling test was carried out in distilled water at room temperature (23±1.5 °C). Swelling was assessed every 20 h for eight days. The degree of swelling of the composites was:

$$S_w = (W_t - W_o) / W_o$$

where S_w is the swelling degree in a given time, W_t the weight of the tested specimens after immersion in water for a time t, and W_o the weight of the tested specimens at the beginning of the test. Each data point represents the mean±SD. Copolymer without ceramic was control.

2.2.2   Hydrolytic Degradation

For studying degradation for two months, 14 specimens per three groups were used in addition to two controls with polymer only, one at pH 3.7 and one at pH 7.3. They were placed into 70 small flasks filled with phosphate buffer. The flasks were allowed to stand in a thermostatic oven for predetermined periods of time. For degradation at 37° C, the pH value 3.7 was selected. Two specimens were withdrawn from the aging media at each degradation time (every ten days), and washed with distilled water. Each data point represents the mean ± SD of two measurements/five experiments. Copolymer without ceramic was control.

2.3   Animals

Male rats of an inbred AO strain were obtained from the Institute for Experimental medicine, Military Medical Academy (Belgrade, Yugoslavia). All animals were used at 3 or 4 months of age, and had ad libitum access to rat chow and water. Mandibles from 20 rats were soaked in butadiene for 15 min to prevent contamination and then rinsed in phosphate-buffered saline (PBS) before being transferred to a sterile environment.

2.3.1   Purified Low Molecular Weight Protein (10kDa) Extracts

About 20 rats were killed by cervical dislocation and the molar tooth buds free of connective tissues were collected in five volumes of cold saline solution. Tissue samples allocated for digestion were treated with collagenase (1 mg/mL) for 12 h in an incubator at 37°C and 5% CO_2 .The suspension was filtrated and washed in PBS. Five mL was plated into 25 cm^2 polystyrene tissue culture flasks and 2 mL was plated onto 35x10 mm^2 Petri dishes for staining purposes. After they were minced in a mixer and centrifuged at 7,000 g for 30 min, supernatants were collected, and then re-centrifuged at 11,000 g for 30 min at 4°C. After lipids were discarded, supernatants were collected and filtered through a 0.45 μm filter. They were used as crude extracts of the tooth buds. Ultrafiltrates of 100 kDa, 30 kDa, and 10 kD (Advantec, Toyo, Ultrafilter) were collected and then sterilized by passage through a 0.22 um filter and stored at -20°C. In all fractions, the concentration of protein was determined by the Lowry[17] method. Only the 10 kDa fraction of ultrafiltrates was lyophilized (Speed Vac Concentrator) and was re-constituted with saline solution at a four-fold higher concentration. Gel-filtration. The 10 kDa fraction was filtered through Sephadex G-25. Experimental conditions were as follows: the column was 35 cm x 2.6 cm, gel 30.9 cm, pressure 34.5 mL/h, and flow rates 35 mL/h. Elution was done with a 0.1 M phosphate buffer at pH 8. After application of the 10 kDa fraction, each 1.16 ml fraction was collected and three tubes were combined. After sterilization by passage through a 0.22 µm filter, fractions were kept at -20°C. RP-HPLC chromatography. All separations were carried out at 30°C and at a constant flow rate of 1 mL/min. Operating parameters were controlled by use of a LKB Ultrachrom GTI System. A LKB-BROMA liquid chromatograph equipped with 2156 solvent conditioner, 2152 LC controller with 2155 HPLC Column Oven, 2151 Variable Wavelength Monitor and Fraction Collector FRAC-100 (Pharmacia Fine Chemicals), were used. The column was Sepherisorb ODS-2 (5 µm), and was developed with a 35 min linear gradient from 0 to 100 per cent B at a flow rate of 1 mL/min, where the primary solvent (A) was water and the secondary solvent (B) was 70% acetonitrile /30% water. The system was fitted with an injection loop for a 1 ml sample. The column elute was monitored by absorbance at 280 nm. Pressure was 134 bar and chart speed 4 mm/min. Each fraction of 1.5 mL per tube was collected automatically.

2.3.2   Cell Cultures

Neonatal AO rat calvaria steoblasts (RCO) were isolated via sequential collagenase digestions. Cells were grown to confluence in standard media (Dulbecco's Modified Eagle's Medium, [DMEM], Life Technologies, Gaitersburg, MD, USA) containing 15% heat inactivated fetal calf serum (FCS; HyClone, Logan; UT, USA) and penicillin and streptomycin (P/S; Life Technologies) on our 3-D scaffolds. All cells were cultured in a humidified 5% CO_2, 37°C incubator, and seeded in flasks. Thirty samples of three composites with various granule sizes were placed in 24-well culture plates, and the cell suspension (1×10^4 cell/mL/100 µL) was applied onto each sample. The control was cells in empty wells. Cells were allowed to attach for 2 h. At the end of the experiment, RCO were characterized for phenotypes: Alkalinephosphatase activity (ALP, Sigma, USA), Lactatedehydrogenase (LDH, Sigma, USA), calcium (Ca, Sigma), phosphorus (P, Sigma), and all tests as indicators of osteoblast activity. The MTT test (Sigma) was performed to assess cell proliferation.

2.3.3   Histology and Scanning Electron Microscopy

Mandibular tooth buds and representative samples of bulk-composites and cells attached were processed for microscopy. Morphological and element analysis (KEVEX) were performed by Etec Autoscan (Etec-Haywood, CA, USA).

2.4   Statistics

All measurements were done in five separate experiments and expressed as mean ±SD of mean. A two-tailed unpaired t-test for the statistical significance was used. The values of cell culture between the groups were compared using analysis of variance (ANOVA) and Scheffé *post hoc* multiple comparison test.

3.   RESULTS

3.1   Physicochemical Characteristics of β-TCP and DLPLG/β-TCP Composites

The characteristics of β-TCP particles were as follows: 25-100 μm , 150-300 μm and >300 μm were 100% pure with 70% interconnected porosity. The Specific Surface Area (SSA) of the particles obtained by BET, yielded 1.07 m^2/g. Electron-microscopy image of bulk-material showed a uniform distribution of the βTCP particles in composites Fig.1.A. On higher magnification porosity and interconnectivity of 20 % are visible Fig.1.B.

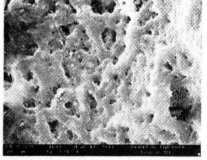

Figure 1. Bioresorbable composites (DLPLG/βTCP) with granule size <100 μm used in experiments. Micrographs of composites with porosity (<20%) and interconnectivity (bar 1μm); A) bar 3 μm, B) bar 1 μm

3.1.1 Mechanical Properties of the Composites

The mechanical properties of composites were affected by the three different sized granules of β-TCP filler. Quantity and size of granules ranging from <100 μm to >300μm, improved effectively elastic modulus of composites formulation (Table 1A). Comparatively, tensile strength and percent of elongation were reduced with increasing the granule size. The tensile tests showed that, although the elastic modulus of the composites was increased by the incorporation of bigger granules, the tensile strength and elongation at break were not changed significantly. Porosity and inter-connectivity of composites processed without any pressure yielded 20%.

3.1.2 Degree of Swelling

The stability of the composite structure in water was studied for eight days. The results are shown in Fig. 2B. All three formulations tested were saturated with solvent for 1 h. In time the swelling degree increased up to the value 0.6%. The composite retained a compact structure and disintegration was not observed, even after experiment.

3.1.3 Hydrolytic Degradation In Vitro

The first signs of degradation were recorded after 10 days Fig. 2A. Hydrolytic degradation generally occurred in two steps. In the first step, the structure became saturated and showed a very slowly increasing concentration of free lactic acid, attributable to the breakage of its chemical bonds, which was particularly marked in the lactic acid part of copolymer. In the second step, the degree of degradation was slightly lower, with no significant difference between the groups and the control.

3.2 Purification of the Molar Tooth Bud Extracts

The crude extracts of the molar tooth germs (Fig. 2) from the AO rats were ultrafiltered and 100 kDa, 30 kDa, and 10 kDa ultrafiltrates were obtained. The protein concentrations of these fractions were markedly decreased as the molecular weight of ultrafiltrates became smaller. The 10 kDa fraction was gelfiltered by Sephadex G-25 chromatography and yielded five fractions. Fraction with highest protein content was further analyzed by RP-HPLC and used to incorporated into composite scaffolds. Fraction with the highest protein content was incorporated into the three composite formulations for cell culturing.

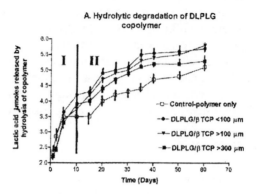

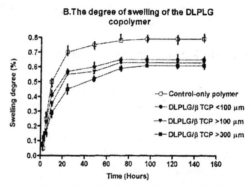

Figure. 2. A) Biodegradation of composites by hydrolysis. The first phase (I) occurred within 10 days and second phase (II) from 10th to 70th day. Control was polymer without any ceramic. Degradation characteristics were measured for two months, 14 specimens per three groups, plus two controls with polymer only. For degradation at 37° C, the pH value 3.7 was selected. Two specimens were withdrawn from the aging media at each degradation time (every ten days), and washed with distilled water. Each data point represents the mean±SD of two measurements per five experiments. B) The degree of swelling of the composites. Note that the degree of swelling nearly reached equilibrium after 24 h immersion in water. For swelling tests six samples/group were cut from hot press sheets with a size of 1×1×0.2 cm³. The swelling test was carried out in distilled water at room temperature (23±1.5 °C). Swelling were assessed every 20 h during eight days. DLPLG copolymer without ceramic was control. Each data point represents the mean±SD of six samples/group/measurements per five experiments.

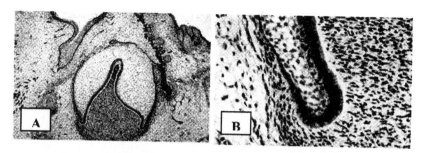

Figure 3. Histological section of molar tooth bud from AO male rat.

3.3 Cell Culture and Biochemical Testing

Results showed that, during the incubation of osteoblast-like cells on the three formulations of composites, no signs of cytotoxicity appeared, (Table 1.B). In fact, the LDH measured in sample cultures of the materials matched those of the control cultures. The ALP values showed that the presence of bone substitute biomaterials did not negatively interfere with osteoblast activity. Ca and P determined in the supernatant were significantly lower than that in the controls. The MTT test revealed that cell proliferation significantly improved in the presence of biomaterials with respect to the control.

3.4 Scanning Electron Microscopy

After culturing on the composites with smaller granule size ($<100\mu m$) RCO osteoblast cells displayed extremely long and thin processes (Fig. 4.A and B). Numerous cells proved to have close relations with the substrate. Additionally, these cells gave rise to many filamentous processes towards an extracellular matrix substrate. The long lamellar-granular structures in the extracellular spaces were closely related to the substrate. The small granular structures in the extracellular spaces were demonstrated to have close relations with the cells. Morphologically, the cells cultured on second size substrate ($>100 \mu m$) were elongated with a large number of processes (Figs. 4.C and D). These cells made numerous contacts with one another and were firmly and strongly attached to the substrates. On the cell surface, a high number of short/thick or long/thin extensions were detected. On these cells, attached particles were observed and appeared to have high affinity towards the third formulation with the largest granules ($>300\mu m$) showed a wide variety of forms and appearance (Figs. 4.E and F).

Table 1. Various mechanical (A) and biochemical (B) characteristics of composites DLPLG/β-TCP with three different granule sizes. A. Mechanical properties in dry state; B. Biochemical values of osteoblast-like cells cultured on composites.

A. Mechanical properties of the Composites* with three different granule size in the dry state (Mean±SD)

COMPOSITES	Elastic modulus (MPa) (n=10)	Tensile strength (MPa) (n=10)	Elongation (%) (n=10)
Polymer only (n=10)	30.5±0.1	7.0±0.2	370±100
Composites (DLPLG/βTCP) <100 μm	49.3±1.3*	6.8±0.5	250±73
Composites >100 μm	56.0±4.7*	6.0±0.2	220±53*
Composites >300 μm	89.6±8.4*	5.1±0.3*	180±24*

B. Biochemical values of osteoblast-like cells after 48 h of culturing with composites DLPLG/β-TCP with three different granule sizes (Mean ± SD)

Osteoblast culture	LDH (U/L) (n=6)	ALP (U/L) (n=6)	Ca (mg/dL) (n=6)	P (mg/dL) (n=6)	MTT OD550nm (n=6)
Control without composites	3.3±0.07	18.5±0.7	7.3±0.26	3.17±0.2	0.3±0.05
Composites (DLPLG/βTCP) <100μm	3,4±0.14	19.3±1.1	3.0±0.4*	1.12±0.1*	0.5±0.03*
Composites >100 μm	3.3±0.01	18.3±0.6	5.1±0.5	1.77±0.1*	0.5±0.03*
Composites >300 μm	3.3±0.07	23.5±1.7*	2.9±0.3*	1.35±0.3*	0.5±0.04*
ANOVA F	0.44, *ns*	1.25, *ns*	113.51, p<0.0005	85.89, p<0.0005	21.27, p<0.0005

Scheffé *post hoc* multiple comparison test: DMEM-control vs other composite materials, $p < 0.001$;

* DLPLG/βTCP; n=experimental samples per group

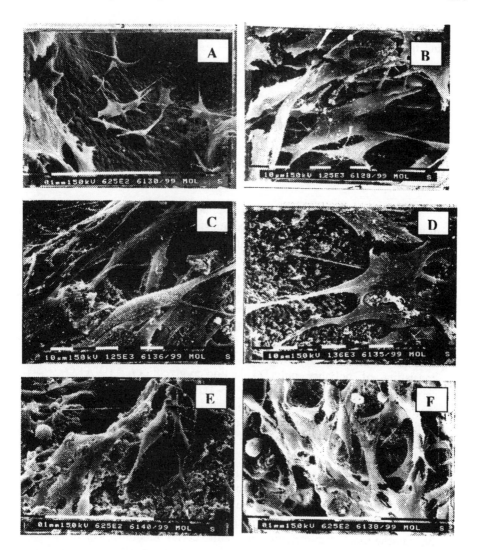

Figure. 4. Electron-micrograph of osteoblast-like cells cultured on DLPLG/βTCP composites
with three granule sizes. A and B) After cultivation on the composites with granules <100 μm
osteoblast-like cells show extremely long and thin cell processes. The numerous cells make
close relations with substrate by cell processes and by cell bodies. Bar: A- 0.1 mm; B -10 μm;
C and D) Cell morphology of osteoblast-like cells cultivated on composites with granules >
100 μm shows great number of short, thick (C) and some very long and very thin (D)
cytoplasmic continuations on cell surface. Bar: C, D - 10 μm. E and F) The osteoblast-like
cells cultivated on composites with granules > 300μm show a great variety of forms and
cellular appearance. On this substrate osteoblasts are abundant, highly proliferated, extremely
connected and firmly attached to the substrate. Significantly increased number of cells (F), on
the small surface area, indicates a high degree of confluence. Bar: E, F - 0.1 mm

RCO osteoblasts were abundant, highly proliferative, closely connected and firmly attached to the substrate. A sharp increase in the number of cells in a small area was indicative of high degree of confluence. These cells were centrally widened and had long cell processes on both poles, although there were cells with numerous processes. A high number of long and short processes were firmly attached to the composite substrate. In extracellular spaces, rough-globular structures of the substrate were attached to or were in close relation with the cells.

4.   DISCUSSION

Tooth loss due to periodontal disease, dental caries, trauma, or a variety of genetic disorders continues to affect most adults adversely at some time in their lives. A biological tooth substitute that could replace lost teeth would provide a vital alternative to currently available clinical treatments. To pursue this goal, we dissociated rat molar tooth buds into single-cell suspensions and seeded them onto biodegradable polymers. The target of the study was to design and develop novel multi-functional composites as a bone alternative for use in three-dimensional bone regeneration thanks to its biocompatibility, mouldability and biodegradability. The research activity included technological · processes for creating resorbable composite consisting of polymer poly(D,L-lactide-co-glycolide) (DLPLG) and ceramic beta-tricalcium phosphate (β-TCP), and then physicochemical, mechanical and biological analyses *in vitro* to evaluate the biocompatibility of the composite as substrates for new bone deposition. We designed and developed a multifunctional bone equivalent by reinforcing poly(D,L-lactide-co-glycolide) (DLPLG) substrate with microparticles of osteoconductive ceramic: β-tricalcium phosphate (β-TCP) with various particle sizes. The performance of the composites were evaluated by physicochemical tests and in vitro culturing with osteoblast-like cells (RCO). The results showed that the composites had excellent mechanical and physicochemical characteristics, including elastic modulus, tensile strength, swelling, biodegradation and had positive effects on RCO. This indicates that the composites are suitable materials as a scaffold for new bone ingrowth, enabling better functional and aesthetic results in patients treated for cranial-facial tumors, malformations and traumas. We decided to design composites of DLPLG/βTCP, because both components are bioresorbable in the shortest time possible and in this case a reconstruction of hard tissue is synchronised with degradation of composite components and the biomaterial starts to be porous during the healing process. Regarding porosity and/or density of composites, reconstruction of hard tissue is synchronised with degradation of the biomaterial and starts to augment its porosity during this process. Therefore, our material being biodegradable does not need to be porous at the beginning of experiment, as it will become porous soon after insertion into dynamic *in vivo* conditions. The question on porosity is crucial only for non-biodegradable biomaterials. DLPLG is the biodegradable polymer with a high starting strength, which lasts for some time, actually a rather short time of up to a few months, and produces nontoxic products during degradation[3, 18]. We tested degradation rate in low pH, which was chosen for *in vitro* "static" conditions where the buffer solution is not replaced over the whole testing period. This results in a dramatic drop in the pH of the solution due to the release of the acids from the device. Although this low pH is not very probable in dynamic *in vivo* conditions in which the products of degradation are transported away from the implantation site, this situation would be possible when large bulky implants degrade at a very high rate, as it is the case with our composites.

While the specific requirements and characteristics of a biodegradable biomaterial depend on a specific application, there is a general set of criteria for "a good biodegradable polymeric biomaterial" and we believe that our composite formulations fulfill these criteria. Further more, DLPLG/β-TCP is excellent due to its good tensile strength and improved elastic modulus at the beginning of experiments. However, diverse factors like the molecular weight, copolymer composition (lactide to glycolide ratio), crystallinity, and β-TCP granule size affect the mechanical strength of the composition[8]. The modulus of elasticity for copolymer is relatively low compared to that of natural cortical bone[12]. To meet that requirement introduction of a certain amount of osteoconductive ceramic such as β-TCP, and change the size and quantity of used particles and the modulus would be improved effectively as was shown in our experiments and others[20]. With our composites formulations, the mechanical properties and degradation time of polymers showed to be adjustable by adding ceramics[11,12,26]. The experimental results of this study demonstrate that all three formulations of β-TCP/DLPLG composites were biocompatible *in vitro*. These formulations have an outstanding propensity to be colonized with osteoblasts, thereby promoting their osteogenic activity. These results are comparable with the others[22, 19]. The results of our *in vitro* study provide strong evidence of good osteocompatibility of the proposed composite formulations. ALP, Ca, P and MTT as indicators of bone cell differentiation seem to be increased when measured in RCO cultured on the composites with the biggest granule size >300 μm. This is in agreement with literature reports on the impact of matrix geometry on osteogenesis[10] and also with our results regarding other biochemical parameters and microscopic observations. Cellular attachment to the implant surface is an important step in the process of tissue-implant interaction and thus long-term successful implantation[21].

Tissue engineered (TE) product formulation proposed in this study will have significant clinical advantages for bone and tooth-like tissue regeneration. The proposed TE formulations (Patent Pending, Number) have a very important potential as scaffolds for cell transfer, thus offering also the possibility of rapid tissue regeneration[5,14-16,23,27]. Also, depending on future developments in TE, it may be multi-functional, i.e. used both as a valuable biomaterial and as an in vitro and in vivo drug delivery device of growth factors. Thus, our future goal is to promote combined clinical, pharmaceutical and TE research efforts to obtain bone equivalent products for successful cranio-facial reconstructions based on tooth-germ epithelial-mesenchymal interactions.

ACKNOWLEDGEMENTS

This investigation was supported by The International Clinic for Neo-Organs Grants's scheme, Rome, Italy.

REFERENCES

1. Alsberg, E., Hill, E.E., and Mooney, D.L., 2001, Craniofacial Tissue Engineering. *Crit Rev Oral Biol Med* 12:64-75.
2. Davin, J.E., Attawis, M.A., and Laurencin, C.T., 1996, Three-dimensional degradable porous polymer-ceramic matrices for use in bone repair. *J Biomater Sci Polym Ed* 7: 661-669.
3. Dunn, A.S., Campbell, P.G., and Marra, K.G., 2001, The influence of polymer blend composition on the degradation of polymer/hydroxyapatite biomaterials. *J Mater Sci: Mater Med* 12: 673-677.
4. Hench, L.L., and Polak, J.M., 2002, Third-Generation Biomedical Materials. *Science* 295: 1014-1017.
5. Young, C.S., Terada, S., Vacanti, J.P., Honda, M., Barlett, J.D., and Yelick, P.C., 2002, Tissue Engineering of Complex Tooth Structures on Biodegradable Polymer Scaffolds. *J Dent Res* 81(10): 695-700.
6. Jain, R.A., 2000, The manufacturing techniques of various drug loaded biodegradable poly(lactide-co-glycolide) (PLGA) devices. *Biomaterials* 21:2475-2490.
7. Glasstone, S., 1967, Development of Teeth in Tissue Culture. *J Dent Res* v. 46; Supp. 5.
8. Kasuga, T., Ota, Y., Nogami, M., and Abe, Y., 2001, Preparation and mechanical properties of polylactic acid composites containing hydroxyapatite fibers. *Biomaterials* 22: 19-23.
9. Kikuchi, M., Suetsugu, Y., Tanaka, J., Akao, M., 1997, Preparation and mechanical properties of calcium phosphate/copoly-L-lactide composites. *J Mater Sci* 8:361-364.
10. Kuboki, Y., Saito, T., Murata, M., Takita, H., Mizuno, M., Inoue, M., Nagai, N., Pool, and R., 1995, Two distinctive BMP-carriers induce zonal chondrogenesis and membranous ossification, respectively; geometrical factors of matrices for cell-differentiation. *Connect Tissue Res* 32: 219-226.
11. Kurashina, K., Kurita, H., Kotani, A., Kobayashi, S., Kyoshima, K., and Hirano, M., 1998, Experimental cranioplasty and skeletal augmentation using an alpha-tricalcium

phosphate/dicalcium phosphate dibasic/tetracalcium phosphate monoxide cement: apreliminary short-term experiment in rabbits. *Biomaterials* 19:701–706.

12. Kurashina, K., Kurita, H., Hirano, M., Kotani, A., Klein, C.P., and de Groot, K., 1997, *In vivo* study of calcium phosphate cements: implantation of an alpha-tricalcium phosphate/dicalcium phosphate dibasic/tetracalcium phosphate monoxide cement paste. *Biomaterials* 18:539–543.

13. Letic-Gavrilovic, A., Scandurra, R., and Abe, K., 2000, Genetic potential of interfacial guided osteogenesis in implant devices. *Dent Mater J* 19(2): 99-132.

14. Letic-Gavrilovic, A., Crudo, V., and Abe, K., 2002, Microspheres-based tissue engineering product for craniofacial bone reconstruction, (abstract). *Tissue Engineeering* 8(6): 1234.

15. Letic-Gavrilovic, A., Crudo, V., and Abe, K., 2002a, Tooth-germ morphogenesis signals for 3D tissue engineering scaffolds in oral reconstructions (abstract). *Tissue Engineeering* 8 (6): 1165.

16. Letic-Gavrilovic, A., Piattelli, A., and Abe, K., 2003, Nerve growth factor β (NGFβ) delivery via a collagen/hydroxyapatite (Col/HAp) composite and its effects on new bone ingrowth. *J Mat Sci: Mater Med* 14: 95-102.

17. Lowry, Q.H., Rosebrough, N.J., Farr, A.L., and Bandall, R.J., 1951, Protein measurements with the Folin Phenol reagent. *J Biol Chem* 193:265-275.

18. Merolli, A., Gabbi, C., Cacchioli, A., Ragionieri, L., Caruso, L., Giannotta, L., Tranquilli Leali, P., 2001, Bone response to polymers based on poly-lactic acid and having different degradation times. *J Mater Sci:Mater Med* 12:775-778.

19. Neo, M., Herbst, H., Voigt, C.F., and Gross, U.M., 1998, Temporal and spatial patterns of osteoblast activation following implantation of beta-TCP particles into bone. *J Biomed Mater Res* 39:71–6.

20. Rejda, B.V., Peelen, J.G., and de Groot, K., 1977, Tri-calcium phosphate as a bone substitute. *J Bioeng* 1: 93–97.

21. Warren, S.M., Fong, K.D., Chen, C.M., Loboa, E.G., Cowan, C.M., Lorenz, H.P., and Longaker, M.T., 2003, Tools and Techniques for Craniofacial Tissue Engineering. *Tissue Engineering* 9(2): 187-200

22. Shikinami, Y., and Okuno, M., 1999, Bioresorbable devices made of forged composites of hydroxyapatite (HA) particles and poly-L-lactide (PLLA): Part I. Basic characteristics. *Biomaterials* 20:859-877.

23. Shikinami, Y., and Okuno, M., 2001, Bioresorbable devices made of forged composites of hydroxyapatite (HA) particles and poly-L-lactide (PLLA): Part II. Practical properties of miniscrews and miniplates. *Biomaterials* 22: 3197–3211.

24. Tada, H., Hatoko, M., Tanaka, A., Kuwahara, M., Mashiba, K., Yurugi, S., Iioka, H., and Niitsuma, K., 2002, Preshaped hydroxyapatite tricalcium-phosphate implant using three-dimensional computed tomography in the reconstruction of bone deformities of craniomaxillofacial region. *J Cranio Surg* 13(2): 287-292.

25. Wang, M., Chen, J., Ni, J., Weng, J., and Yue, C.Y., 2001, Manufacture and evaluation of bioactive and biodegradable materials and scaffolds for tissue engineering. *J Mater Sci: Mater Med* 12: 855-860.

26. Wiltfang, J., Merten, H.A., Schleger, K.A., Schultze-Mosgau, S., Kloss, F.R., Rupprecht, S., and Kessler, P., 2002, Degradation characteristics of alpha and beta tri-calcium-phosphate (TCP) in minipigs. *J Biomed Mater Res* 63: 115-121.

27. Zerbo, I.R., Bronckers, A.L.J.J., de Lange, G.L., van Beek, G.J., Burger, E.H., 2001, Histology of human alveolar bone regeneration with a porous tricalcium phosphate. A report of two cases. *Clin Oral Impl Res* 12: 379-384.

Microparticulate Release Systems Based on Natural Origin Materials

GABRIELA A. SILVA[*,#], FILIPA J. COSTA[*], NUNO M. NEVES[*,#] and RUI L. REIS[*,#]

[*]3B's Research Group – Biomaterials, Biodegradables, Biomimetics, University of Minho, Campus de Gualtar, 4710-057 Braga, Portugal; [#]Department of Polymer Engineering, University of Minho, Campus de Azurém, 4800-058 Guimarães, PORTUGAL

1. PARTICULATE SYSTEMS: FUNDAMENTALS

It has been a long path for particulate systems in biomedical applications. Since long time ago that these micron and nanosize systems synthesised from the most varied materials find application in the biomaterials field, mainly as drug delivery carrier systems. The aim of drug delivery systems is to facilitate the dosage and duration of the drug effect, causing the minimal harm to the patient and improving human health[1,2]. Typically, they allow for the reduction of the dosage frequency[3] and are non-toxic[4]. Back in time, these systems designed for the controlled delivery of drugs were found extremely promising for several applications, such as the delivery of insulin[5-8], contraceptives[9-12], cancer therapeutics[13-16] among others[17].

Polymer microparticles have attracted attention as carrier matrices in a wide variety of medical and biological applications, such as affinity chromatography[18-20], immobilization technologies[20-23], drug delivery systems[4, 7, 24-30] and cell culturing[31].

Various parameters including particle size and size distribution, porosity and pore structure and surface area are considered to describe the overall performance of polymeric microspheres in these applications[32].

Biomaterials: From Molecules to Engineered Tissues, edited by
N. Hasırcı and V. Hasırcı, Kluwer Academic/Plenum Publishers, 2004

At the present, the applications are diverse, ranging from drug delivery devices – where these particulate systems find their primary field of action – to other biomedical applications, namely those related with detection and analysis applications[20,33-35]. These systems could also be used in the Tissue Engineering field, has it has been reported by several authors[36-42].

1.1   Natural Origin Materials

The development of particulate systems is in constant evolution, and every year more and more new materials are used to synthesise them. PLA, PGA and their co-polymers have been widely used, but it is well known the problems associated with these materials[43-47]. Many other synthetic polymers have been studied. These polymers have found several applications, but are clearly not adequate for others.

This led the scientific community to turn to natural origin materials, as an alternative group of materials. The main characteristics of useful natural origin materials are their availability/renewability, low cost, biodegradable character, easy of processing, among others. These materials can be processed alone or in combination with other polymers or with other groups of materials (ceramics, metals).

Polymers from crab shells, such as chitin and its deacetylated derivative, chitosan, proteins from plant origin, such as soybean protein, starch from corn, potato, cassava, casein, albumin, collagen, gelatin among others are currently being studied within the biomaterials field.

Proteins are a diverse group of biopolymers widely studied for biomedical applications. Albumin, collagen and gelatin are three of the most studied proteins to form microcapsules/spheres. Their main disadvantages are their susceptibility of denaturation in adverse environments, the risk of allergy, and potential disease transmission.

Chitin and its derivative chitosan are now being widely used for several drug delivery applications. Chitin is, after cellulose, the most abundant polymer found in nature[48]. Chitin can be complexed with iron to form microspheres that can be used as possible drug delivery systems targeted for liver, due to their bioavailability[49].

Chitosan [β(1-4)2-amino-2-deoxy-D-glucose] is a cationic polysaccharide derived by hydrolysing the aminoacetyl groups of chitin, hydrophilic, biocompatible and biodegradable polymer[11,48,50]. Chitosan has been used for several applications, namely as delivery systems for anticancer drugs, such as doxorubicin[11], 5-fluorouracil[50], nucleic acids carriers, such as DNA-chitosan nanoparticle complexes[51,52]. This material can also be modified to accommodate other bioactive agents such as in chitosan succinate and hydroxamated chitosan succinate (both crosslinked with iron) that were found successful in the encapsulation and prolonged release of theophylline[53].

Works on soy and casein proteins has revealed the potential of these materials for biomedical applications. Soy thermoplastics have been developed so as to constitute dual release systems when processed by a double-layer co-injection moulding method[54], as well as their ability to incorporate and release theophilline in a two stage process[55,56]. When processed as transdermal delivery systems, such as membranes, these were found cytocompatible when tested in vitro[57-59].

Also starch has been widely used, usually modified, either by blending with other polymers[60-64] or by chemical modification of its structure. Several works performed with starch, namely in the microspheres form have shown to be nasally inhaled to deliver proteins and drugs[60,65-68], as well as its processing as porous scaffolds, that could be combined with other materials to yield materials with adequate mechanical, chemical and biocompatible character.

Other materials have been used, but their description would become this chapter too extensive.

1.2 Blends and Composite Materials

The development of polymeric matrix composite materials aims to combine the most desired properties of two or more materials[36], either between the same group of materials (polymers) or between different classes of materials (polymer-ceramic). Composite materials aim to combine the best properties of each material. Depending on the application, one should choose the materials that may bring together the desired properties, and to tailor them for the foreseen application.

In the polymer field, there are frequently examples of the combination of two different polymers to yield a final material that has enhanced properties compared to the ones displayed by each of the individual materials. Examples are found in the literature, such as the combination of chitosan with xanthan[69,70], two polysaccharides that when complexed in the form of microspheres presented a slower degradation profile when compared with chitosan microspheres alone. Hyaluronic acid was also blended with chitosan to create microspheres with mucoadhesive properties[50], that can be very useful in intranasally administered drugs. Also natural and synthetic polymers can be combined, such as chitin and PLGA, that were blended to form biodegradable microspheres for protein delivery[25].

Chitosan can also be combined with ceramic materials, such as coralline hydroxyapatite[71], that were found to be effective in the entrapment and zero-order release kinetics of gentamicin, thus presenting a great potential for bone and dental applications.

Hydroxyapatite is a ceramic material widely used for creating composite polymer-ceramic materials. For instance, embedding hydroxyapatite particles into reconstituted fibrous collagen was found capable of promoting osteoblast attachment and growth, thus rendering it suitable for clinical applications[72].

Other combination of polymers with ceramic materials, namely starch-based materials and Bioactive Glass 45S5, has shown that microparticles with spherical shape were formed, entrapping Bioactive Glass 45S5[36].

2.   APPLICATIONS OF PARTICULATE SYSTEMS IN TISSUE ENGINEERING

Particulate systems find their application in the biomedical field mainly as drug delivery vehicles. In the field of Tissue Engineering, particulate systems are not used *per se*, but usually produced and then moulded into porous scaffolds[73]. This can be a strategy to combine scaffolding and drug delivery, but work still has to be done in order to optimise properties such as pore dimension or mechanical properties namely for bone related applications. The advantages of these systems for Tissue Engineering applications comes from their ability to be targeted to specific sites of action, and their biodegradability, that will allow them to disappear when they are no longer required. These properties can be advantageous when applications like drug or cell delivery are foreseen, where drugs or cells will be released in particular time points. For instance, a dual release system that may entrap cells and its differentiation agents, cells that could only be released as the material degrades, and during that time the differentiation agents could act in the cells, could be created.

The fast development of Nanotechnology has pushed forward the interest in the small-scale materials. One can easily predict that the Nano era, that is just now beginning, will allow for enormous developments in the applications of these micron and submicron size systems.

2.1   Bone and Cartilage

Concerning hard tissues such as bone, not many advances have been made in the use of particulate systems in tissue engineering. Ceramic materials are sometimes used in the form of particulate materials for filling of small defects, especially in maxillofacial and dental applications[74-76]. As for polymers, they are so far restricted to be used as porous scaffolds, either by polymer processing or "double-processing", where particles are produced and then sintered to form 3D scaffolds[40].

Thus, a strategy that would combine 3D-porous scaffolds with micron size particle systems acting as drug delivery/cell encapsulation systems creating the perfect scaffold, could constitute a leap forward in bone and cartilage tissue engineering.

Bearing this in mind, starch-based materials present themselves as a good alternative, since they can be processed into porous scaffolds with interesting properties[62,63,77-81], and as it will be described, into particulate systems that can be used as delivery systems or as composite materials (with ceramics) for reinforcement.

3.   STARCH-BASED MATERIALS

Reis and co-workers[61,82-91] have previously demonstrated the potential starch-based materials for several biomedical applications. These materials have been proving their potential for biomedical applications such as bone cements[86], drug delivery systems[85,92], bone tissue engineering scaffolds[83, 84], among other possible applications.

Their adequate profile can be evidenced by their easiness of processing, their ability to be combined with other polymers or ceramic materials (such as hidroxylapatite or bioactive glasses), their renewable and natural origin nature, their enzymatic degradation[93], water uptake ability[94,95], their physical and chemical properties that can be tailored and modified according to the foreseen application[62,96-98] as well as its unusual (for a biodegradable material) good biological performance[91,99].

Different blends of starch with other polymers, such as polylactic acid, polycaprolactone, ethylene vinyl alcohol and cellulose acetate have been studied in order to obtain materials that can have a combination of suitable properties, can be easily processed and are biocompatible[88,91,100,101].

3.1   SPLA and Composite Particulate Systems

In this work, a blend of corn starch and polylactic acid (PLA) was used to synthesise particles that can be applied as drug delivery carriers, both alone or in combination with scaffolds. In the latest case the aim is to achieve a release profile closer to the desired, since release from three-dimensional porous scaffolds occurs too fast in the majority of the cases. Thus, by combining micron size delivery systems and scaffolds, it would be possible to tailor the release profile of growth factors or other bioactive agents that should act in the implantation site.

A simple method were the blend of starch and polylactic acid was dissolved in an organic solvent and then emulsified with a stirring aqueous solution was used to prepare the particles. This method yielded particles of a size of less than 10<μm up to 350 μm (Figure 1), with a spherical morphology, whose surface alternates between smooth areas (attributed to PLA) and porous areas (starch phase).

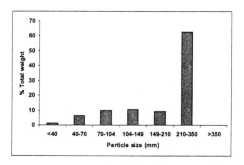

Figure 1. Size distribution of the developed SPLA particles as determined by sieving. The size distribution shows that the particle size between 210-350 μm predominates over all other sizes.

Particles composed of SPLA and 30% (w/w) of Bioactive Glass 45S5 (BG 45S5) were also synthesised, by mixing BG 45S5 with SPLA and then forming the particles as described for SPLA. No differences on both size distribution and morphology (Figures 1 and 2) could be found as compared to polymeric SPLA particles.

Figure 2. Scanning electron microscopy image representative for SPLA and SPLA-BG particles. No differences were found between polymer and composite particles. Regarding the morphology, smooth areas and porous areas are observed in the particles.

However, when evaluating their water uptake ability (Figure 3) and degradation profile (through weight loss and reducing sugars quantification) it was found that the composite particles degrade slower than the polymeric ones (Figure 4).

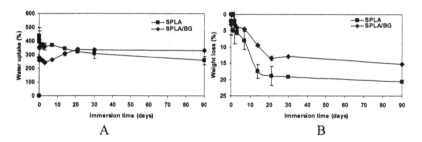

Figure 3. Water uptake (A) and weight loss (B) profiles for SPLA and SPLA-BG particles. Composite particles have a similar water uptake profile as SPLA particles, differing in the weight loss, which is lower than the one observed for the polymeric particles.

It was also found that the composite materials are bioactive upon immersion in a solution simulating the human blood ionic composition, as seen by the formation of a calcium-phosphate layer at the surface of the particles[36].

SPLA particles have already been studied for their ability to encapsulate and release bioactive agents, being proved their potential to be used as drug delivery carriers[28], namely of growth factors or other bioactive agents.

3.2   Soluble Potato Starch and Composite Particulate Systems

Paselli II starch is a potato starch chemically modified by ATO (Wagningen, The Netherlands) as to become soluble in water. This material was used with an emulsion crosslinking method based on the creation of a water-in-oil emulsion. Starch is crosslinked with trisodium trimetaphosphate (TSTP). Briefly, this method can be described as the dissolution of starch and TSTP in water, which is then emulsified with paraffin and a surfactant for emulsion formation. When the reaction is completed, the particles are extensively washed with water and ethanol and then freeze-dried. Due to particle aggregation, no size separation was performed for this material, but sizes up to 1 mm were observed, in particular for composite PaII/BG 45S5 particles, synthesised as described for polymer particles, by mixing starch and BG 45S5 (30% w/w) in the initial step of the method.

In Figures 4 and 5, representative images of PaII (Figures 4A, 4B) and of PaII/BG 45S5 (Figures 5A and 5B) particles are shown.

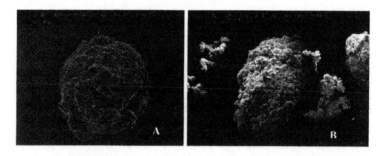

Figure 4. Scanning electron microscopy images representative of PaII particles (A, B). Regarding the size distribution, particle size ranges from few microns (A) to hundreds of microns (B). Morphological evaluation shows that the particles seem to be composed of smaller particles, all brought together. This feature creates pores in the particles, but they do possess a dense matrix with only very small pore sizes.

The size range of these particles goes from a few microns (Fig. 4A) to hundreds of microns (Fig. 4B). The polymer particles present a spherical surface, which seems to be composed from several smaller particles, thus rendering this material porous, but with a dense matrix were only some very small pores observed (data not shown).

When analysing the composite materials (Figs. 5A and 5B), the size range is similar to the polymeric particles. However, when comparing the morphology, it is clearly different from the polymeric ones.

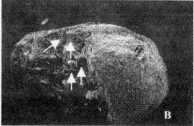

Figure 5. Scanning electron microscopy images representative of PaII-BG composite particles (A, B). Size range is the same as compared with polymeric particles, however differences in the morphology do occur. Composite particles are more regularly spherical, with a smoother surface. In Pa-BG particles the cross-section of a composite particle cut to expose its interior shows the BG 45S5 granules dispersed in the starch network.

Composite Pa-II-BG particles present a smoother surface when compared to PaII particles. The presence of BG45S5 in the particles is shown dispersed in the particles matrix, as seen in Figure 5B (BG 45S5 granules pointed by arrows).

The bioactivity of both PaII polymeric and PaII-BG composite particles is being evaluated[102]. Regarding their water uptake (Figure 6A) and weight loss (Figure 6B) profiles, as for water uptake the values are higher for composite particles. This feature can be explained by high water uptake at the interaction between polymer and ceramic phases[64]. PaII polymer particles have a higher loss of weight when immersed in a phosphate buffered saline solution, proved by the higher amount of quantified reducing sugars, that measure degradation of small weight starch chains.

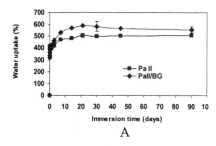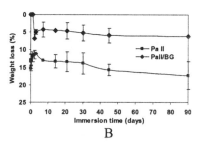

Figure 6. Water uptake (A) and weight loss (B) profiles for PaII and PaII-BG particles. Composite particles have a slight higher water uptake profile as PaII particles, differing in the weight loss, which is lower than the one of polymeric particles.

3.3   Other Starch-Based Particle Systems: SCA, SEVA-C

Other starch-based materials were used for producing micron size particles. For example, a blend of starch with cellulose acetate (SCA) was used for the synthesis of particles. The method chosen was as described for Paselli II (soluble starch) particles. The morphology was evaluated prior to any other characterization and it shows (Figure 7A) a spherical morphology of the particles, with an average size of 160 μm, and with a surface alternating between smoother areas and rougher areas (B), with a dense matrix.

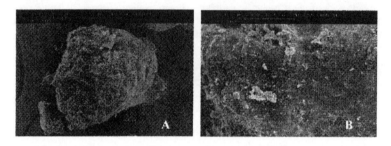

Figure 7. Scanning electron microscopy images of a representative SCA particle (A) and detail of SCA particles' surface (B).

When evaluating the ability of SEVA-C, a blend of starch and ethylene vinyl alcohol to be processed as particulate systems, using the method described above for Paselli II, this material has shown to form microparticles with spherical shape (Figure 8A), with sizes up to 700 μm. These particles present a rough surface, with evident pores (Figure 8B).

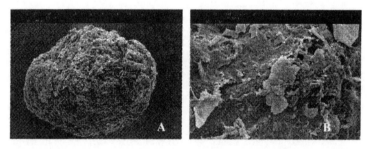

Figure 8. Scanning electron microscopy images of SEVA-C particles, showing its spherical morphology and average size (A) and detail of the surface of a SEVA-C particle (B).

4. CONCLUSION

Natural origin materials are in the biomedical field to stay. Their potential for biomedical applications has been coming to scene, and scientists are realising their potential.

From crab shells to crop reserves, from plant proteins to coralline materials, from animal to plant by-products, there is a vast panorama from where to choose, and many more will be continuously brought front as the search for the perfect material continues. Combining expertise from different fields such as chemistry, materials science, biology and other sciences can lead the scientific community to develop adequate materials.

For some biomedical applications, natural origin materials cannot alone suffice the needs. However, their combination with other materials, either of natural or synthetic origin can create new materials that can in fact combine the best properties from each material, pushing development forward.

In this context, starch-based polymers are one of the good alternatives, either alone or in combination with other materials – for instance, ceramic materials such as hydroxyapatite or bioactive glasses. The ability of starch-based materials to be tailored, modified, as well as their biological performance and variety of ways of processing them, has brought up their potential to be used as scaffolds, drug carrier systems, bone cements, among others.

It is our firm belief that the future lies in natural origin materials, and that by learning with Nature we can actually create materials that can perform what they were designed to: improve, regenerate, heal, ultimately, give hope for a better quality of life of patients worldwide.

ACKNOWLEDGEMENTS

G. A. Silva is recipient of a grant from Fundação Ciência e Tecnologia (FCT, Portugal) (reference SFRH/BD/4648/2001). This work was partially supported by FCT Foundation for Science and Technology, through funds from the POCTI and/or FEDER programmes.

REFERENCES

1. Langer, R., 1991, Drug Delivery Systems. *Mrs Bulletin* 16:47-49
2. Pillai, O., Dhanikula, A. B. and Panchagnula, R., 2001, Drug delivery: an odyssey of 100 years. *Current Opinion in Chemical Biology* 5:439-446
3. Pillai, O. and Panchagnula, R., 2001, Polymers in drug delivery. *Current Opinion in Chemical Biology* 5:447-451

4. Kumar, M., 2000, Nano and microparticles as controlled drug delivery devices. *Journal of Pharmacy and Pharmaceutical Sciences* 3:234-258

5. Belmin, J. and Valensi, P., 2003, Novel drug delivery systems for insulin - Clinical potential for use in the elderly. *Drugs & Aging* 20:303-312

6. Victor, S. P. and Sharma, C. P., 2002, Stimuli sensitive polymethacrylic acid microparticles (PMAA) - Oral insulin delivery. *Journal of Biomaterials Applications* 17:125-134

7. Ramadas, M., Paul, W., Dileep, K. J., Anitha, Y. and Sharma, C. P., 2000, Lipoinsulin encapsulated alginate-chitosan capsules: intestinal delivery in diabetic rats. *Journal of Microencapsulation* 17:405-411

8. Carino, G. P., Jacob, J. S. and Mathiowitz, E., 2000, Nanosphere based oral insulin delivery. *Journal of Controlled Release* 65:261-269

9. Kost, J., Liu, L. S., Gabelnick, H. and Langer, R., 1994, Ultrasound as a Potential Trigger to Terminate the Activity of Contraceptive Delivery Implants. *Journal of Controlled Release* 30:77-81

10. Dasaratha Dhanaraju, M., Vema, K., Jayakumar, R. and Vamsadhara, C., 2003, Preparation and characterization of injectable microspheres of contraceptive hormones. *International Journal of Pharmaceutics* 268:23-29

11. Janes K. A., F. M. P., Marazuela A., Fabra A., Alonso M. J., 2001, Chitosan nanoparticles as delivery systems for doxorubicin. *Journal of Controlled Release* 73:255-267

12. Dhanaraju, M. D., Vema, K., Jayakumar, R. and Vamsadhara, C., 2003, Preparation and characterization of injectable microspheres of contraceptive hormones. *International Journal of Pharmaceutics* 268:23-29

13. Ehrhart, N., Dernell, W. S., Ehrhart, E. J., Hutchison, J. M., Douple, E. B., Brekke, J. H., Straw, R. C. and Withrow, S. J., 1999, Effects of a controlled-release cisplatin delivery system used after resection of mammary carcinoma in mice. *American Journal of Veterinary Research* 60:1347-1351

14. Yapp, D. T. T., Lloyd, D. K., Zhu, J. and Lehnert, S. M., 1998, Cisplatin delivery by biodegradable polymer implant is superior to systemic delivery by osmotic pump or ip injection in tumor-bearing mice. *Anti-Cancer Drugs* 9:791-796

15. Ike, O., Shimizu, Y., Wada, R., Hyon, S. H. and Ikada, Y., 1992, Controlled Cisplatin Delivery System Using Poly(D,L-Lactic Acid). *Biomaterials* 13:230-235

16. Arica, B., Calis, S., Kas, H. S., Sargon, M. F. and Hincal, A. A., 2002, 5-Fluorouracil encapsulated alginate beads for the treatment of breast cancer. *International Journal of Pharmaceutics* 242:267-269

17. Schlapp, M. and Friess, W., 2003, Collagen/PLGA microparticle controlled composites for local delivery of gentamicin. *Journal of Pharmaceutical Sciences* 92:2145-2151

18. Mateo, C., Fernandez-Lorente, G., Pessela, B. C. C., Vian, A., Carrascosa, A. V., Garcia, J. L., Fernandez-Lafuente, R. and Guisan, J. M., 2001, Affinity chromatography of polyhistidine tagged enzymes - New dextran-coated immobilized metal ion affinity chromatography matrices for prevention of undesired multipoint adsorptions. *Journal of Chromatography A* 915:97-106

19. Mao, Q. M., Johnston, A., Prince, I. G. and Hearn, M. T. W., 1991, High-Performance Liquid-Chromatography of Amino-Acids, Peptides and Proteins .113. Predicting the Performance of Nonporous Particles in Affinity-Chromatography of Proteins. *Journal of Chromatography* 548:147-163

20. An, X. N., Su, Z. X. and Zeng, H. M., 2003, Preparation of highly magnetic chitosan particles and their use for affinity purification of enzymes. *Journal of Chemical Technology and Biotechnology* 78:596-600

21. Sun, Y. M., Yu, C. W., Liang, H. C. and Chen, J. P., 1999, Temperature-sensitive latex particles for immobilization of alpha-amylase. *Journal of Dispersion Science and Technology* 20:907-920

22. Dasilva, M. A., Burrows, H. D., Formosinho, S. J., Gil, M. H., Lourenco, A. R., Paula, F. J. A. and Piedade, A. P., 1991, Photopolymerization of Acrylamide onto Magnetite Particles - Preparation of Magnetic Supports for Enzyme Immobilization. *Materials Letters* 11:96-100

23. Safarikova, M., Roy, I., Gupta, M. N. and Safarik, I., 2003, Magnetic alginate microparticles for purification of alpha-amylases. *Journal of Biotechnology* 105:255-260

24. Lee, T. H., Wang, J. and Wang, C.-H., 2002, Double-walled microspheres for the sustained release of a highly water soluble drug: characterization and irradiation studies. *Journal of Controlled Release* 83:437-452

25. Mi F.-L., S. S.-S., Lin Y.-M., Wu Y.-B., Peng C.-K. and Tsai Y.-H., 2003, Chitin/PLGA blend microspheres as a biodegradable drug delivery system: a new delivery system for protein. *Biomaterials* 24:5023-5036

26. Rongved, P., Klaveness, J. and Strande, P., 1997, Starch microspheres as carriers for X-ray imaging contrast agents: Synthesis and stability of new amino-acid linker derivatives. *Carbohydrate Research* 297:325-331

27. Sahin, S., Selek, H., Ponchel, G., Ercan, M. T., Sargon, M., Hincal, A. A. and Kas, H. S., 2002, Preparation, characterization and in vivo distribution of terbutaline sulfate loaded albumin microspheres. *Journal of Controlled Release* 82:345-358

28. Silva G. A., D. A. C. P., Coutinho O. P., Reis R. L., 2004, Evaluation of the encapsulation ability and release profile of starch-based particles using glucocorticoids as model drugs. *in final stage of preparation*

29. Sinha, V. R. and Trehan, A., 2003, Biodegradable microspheres for protein delivery. *Journal of Controlled Release* 90:261-280

30. Soriano, I., Llabres, M. and Evora, C., 1995, Release control of albumin from polylactic acid microspheres. *International Journal of Pharmaceutics* 125:223-230

31. Takagi, M., Hayashi, H. and Yoshida, T., 1999, Starch particles modified with gelatin as novel small carriers for mammalian cells. *Journal of Bioscience and Bioengineering* 88:693-695

32. Tuncel, A. and Piskin, E., 1996, Nonswellable and swellable poly(EGDMA) microspheres. *Journal of Applied Polymer Science* 62:789-798

33. Czejka, M. J., Jager, W. and Schuller, J., 1989, Mitomycin-C Determination Using Loop-Column Extraction - a Rapid and Sensitive High-Performance Liquid-Chromatographic Assay for Pharmacokinetic Studies with Spherex Starch Particles. *Journal of Chromatography-Biomedical Applications* 497:336-341

34. Erdogan, S., Ozer, A. Y., Volkan, B., Caner, B. and Bilgili, H., 2003, Particulate drug delivery systems in diagnosis of venous thrombosis. *Journal of Controlled Release* 87:238-240

35. Genc, O., Arpa, C., Bayramoglu, G., Arica, M. Y. and Bektas, S., 2002, Selective recovery of mercury by Procion Brown MX 5BR immobilized

poly(hydroxyethylmethacrylate/chitosan) composite membranes. *Hydrometallurgy* 67:53-62

36. Silva G.A., C. F. J., Coutinho O. P., Radin S., Ducheyne P., Reis R. L., 2003, Synthesis and evaluation of novel bioactive starch/Bioactive Glass microparticles. *Journal of Biomedical Materials Research: Part A*

37. Yenice, I., Calis, S., Atilla, B., Kas, H. S., Ozalp, M., Ekizoglu, M., Bilgili, H. and Hincal, A. A., 2003, In vitro/in vivo evaluation of the efficiency of teicoplanin-loaded biodegradable microparticles formulated for implantation to infected bone defects. *Journal of Microencapsulation* 20:705-717

38. Qu, Q. Q., Ducheyne, P. and Ayyaswamy, P. S., 2002, Bioactive, degradable composite microspheres - Effect of filler material on surface reactivity. In *Microgravity Transport Processes in Fluid, Thermal, Biological, and Materials Sciences* pp. 556-564

39. Schepers, E., Barbier, L., van Steenberghe, D. and Ducheyne, P., 2001, Guided tissue regeneration versus two types of bioactive glass particles in the treatment of furcation type II defects in the beagle dog. *Journal of Dental Research* 80:1215-1215

40. Qiu, Q. Q., Ducheyne, P. and Ayyaswamy, P. S., 2001, 3D Bone tissue engineered with bioactive microspheres in simulated microgravity. *In Vitro Cellular & Developmental Biology-Animal* 37:157-165

41. Qiu, Q. Q., Ducheyne, P. and Ayyaswamy, P. S., 2000, New bioactive, degradable composite microspheres as tissue engineering substrates. *Journal of Biomedical Materials Research* 52:66-76

42. Qiu, Q. Q., Ducheyne, P. and Ayyaswamy, P. S., 1999, Fabrication, characterization and evaluation of bioceramic hollow microspheres used as microcarriers for 3-D bone tissue formation in rotating bioreactors. *Biomaterials* 20:989-1001

43. Hutmacher, D. W., 2000, Scaffolds in tissue engineering bone and cartilage. *Biomaterials* 21:2529-2543

44. Santavirta, S., Konttinen, Y. T., Saito, T., Gronblad, M., Partio, E., Kemppinen, P. and Rokkanen, P., 1990, Immune-Response to Polyglycolic Acid Implants. *Journal of Bone and Joint Surgery-British Volume* 72:597-600

45. Paivarinta, U., Bostman, O., Majola, A., Toivonen, T., Tormala, P. and Rokkanen, P., 1993, Intraosseous Cellular-Response to Biodegradable Fracture Fixation Screws Made of Polyglycolide or Polylactide. *Archives of Orthopaedic and Trauma Surgery* 112:71-74

46. Ignatius, A. A., Betz, O., Augat, P. and Claes, L. E., 2001, In vivo investigations on composites made of resorbable ceramics and poly(lactide) used as bone graft substitutes. *Journal of Biomedical Materials Research* 58:701-709

47. Toth, J. M., Wang, M., Scifert, J. L., Cornwall, G. B., Estes, B. T., Seim, H. B. and Turner, A. S., 2002, Evaluation of 70/30 D,L-PLa for use as a resorbable interbody fusion cage. *Orthopedics* 25:S1131-S1140

48. Filipovic-Grcic J., V. D., Moneghini M., Becirevic-Lacan M., Magarotto L., Jalsenjak I., 2000, Chitosan microspheres with hydrocortisone and hydrocortisone-hydroxypropyl-B-cyclodextrin inclusion complex. *European Journal of Pharmaceutical Sciences* 9:373-379

49. Hata H., O. H., Machida Y., 2000, Preparation of CM-chitin microspheres by complexation with iron(II) in w/o emulsion and their biodisposition characteristics in mice. *Biomaterials* 21:1779-1788

50. Lim S. T., M. G. P., Berry D. J., and Brown M. B., 2000, Preparation and evaluation of the in vitro drug release properties and mucoadhesion of novel microspheres of hyaluronic acid and chitosan. *Journal of Controlled Release* 66:281-292

51. Mao H-Q., R. K., Troung-Le V. L., Janes K. A., Lin K. Y., Wang Y., August J. T., Leong K. W., 2001, Chitosan-DNA nanoparticles as gene carriers: synthesis, characterization and transfection efficiency. *Journal of Controlled Release* 70:399-421

52. Cui Z., M. R. J., 2001, Chitosan-based nanoparticles for topical genetic immunization. *Journal of Controlled Release* 75:409-419

53. Aiedeh K. and Taha, M. O., 2001, Synthesis of iron-crosslinked succinate and iron-crosslinked hydroxamated chitosan succinate and their in vitro evaluation as potential matrix materials for oral theophylline sustained-release beads. *European Journal of Pharmaceutical Sciences* 13:159-168

54. Vaz, C. M., Fossen, M., van Tuil, R. F., de Graaf, L. A., Reis, R. L. and Cunha, A. M., 2003, Casein and soybean protein-based thermoplastics and composites as alternative biodegradable polymers for biomedical applications. *Journal of Biomedical Materials Research Part A* 65A:60-70

55. Vaz, C. M., van Doeveren, P., Reis, R. L. and Cunha, A. M., 2003, Soy matrix drug delivery systems obtained by melt-processing techniques. *Biomacromolecules* 4:1520-1529

56. Vaz, C. M., van Doeveren, P., Reis, R. L. and Cunha, A. M., 2003, Development and design of double-layer co-injection moulded soy protein based drug delivery devices. *Polymer* 44:5983-5992

57. Vaz, C. M., De Graaf, L. A., Reis, R. L. and Cunha, A. M., 2003, Effect of crosslinking, thermal treatment and UV irradiation on the mechanical properties and in vitro degradation behavior of several natural proteins aimed to be used in the biomedical field. *Journal of Materials Science-Materials in Medicine* 14:789-796

58. Vaz, C. M., de Graaf, L. A., Reis, R. L. and Cunha, A. M., 2003, In vitro degradation behaviour of biodegradable soy plastics: effects of crosslinking with glyoxal and thermal treatment. *Polymer Degradation and Stability* 81:65-74

59. Silva, G. A., Vaz, C. M., Coutinho, O. P., Cunha, A. M. and Reis, R. L., 2003, In vitro degradation and cytocompatibility evaluation of novel soy and sodium caseinate-based membrane biomaterials. *Journal of Materials Science-Materials in Medicine* 14:1055-1066

60. Osth, K., Strindelius, L., Larhed, A., Ahlander, A., Roomans, G. M., Sjoholm, I. and Bjork, E., 2003, Uptake of ovalbumin-conjugated starch microparticles by pig respiratory nasal mucosa in vitro. *Journal of Drug Targeting* 11:75-82

61. Sousa, R. A., Mano, J. F., Reis, R. L., Cunha, A. M. and Bevis, M. J., 2002, Mechanical Performance of Starch Based Bioactive Composite Biomaterials Molded with Preferred Orientation for Potential Medical Applications. *Polym Eng & Sci* 42:1032-1045

62. Reis, R. L. and Cunha, A. M., 1995, Characterization of two biodegradable polymers of potential application within the biomaterials field. *Journal of Materials Science-Materials in Medicine* 6:786-792

63. Reis, R. L., Mendes, S. C., Cunha, A. M. and Bevis, M. J., 1997, Processing and in vitro degradation of starch/EVOH thermoplastic blends. *Polymer International* 43:347-352

64. Vaz, C. M., Reis, R. L. and Cunha, A. M., 2001, Degradation model of starch-EVOH plus HA composites. *Materials Research Innovations* 4:375-380

65. Borissova, R., Lammek, B., Stjarnkvist, P. and Sjoholm, I., 1995, Biodegradable Microspheres .16. Synthesis of Primaquine-Peptide Spacers for Lysosomal Release from Starch Microparticles. *Journal of Pharmaceutical Sciences* 84:249-255

66. Laakso, T., 1987, Preparation and Properties of Polyacryl Starch Microparticles - Potential-Drug Carriers in the Treatment of Lysosomal Parasitic Diseases. *Acta Pharmaceutica Suecica* 24:208-208

67. Baillie, A. J., Coombs, G. H., Dolan, T. F., Hunter, C. A., Laakso, T., Sjoholm, I. and Stjarnkvist, P., 1987, Biodegradable Microspheres .9. Polyacryl Starch Microparticles as a Delivery System for the Antileishmanial Drug, Sodium Stibogluconate. *Journal of Pharmacy and Pharmacology* 39:832-835

68. Artursson, P., Edman, P., Laakso, T. and Sjoholm, I., 1984, Characterization of Polyacryl Starch Microparticles as Carriers for Proteins and Drugs. *Journal of Pharmaceutical Sciences* 73:1507-1513

69. Chellat, F., Tabrizian, M., Dumitriu, S., Chornet, E., Rivard, C. H. and Yahia, L., 2000, Study of biodegradation behavior of chitosan-xanthan microspheres in simulated physiological media. *J Biomed Mater Res* 53:592-599

70. Chellat, F., Tabrizian, M., Dumitriu, S., Chornet, E., Magny, P., Rivard, C. H. and Yahia, L., 2000, In vitro and in vivo biocompatibility of chitosan-xanthan polyionic complex. *J Biomed Mater Res* 51:107-116

71. Sivakumar, M., Manjubala, I. and Panduranga Rao, K., 2002, Preparation, characterization and in-vitro release of gentamicin from coralline hydroxyapatite-chitosan composite microspheres. *Carbohydrate Polymers* 49:281-288

72. Wu, T.-J., Huang, H.-H., Lan, C.-W., Lin, C.-H., Hsu, F.-Y. and Wang, Y.-J., 2004, Studies on the microspheres comprised of reconstituted collagen and hydroxyapatite. *Biomaterials* 25:651-658

73. Borden, M., Attawia, M., Khan, Y. and Laurencin, C. T., 2002, Tissue engineered microsphere-based matrices for bone repair: : design and evaluation. *Biomaterials* 23:551-559

74. Wilson, J. and Low, S. B., 1992, Bioactive Ceramics for periodontal treatment: comparative studies in the patus monkey. *J Applied Biomaterials* 3:123-129

75. Stanley, H. R., Hall, M. B., Colaizzi, F. and Clark, A. E., 1987, Residual alveolar rige maintenance with a new endosseous implant material. *J Prosthetic Dentristy* 58:607-613

76. Wilson, J. and Merwin, G. E., 1988, Biomaterials for facial bone augmentation: comparative studies. *J Applied Biomaterials* 22:159-177

77. Reis, R. L., Cunha, A. M., Allan, P. S. and Bevis, M. J., 1996, Mechanical behavior of injection-molded starch-based polymers. *Polymers for Advanced Technologies* 7:784-790

78. Reis, R. L., Cunha, A. M., Fernandes, M. H. and Correia, R. N., 1997, Treatments to induce the nucleation and growth of apatite-like layers on polymeric surfaces and foams. *Journal of Materials Science-Materials in Medicine* 8:897-905

79. Reis, R. L., Cunha, A. M., Allan, P. S. and Bevis, M. J., 1997, Structure development and control of injection-molded hydroxylapatite-reinforced starch/EVOH composites. *Advances in Polymer Technology* 16:263-277

80. Reis, R. L., Cunha, A. M. and Bevis, M. J., 1999, Oriented composites meet tough orthopedic demands. *Modern Plastics* 76:73-75

81. Sousa, R. A., Kalay, G., Reis, R. L., Cunha, A. M. and Bevis, M. J., 2000, Injection molding of a starch/EVOH blend aimed as an alternative biomaterial for temporary applications. *Journal of Applied Polymer Science* 77:1303-1315

82. Reis, R. L., Cunha, A. M., Allan, P. S. and Bevis, M. J., 1996, Mechanical Behaviour of Injection Moulded Starch Based Polymers. *J Polym Adv Tech* 7:784-790

83. Gomes, M. E., Ribeiro, A. S., Malafaya, P. B., Reis, R. L. and Cunha, A. M., 2001, A new approach based on injection moulding to produce biodegradable starch-based polymeric scaffolds: morphology, mechanical and degradation behaviour. *Biomaterials* 22:883-889

84. Gomes, M. E., Godinho, J. S., Tchalamov, D., Cunha, A. M. and Reis, R. L., 2002, Alternative tissue engineering scaffolds based on starch: processing methodologies, morphology, degradation, mechanical properties and biological response. *Materials Science and Engineering: C* 20:19-26

85. Malafaya, P. B., Elvira, C., Gallardo, A., San Roman, J. and Reis, R. L., 2001, Porous starch-based drug delivery systems processed by a microwave route. *J Biomater Sci Polym Ed* 12:1227-1241

86. Boesel, L. F., Mano, J. F., Elvira, C. and Reis, R. L., 2003, Hydrogels and hydrophilic partially degradable bone cements based on biodegradable blends incorporating starch. In *Biodegradable polymers and plastics* (E., C.).Kluwer Academic, Drodrecht, pp.

87. Mendes, S. C., Reis, R. L., Bovell, Y. P., Cunha, A. M., van Blitterswijk, C. A. and de Bruijn, J. D., 2001, Biocompatibility testing of novel starch-based materials with potential application in orthopaedic surgery: a preliminary study. *Biomaterials* 22:2057-2064

88. Mendes, S. C., Bezemer, J., Claase, M. B., Grijpma, D. W., Bellia, G., Degli-Innocenti, F., Reis, R. L., De Groot, K., Van Blitterswijk, C. A. and De Bruijn, J. D., 2003, Evaluation of two biodegradable polymeric systems as substrates for bone tissue engineering. *Tissue Engineering* 9:S91-S101

89. Gomes, M. E., Reis, R. L., Cunha, A. M., Blitterswijk, C. A. and de Bruijn, J. D., 2001, Cytocompatibility and response of osteoblastic-like cells to starch-based polymers: effect of several additives and processing conditions. *Biomaterials* 22:1911-1917

90. Marques, A. P., Reis, R. L. and Hunt, J. A., 2002, The biocompatibility of novel starch-based polymers and composites: in vitro studies. *Biomaterials* 23:1471-1478

91. Salgado, A. J., Gomes, M. E., Chou, A., Coutinho, O. P., Reis, R. L. and Hutmacher, D. W., 2002, Preliminary study on the adhesion and proliferation of human osteoblasts on starch-based scaffolds. *Materials Science & Engineering C-Biomimetic and Supramolecular Systems* 20:27-33

92. Elvira, C., Mano, J. F., San Roman, J. and Reis, R. L., 2002, Starch-based biodegradable hydrogels with potential biomedical applications as drug delivery systems. *Biomaterials* 23:1955-1966

93. Azevedo, H. S., Gama, F. M. and Reis, R. L., 2003, In vitro assessment of the enzymatic degradation of several starch based biomaterials. *Biomacromolecules* 4:1703-1712

94. Elvira, C., Mano, J. F., San Roman, J. and Reis, R. L., 2002, Starch-based biodegradable hydrogels with potential biomedical applications as drug delivery systems. *Biomaterials* 23:1955-1966

95. Pereira, C. S., Cunha, A. M., Reis, R. L., Vazquez, B. and San Roman, J., 1998, New starch-based thermoplastic hydrogels for use as bone cements or drug-delivery carriers. *Journal of Materials Science-Materials in Medicine* 9:825-833

96. Leonor, I. B. and Reis, R. L., 2003, An innovative auto-catalytic deposition route to produce calcium-phosphate coatings on polymeric biomaterials. *Journal of Materials Science-Materials in Medicine* 14:435-441

97. Oliveira, A. L., Elvira, C., Reis, R. L., Vazquez, B. and San Roman, J., 1999, Surface modification tailors the characteristics of biomimetic coatings nucleated on starch-based polymers. *Journal of Materials Science-Materials in Medicine* 10:827-835

98. Demirgoz, D., Elvira, C., Mano, J. F., Cunha, A. M., Piskin, E. and Reis, R. L., 2000, Chemical modification of starch based biodegradable polymeric blends: effects on water uptake, degradation behaviour and mechanical properties. *Polymer Degradation and Stability* 70:161-170

99. Gomes, M. E., Sikavitsas, V. I., Behravesh, E., Reis, R. L. and Mikos, A. G., 2003, Effect of flow perfusion on the osteogenic differentiation of bone marrow stromal cells cultured on starch-based three-dimensional scaffolds. *Journal of Biomedical Materials Research Part A* 67A:87-95

100. Marques, A. P., Reis, R. L. and Hunt, J. A., 2002, The biocompatibility of novel starch-based polymers and composites: in vitro studies. *Biomaterials* 23:1471-1478

101. Mendes, S. C., Reis, R. L., Bovell, Y. P., Cunha, A. M., van Blitterswijk, C. A. and de Bruijn, J. D., 2001, Biocompatibility testing of novel starch-based materials with potential application in orthopaedic surgery: a preliminary study. *Biomaterials* 22:2057-2064

102. Silva, G. A., Costa, F. J., Pedro, A., Coutinho, O. P., Neves, N. M. and Reis, R. L., 2004, The response of starch-based particles to in vitro bioactivity and biocompatibility testing. *in preparation*

Stem Cells and Tissue Engineering

Y. MURAT ELÇİN
Ankara University, Faculty of Science and Biotechnology Institute, Tissue Engineering and Biomaterials Laboratory, Ankara 06100, Turkey

1.   A POTENTIAL CELL SOURCE FOR ENGINEERING TISSUES

Tissue engineering has emerged as an alternative approach for the treatment of the loss or malfunction of a tissue or organ with the advantage of not having the limitations of the current orthodox therapies. Basically, the concept of this technology has been the transplantation of constructs consisting of cells grown *ex vivo* within predesigned scaffolds made up of exogenous three-dimensional extracellular matrices (ECMs). The scaffolds employed to guide the functional tissue development, eventually break down leaving only the cells and the stroma that they produce in the body[1-3].

The availability of a reliable and unlimited source of cells has always been a major concern in cell-based technologies[1-3]. Thus, stem cell research has been the recipient of increasing attention in the context of therapeutic transplantation, and the emerging field of tissue engineering. Stem cells by definition are self-renewing cells with the potential to differentiate into specialized cells[4-6] (Figure 1). The unique properties of human embryonic (hESCs) and adult stem cells (ASCs) have opened up the possibility of generating organs and tissues *ex vivo,* from an unlimited supply of pluri-(multi) potent, self-renewable cells. This remarkable plasticity suggests that endogenous or transplanted stem cells can be tweaked in ways that allow them to replace lost or dysfunctional cell populations in numerous pathologies ranging from neurodegenerative and haematopoietic disorders to

Biomaterials*: From Molecules to Engineered Tissues,* edited by
N. Hasırcı and V. Hasırcı, Kluwer Academic/Plenum Publishers, 2004

diabetes, liver and cardiovascular disease[5-7]. Some stem cell-based therapies have already entered routine clinical practice. Among them, bone marrow transplantation and skin grafting can be mentioned.

Furthermore, stem cells can be used as vehicles for gene therapy. It has now become possible to isolate and expand human embryonic- and various types of somatic- stem cells, including haematopoietic and mesenchymal stem cells. The brain, liver, skin, adipose tissue, intestinal crypts, and pancreatic islets are other tissues that harbor stem cell populations.

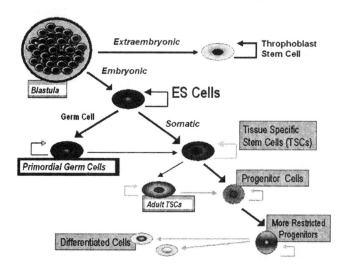

Figure 1. Progeny of stem cells during development (*redrawn from reference 7*).

2. EMBRYONIC STEM CELLS (ESCS)

Transplantation medicine is confronted by the problem of limited availability of donor tissues or organs. Current therapies for these conditions are also subject to problems of acute and chronic rejections after allogenic transplantations. ESCs are pluripotent cell lines established from the inner cell mass of mammalian blastocyst characterized by nearly unlimited self-renewal and differentiation capacity[8,9]. Human ESCs express markers, such as stage-specific embryonic antigens: SSEA-3 and SSEA-4, the germ line transcription factor Oct-4, Tra-1-60, Tra-1-81 and high telomerase and alkaline phosphatase activities[9-11]. hES cells can be co-cultured on inactivated mouse embryonal fibroblast (MEFs) feeder cells or in a medium conditioned by prior incubation with MEF cells[12]; supplementation with basic fibroblast growth factor (bFGF) is also preferred[13] (Figure 2). Recently, it has been shown that human fetal and adult fibroblasts can also support the undifferentiated growth of hES cells[14].

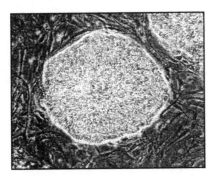

Figure 2. Embryoid bodies generated from human embryonic stem cells on murine embryonic fibroblast feeder layer.

The establishment of human ESC lines has allowed the transfer of the knowledge of mouse (m) ESC differentiation into human biology and medical therapy, especially in areas, such as the *in vitro* study of human embryonic development, pharmacology for drug testing, reproductive toxicology, and potentially most important, (by the use of ESC-generated donor cells) in regenerative therapies[5,11]. Recent studies involving the differentiation of human (h) ESCs show the possibility of using this unique cell source for the reconstruction of a variety of tissue types. Using specific culture conditions, hESCs can differentiate very efficiently *in vitro* into neural progenitor cells that subsequently give rise to mature, functional neurons[15]. Moreover, functional cardiomyocytes can be differentiated from hES cells and enriched by mechanical separation[16]. It has also been demonstrated that, hES cells can form vessel-like structures when cultured on special substrates[17,18]. More recently, hESC-derived hepatocyte cultures expressing albumin and other liver-specific markers have been reported[19].

However, before hESCs may be used in cell and tissue therapy, the following prerequisites have to be fulfilled[5]: *i.* Methods for directed differentiation and selective isolation of pure populations have to be established; *ii.* Non-tumorigenicity of ESC-derived somatic donor cells need to be shown; *iii.* Their integration and tissue-specific function have to be demonstrated; *iv.* Immunological incompatibility between the ESC-generated donor cells and the specific organism after transplantation should be solved by using strategies, such as genetic manipulation of histocompatibility genes in donor cells[20], immunosuppression[21], or therapeutic cloning[22] which allows the derivation of autologous donor cells for transplantation.

3.   ADULT STEM CELLS (ASCS)

The use of ASCs of the adult organism has been proposed as another alternative strategy to generate donor cells for tissue transplantations. The research of the recent years has raised the possibility for future cell and tissue therapies. It is well known that not only fetal, but even adult tissues with high self-renewal capacity do contain stem cell populations, as for example, blood, skin and gut tissues. Adult bone marrow stem cells, such as haematopoietic and mesenchymal stem cells, are already in clinical use for stem cell therapies of leukemia patients and for skeletal regeneration, respectively[23].

3.1   Biology of Adult Stem Cell Renewal

The successful use of stem cells for therapeutic applications will depend on a better understanding of the biology of stem cell differentiation. At present, there are two major models that explain stem cell renewal: *i.* the **stochastic model** that postulates that stem cells in tissue units exist in stable pools from which some remain as stem cells, whereas others differentiate to form specialized tissue cells[24,25] and *ii.* the **transdifferentiation model** that postulates that stem cells begin to produce differentiated cells when transferred to a heterologous tissue site, and that the generated cells are unique to the recipient tissue[23,26].

According to the second model, there may be a restriction of stem cell differentiation by the local tissue factors yielding only the mature cell types of a stem cell's current tissue of residence. For some cases, transit cell differentiation could be responsible for transdifferentiation. It is possible that all adult stem cells might share the property of generating transit cells that are subsequently instructed by local microenvironment factors to adopt specific differentiation lineages.

Although possessing a developmental potential, stem cells from adults may have limitations: They may not have the same proliferative and developmental capacity as embryonic or fetal cells[5]. Asymmetric cell kinetics is a prerequisite for ASC function *in vivo*. *Ex vivo*, the same cell kinetics presents a major barrier to the propagation of ASCs in culture[27]. Conversion of ASCs into symmetric cell kinetics, producing two daughter stem cells, is predicted to promote their exponential expansion in culture. Recently, **Suppression of Asymmetric Cell Kinetics (SACK)** agents that reversibly convert asymmetric cell kinetics to symmetric cell kinetics were identified, consistent with the idea that ASCs are transpotent (*section 3.3.; Liver Stem Cells*). However, the direct identification of the stable undifferentiated stem cells postulated to exist amidst differentiating transit cells in different culture conditions has not been made yet.

3.2 Bone Marrow Stem Cells

The marrow seems to be at the centre stage for future biotechnological developments in tissue engineering, not only as the only organ in which at least two types of stem cells (HSCs and MSCs) reside, but also as the organ in which progenitors for a number of distant tissues can be found[28]. The traditional wall separating the haematopoietic and mesodermal tissue systems and lineages is being demolished by recent data from several groups. Cells capable of regenerating blood vessels, cardiac muscle and skeletal muscle are found in the marrow[29-31]. The unexpected potential for myogenesis and cardiomyogenesis has been ascribed to both haematopoietic and mesenchymal stem cells in the marrow[32-34].

3.2.1 Haematopoietic Stem Cells (HSCs)

Recent data suggest that a group of cells with unknown phenotype in the bone marrow (and in circulating blood) have the ability to differentiate not only into blood cells, but also into a multide of other cell types of the body. It is possible that cells which are partially committed to haematopoiesis may be able to be programmed to differentiate into other cell types. There is a possibility that bone marrow subpopulations enriched for HSC activity (*e.g.* $CD34^+CD38^-$ for humans) may actually contain cells that are less mature than HSCs, and are not yet 'committed' to the haematopoietic lineage. Other subsets characterized either as angioblasts ($CD34^+133^+$) or endothelial progenitor cells ($CD34^+KDR^+$), are both involved in the angiogenesis process, and are myocyte progenitors as well[35]. Once reinjected into the myocardic ischemic lesion, one can then expect that these $CD34^+$ subsets will reconstitute at least partially, a functional myocardiac tissue[36,37].

3.2.2 Stromal Cells

This is a diverse population consisting of fibroblasts, endothelial cells, muscle cells and others, that provide a scaffold to developing stem and progenitor cells. Stromal cells also produce extracellular matrix components and soluble proteins. Within this population are **Mesenchymal Stem Cells (MSCs)**, that are capable of self-renewal and that can differentiate into many mesenchymal-derived tissues[38]. MSCs can be purified from bone marrow, on the basis of their ability to attach to plastic substrates. Flow cytometry using several surface markers has demonstrated that expanded human MSCs are

>98% homogenous and can give rise to several tissues[4,38], such as bone[39], cartilage[40], and adipose[41] when exposed to appropriate stimuli (*e.g.* cytokines and biomechanical forces).

Marrow-derived hMSCs represent a useful, characterized cell population to explore mesenchymal tissue regeneration, and there is ample evidence to suggest that the cells can be used allogeneically. hMSCs express small amounts of the major histocompatibility complex (MHC) class I molecule, but express little or no MHC class II or B7 costimulatory molecules[38]. Clinical trials for the use of allogeneic hMSCs to aid engraftment of matched bone marrow or mobilized periphereal blood progenitor cells have been undertaken by Osiris Therapeutics. Phase I results suggest that the matched hMSCs are well tolerated, and ongoing Phase II studies should provide the appropriate dosage data[38].

Recently, a population of highly plastic adult-derived marrow cells (from postnatal mice bone marrow stroma) has been identified[42]. This population, the so called **Multipotent Adult Progenitor Cells (MAPCs)** grows as an adherent layer and can be cultured indefinitely in a relatively nutrient poor medium without senescence. MAPCs can differentiate extensively into derivatives of the three germ layers, both in culture and following microinjection into host blastocysts or transplantations onto irradiated and nonirradiated recipients[42]. Unlike what is commonly observed in ES cells, no tumorigenesis is observed in the latter setting. It has also been shown that MAPCs are progenitors for angioblasts[43] and these can also differentiate into functional hepatocyte-like cells[44]. However, MAPCs do not contribute to the development of skeleton, heart or brain, at least in physiological situation[35].

Recent limited data suggest that there are both similarities and important differences between MAPCs and ESCs. MAPCs might be very rare pluripotent stem cells that persist from the embryo into adult life[45]. To prove this it would be necessary to identify these cells prospectively *in vivo*, by the marker proteins they express, and to purify them without an intervening culture step. As an alternative, MAPCs may not actually exist *in vivo*, thus the extended period of culture may be triggering certain bone marrow cells to regress to a more primitive state, just as primordial germ cells can be reprogrammed in culture to acquire properties like those of ESCs[45].

Potential clinical uses of MAPCs include autologous transplantation into healing or growing tissues or use as stromal support for haematopoietic cells. Allogeneic transplantation may also be a potential application of MAPCs, since these cells do not express HLA antigens and therefore may elude immune rejection.

3.3   Liver Stem Cells

As a consequence of the increase in liver disease incidents, the shortage of donor organs, and the demand for sophisticated transplant teams, interest in hepatocellular therapies is gradually increasing[46,47]. These therapies can be categorized under three headings, namely extracorporeal bioartificial liver devices, cell transplantation, and tissue engineering. The choice of the cell type in hepatocellular therapies is of great importance. On the other hand, full complement of cellular functions required to replace the liver and positively affect clinical outcomes has not yet been determined. The use of primary hepatocytes does not seem to be the practical choice for cellular therapies, due to their low stability under culture conditions, and during cryopreservation, thus they are in limited supply[46-50]. Immortalized hepatocyte cell lines carry the risk of oncogenic factor transmission to the host, especially in cell transplantation and tissue engineering therapies. Potential stem cell sources for use in hepatocellular therapies are the embryonic stem cells[19], liver stem cells (adult liver progenitors), and transdifferentiated non-hepatic cells[51,52].

Liver stem cells are the precursors of the two epithelial liver cell types, the hepatocytes and the bile-duct epithelial cells. Oval Cells are the descendants of the stem cells and are found in the portal and periportal regions in experimental animals within days of the liver injury and blockade of hepatocyte proliferation[47,53,54]. It has been shown that oval cells can transdifferentiate *in vitro* into immature pancreatic endocrine-like cells. Hepatic progenitors are proposed as ideal cells for use in liver cell therapies given their ability to expand under certain conditions; they possess the capacity to differentiate into all mature liver cells. Moreover, hepatic progenitors exhibit minimal immunogenicity, and can be cryopreservable, and are able to reconstitute liver tissue when transplanted[54].

Although strategies are available to purify and culture hepatic oval cells[54], a detailed understanding of the growth factors and cytokines that are important in their survival and differentiation is yet to be reached. Thus, the propagation and expansion of many normal mature cells, including liver stem cells in sufficient quantities for therapeutic applications is still impractical. Expanding somatic stem cells programmed for indefinite division, would provide a solution for this problem. A general approach based on suppression of asymmetric cell kinetics (SACK) was described for clonal expansion of adult somatic stem cells[24,55], and has recently been used to derive hepatic stem cells from adult rat liver[56]. The SACK nucleoside xanthosine (Xs) was used to derive these stem cell lines which exhibited Xs-dependent asymmetric stem cell kinetics, expressed alpha-fetoprotein, and produced progeny expressing indicators of mature hepatocytes (*i.e.* self-adhesion, hepatocyte morphology, expression of albumin, H4 antigen and cytochrome p450 activity)[56].

3.4   Insulin-Producing Stem Cells

The use of either ESCs or ASCs to differentiate and expand fully functional, insulin-producing beta cells is a novel approach to overcome the problems of limited tissue supply for clinical islet transplantation in diabetes mellitus[57].

Recent reports on the *in vitro* differentiation and expansion of mESCs and hESCs into insulin-producing beta cells are very promising[58,59]. When transplanted, these cells were able to normalize blood glucose of the diabetic recipients[58]. ASCs may be differentiated into insulin-producing cells as well. Nestin-positive cells in the pancreatic duct or within the pancreatic islets may be relevant progenitor cells[60]. Ramiya and co-workers[61] were able to obtain islets from ductal epithelial cells of adult mice, which responded to glucose stimulation *in vitro* and reversed diabetes when transplanted into diabetic NOD mice. Furthermore, Bonner-Weir *et al.*[62] have developed a method to yield islets through *in vitro* cultivation of adult human pancreatic ductal cells.

Engineering pancreatic islets is another potential alternative to overcome the problems with limited donor tissue supply[63]. Since autologous tissue and cells can be used in this approach, graft rejection and disease recurrence will probably no longer exist as problems. Several groups have been able to target the cells of the gut, the liver and the muscle for transfection with the insulin-/pro-insulin gene[64-67]. Ferber *et al.*[68] have recently described the expression of insulin genes in the liver and amelioration of hyperglycemia in mice after viral transfection of the hepatocytes with the pancreatic and duodenal homeobox (PDX)-1 gene, which is a valid proof for converting hepatocytes into beta cells[69].

3.5   Neural Stem Cells

Neural Stem Cells are a subtype of progenitor cells in the nervous system that can self-renew and generate both neurons and glia[70]. Initially, stem-like cells have been isolated from the embryonic mammalian central nervous system (CNS)[71,72], and from the peripheral nervous system (PNS)[73].

Since then, stem cells have been isolated from many regions of the embryonic nervous system, indicating their ubiquity. However, isolation of stem-like cells from adult brain has opened another chapter in neuroscience[74]. Adult neural stem cells have now been found in the two principal adult neurogenic regions, *i.e.* the hippocampus and the subventricular zone, and in some non-neurogenic regions, including the spinal cord[75-77].

CNS stem cells produce large clones containing neurons, glia and more stem cells in adherent cultures. This population can also be cultured as floating, multicellular neurospheres[75-77]. The use of CNS-derived stem cells in regenerative medicine seems unlikely at the time, but may prove to be useful by the developments in acellular tissue engineering technologies in the future.

Adult neural stem cells are not highly plastic. They have a limited potential to generate different somatic cell types which seems to be restricted to rare events or rare cells[70,78]. It is possible that most neural stem cells might be regionally and temporarily specified. There may also be rare stem cells present in the nervous system, perhaps not even of neural origin, that have greater plasticity, at least in terms of producing diverse somatic cell types[78].

4. TISSUE ENGINEERING STUDIES WITH HUMANS

The use of the appropriate scaffold design, the precise control of the physiological milieu, and the biochemical and physical factors have been the major concerns of tissue engineering during the *ex vivo* culture step (Figure 3). Tissues that can now be engineered comprise a diverse range from epithelial surfaces, such as skin, cornea and mucosal membranes to skeletal tissues, that have different self-renewal potential and physical structure[79]. Some examples of the clinical application of the tissue engineering approach are given below.

4.1 Engineering of the Skin

Skin has been the first engineered tissue which has progressed from experimental research to accepted patient care[80,81]. This can be related to the essentially two-dimensional organization of skin, which is in contrast with systems having more complex shapes and architectures (*e.g.* bone). LifeCell has developed a dermal skin substitute (AlloDerm®) composed of a dermal matrix lacking immunogenic cells. Integra® is the product of Integra which is comprised of dermal fibroblasts and bovine collagen. Dermagraft® of Applied Tissue Sciences consists of non-immunogenic neonatal fibroblasts cultured on a polyglactin mesh and has been used to treat burns and diabetic foot ulcers[82]. The first mass-produced skin product has been developed by Organogenesis (Apligraf®), and is composed of human neonatal foreskin fibroblasts and keratinocytes in a bovine collagen (type I) ECM. Apligraf®

is currently being used for the treatment of partial and full thickness skin loss related to ulcers of venous stasis[83]. Pellegrini *et al.*[84] have used autologous keratinocytes that were cultured on a fibrin matrix to treat massive full-thickness burns by controlling epidermal stem cells (holoclones), in humans.

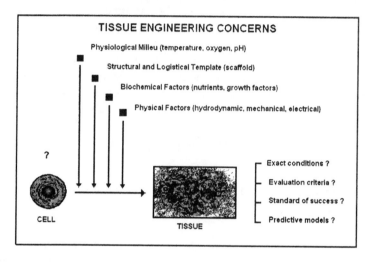

Figure 3. Tissue engineering concerns are summarized above.

4.2   Engineering of the Bone

MSCs, the skeletal stem cells are found in the subset of clonogenic adherent marrow-derived cells, which can undergo extensive replication in culture. Engineering the bone requires the use of appropriate three-dimensional osteoconductive carriers made up of, for example hydroxyapatite/tricalcium phosphates alone or together with polyglycolic and polylactic acids for the attachment of MSCs, before transplantation. Such scaffolds provide mechanical stability to the newly-forming tissue in which a vascular bed can be established, and transplanted progenitor cells can differentiate and form a bone/marrow organ. A convincing preclinical study adopting this approach has been performed in humans[85]; thus, clinical trials have recently started in a number of centres.

4.3   Other Tissues

The major challenge to restore corneal epithelial surfaces consists of the definition of culture conditions and of carriers that are designed to specifically to maintain an adequate stem cell compartment in the engineered graft. This dependence on relevant stem cell compartments is further highlighted by the successful reconstruction of damaged human corneas by using stem cells derived from the autologous limbal epithelial cells[86,87] attached to an amniotic membrane.

Cell-based therapies have emerged as a promising approach for restoration of function in neurodegenerative diseases, particularly Parkinson's and Huntington's disease. Besides the ongoing experimental studies which share the aim to upregulate the efficiency of cells, that are of clinical relevance, by using 3D supports or immunoisolation devices[88-90], cell replacement therapy for neurodegenerative disorders has already been applied to humans[91].

5.   LIMITATIONS AND FUTURE ASPECTS

Despite the knowledge on the plasticity of stem cells of various origins, our understanding of the fundamental mechanisms of differentiation and organogenesis *in vivo* is still very limited; and still, many practical obstacles stand in the way to the routine application of stem cell technologies in medicine. For the time being, -except for certain types of tissues including the skin, the bone, and some others-, engineering of implantable tissue constructs is largely experimental and is still to overcome significant hurdles before it becomes a viable clinical modality. Tissue engineering and regenerative medicine will eventually benefit from the emerging science of stem cell biology and signal transduction. Acellular tissue engineering approach using native biomaterials maintaining cell-signalling motifs to mobilize the stem/progenitor cell reservoirs of the body (*e.g.* from the circulating blood or from the implantation site) presents potential for engineering functional tissue.

Concerning hESC research, there are scientific and ethical obstacles that need to be overcome, before these cells can be accepted as a reliable source for tissue engineering. Results from adult stem cell research studies elucidate the nature of stem cell differentiation *in vitro* and *in vivo*. While most of the current *ex vivo* culture and expansion protocols are for the mixed cell populations at distinct maturational stages, recent studies have started to provide methods for the routine expansion of clinically-desirable pure populations. Increased attention to the heterogenous cell kinetics architecture of stem cell-based tissue units in adult tissues should better inform current efforts to improve existing stem cell therapies (*e.g.* bone marrow transplant) and promote an acceleration of development of new cell-based treatments, including tissue engineering.

ACKNOWLEDGEMENTS

The support of Ankara University Biotechnology Institute, and the Turkish Academy of Sciences (EA-TÜBA-GEBİP/2001-1-1), Ankara, Turkey is acknowledged.

REFERENCES

1. Langer, R., and Vacanti, J.P., 1993, Tissue engineering. *Science* **260**, 920-926.
2. Lanza, R.P., Langer, R, and Chick, W.L., (Eds.), 1997, *Principles of Tissue Engineering*, Academic Press, San Diego and London, 808 pages.
3. Elçin, Y.M., (Ed.), 2003, *Tissue Engineering, Stem Cells and Gene Therapies*, AEMB Series, Vol. 534, Kluwer Academic-Plenum Publishers, New York, Boston, Dordrecht, London, Moscow, 350 pages.
4. Pittenger, M.F., *et al.*, 1999, Multilineage potential of adult human mesenchymal stem cells. *Science* **284**, 143-147.
5. Wobus, A.M., 2001, Potential of embryonic stem cells. *Mol Asp Med* **22**, 149-164.
6. Krause, D.S., 2002, Plasticity of marrow-derived stem cells. *Gene Therapy* **9**, 754-758.
7. Rao, M.S., and Mattson, M.P., 2001, Stem cells and aging: expanding the possibilities. *Mech Age Dev* **122**, 713-734.
8. Evans, M.J., and Kaufman, M., 1981, Establishment in culture of pluripotential cells from mouse embryos. *Nature* **292**, 154-156.
9. Thomson, J.A., Itskovitz-Eldor, J., Shapiro, S.S., Waknitz, M.A., Swiergiel, J.J., Marshall, V.S., and Jones, J.M., 1998, Embryonic stem cell lines derived from human blastocysts. *Science* **282**, 1145-1147.
10. Reubinoff, B.E., Pera, M.F., Fong, C., Trounson, A., and Bongso, A., 2000, Embryonic stem cell lines from human blastocysts: somatic differentiation *in vitro*. *Nature Biotechnology* **18**, 399-404.
11. McWhir, J., Thomson, A., and Sottile, V., 2003, *Tissue Engineering, Stem Cells and Gene Therapies*, (Y.M. Elçin, Ed.), AEMB Series, Vol. 534, Kluwer Academic-Plenum Publishers, New York, Boston, Dordrecht, London, Moscow, pp. 11-26.
12. Xu, C., Inokuma, M.S., Denham, J., Golds, K., Kundu, P., Gold, J.D., and Carpenter, M., 2001, Feeder-free growth of undifferentiated human embryonic stem cells. *Nature Biotechnology* **19**, 971-974.
13. Lebkowski, J.S., Gold, J., Xu, C., Funk, W., Chiu, C., and Carpenter, M., 2001, Human embryonic stem cells: culture differentiation, and genetic modification for regenerative medicine applications. *The Cancer J* **7** (suppl.2), S83-S93.
14. Richards, M., Fong, C., Chan, W., Wong, P., and Bongso, A., 2002, Human feeders support prolonged undifferentiated growth of human inner cell masses and embryonic stem cells. *Nature Biotechnology* **20** (9), 933-936.
15. Zhang S., Wernig, M., Duncan, I.D., Brustle, O., and Thomson, J.A., 2001, *In vitro* differentiation of transplantable neural precursors from human embryonic stem cells. *Nature Biotechnology* **19**, 1129-1133.
16. Kehat, I., Kenyagin-Karsenti, D., Snir, M., Segev, H., Amir, M., Gepstein, A., Livne, E., Binah, O., Itskovitz-Eldor, J., and Gepstein. L., 2002, Human embryonic stem cells can differentiate into myocytes with structural and functional properties of cardiomyocytes. *J Clin Invest* **108**, 407-414.

17. Elçin, A.E., *et al.*, 2002, Formation of vessel-like structures from human embryonic stem cells in culture: preliminary findings. *Proc BIOMED 2002*, Antalya, Turkey, P14, 049.

18. Levenberg, S., Golub, J.S., Amit, M., Itskovitz-Eldor, J., and Langer, R., 2002, Endothelial cells derived from human embryonic stem cells. *Proc Natl Acad Sci USA* **99** (7), 4391-4396.

19. Rambhatla, L., Chui, C-P., Kundu, P., Peng, Y., and Carpenter, M.K., 2003, Generation of hepatocyte-like cells from human embryonic stem cells. *Cell Transplant* **12** (1), 1-11.

20. Rathjen, P.D., Lake, J., Whyatt, L.M., Bettess, M.D., and Rathjen, J., 1998, Properties and uses of embryonic stem cells: prospects for application to human biology and gene therapy. *Reprod Fertil Dev* **10**, 31-47.

21. Gage, F.H., 1998, Cell therapy. *Nature* **392**, 18-24.

22. Lanza, R.P., Cibelli, J.B., and West, M.D., 1999, Human therapeutic cloning. *Nature Med* **5**, 975-976.

23. Fuchs, E., and Segre, J.A., 2000, Stem cells: a new lease on life. *Cell* **100**, 143-155.

24. Sherley, J.L., 2002, Asymmetric cell kinetics genes: the key to expansion of adult stem cells in culture. *Stem Cells* **20** (6). 561-572.

25. Loeffler, M., and Potten, C.S., Stem cells and cellular pedigrees – a conceptual introduction. In Stem Cells, C.S. Potten, Ed. San Diego, Harcourt Brace & Co. 1997, pp.1-28.

26. Clark, D., and Frisen, J., 2001, Differentiation potential of adult stem cells. *Curr Opin Genet & Devel* **11**, 575-580.

27. Rambhatla, L., *et al.*, 2001, Cellular senescence: *Ex vivo* p53-dependent asymmetric cell kinetics. *J Biomed Biotech* **1**, 28-37.

28. Bianco,P., and Robey, P.G., 2001, Stem cells in tissue engineering. *Nature* **414**, 118-121.

29. Kocher, A.A., *et al.*, 2001, Neovascularization of ischemic myocardium by human bone-marrow-derived angioblasts prevents cardiomyocyte apoptosis, reduces remodelling and improves cardiac function. *Nature Med* **7**, 430-436.

30. Ferrari, G., *et al.*, 1998, Muscle regeneration by bone-marrow-derived myogenic progenitors. *Science* **279**, 1528-1530.

31. Orlic, D., *et al.*, 2001, Bone marrow cells regenerate infarcted myocardium. *Nature* **410**, 701-705.

32. Wakitani, S., Saito, T., and Caplan, A.I., 1995, Myogenic cells derived from rat bone marrow mesenchymal stem cells exposed to 5-azacytidine. *Muscle Nerve* **18**, 1417-1426.

33. Makino, S., *et al.*, 1999, Cardiomyocytes can be generated from marrow stromal cells *in vitro. J Clin Invest* **103**, 697-705.

34. Jackson, K.A., *et al.*, 2001, Regeneration of ischemic cardiac muscle and vascular endothelium by adult stem cells. *J Clin Invest* **107**, 1395-1402.

35. Henon, P., 2003, Human embryonic or adult stem cells: an overview on ethics and perspectives for tissue engineering. In *Tissue Engineering, Stem Cells and Gene Therapies*, (Y.M. Elçin, Ed.), AEMB Series, Vol. 534, Kluwer Academic-Plenum Publishers, New York, Boston, Dordrecht, London, Moscow, pp. 27-46.

36. Hamano, K., Nishida, M., Hirata, K. *et al.*, 2001, Local implantation of autologous bone marrow cells for therapeutic angiogenesis in patients with ischemic heart disease: clinical trial and preliminary results. *Jpn Circ J* **65**, 845-847.

37. Strauer, B.E., Brehm, M., Zeus, T. *et al.*, 2002, Repair of infarcted myocardium by autologous intracoronary mononuclear bone marrow cell transplantation in humans. *Circulation* **106**, 1913-1918.

38. Pittenger, M.F., and Marshak, D.R., 2001, Mesenchymal stem cells of human adult bone marrow, in *Stem Cell Biology*, Ed. D.R. Marshak, pp. 349-373, Cold Spring Harbor Laboratory Press, Woodbury, New York.

39. Kim, C.-H., Cheng, S.-L., and Kim, G.S., 1999, Effects of dexamethasone on proliferation, activity and cytokine secretion of normal human bone marrow stromal cells:possible mechanisms of glucocorticoid-induced bone loss. *J Endocrin* **162**, 371-379.

40. Mackay, A.M., Beck, S.C., Murphy, J.M., Barry, F.P., Chichester, C.O., and Pittenger, M.F., 1998, Chondrogenic differentiation of cultured human mesenchymal stem cells from marrow. *Tissue Eng* **4**, 415-428.

41. Pittenger, M.F., 1998, Adipogenic differentiation of human mesenchymal stem cells. U.S. Patent #5,827,740.

42. Jiang, Y., Jahagirdar, B.N., Reinhardt, R.L. *et al.*, 2002, Pluripotency of mesenchymal stem cells derived from adult marrow. *Nature* **418**, 41-49.

43. Reyes, M., Dudek, B., Jahagirdar, B., Koodie, K., Marker, Ph., and Verfaillie, C.M., 2002, Origin of endothelial progenitors in human postnatal bone marrow. *J. Clin. Invest.* **109**, 337-346.

44. Schwartz, R.E., Reyes, M., Koodie, L. *et al.*, 2002, Multipotent adult progenitor cells from bone marrow differentiate into functional hepatocyte-like cells. *J.Clin.Invest.* **109**, 1291-1302.

45. Orkin, S.H., and Morrison, S.J., 2002, Stem-cell competition. *Nature* **418**, 25-27.

46. Elçin, Y.M., 1998, Tissue engineering of liver, review. In *Biomedical Science and Technology: Recent Developments In The Pharmaceutical and Medical Sciences*, (A.A.Hıncal and H.S. Kaş, Eds.), Plenum Press, New York and London, pp. 109-116.

47. Elçin, Y.M., 2004, Tissue engineering of the liver, review. In *Biodegradable Systems in Medical Functions: Design, Processing, Testing and Applications*, (R.Reis and J.S. Roman, Eds.), CRC Press, Boca Raton, in press.

48. Elçin, Y.M., Dixit, V., and Gitnick, G., 1998, Hepatocyte attachment on modified chitosan membranes: *In vitro* evaluation for the development of liver organoids. *Artif Organs* **22**, 837-846.

49. Elçin, Y.M., Dixit, V., Lewin K., and Gitnick, G., 1999, Xenotransplantation of fetal porcine hepatocytes in rats using a tissue engineering approach. *Artif Organs* **23**(2), 146-152.

50. Dixit, V, and Elçin, Y.M., 2003, Liver tissue engineering: successes and limitations. In *Tissue Engineering, Stem Cells and Gene Therapies*, (Y.M. Elçin, Ed.), AEMB Series, Vol. 534, Kluwer Academic-Plenum Publishers, New York, Boston, Dordrecht, London, Moscow, pp. 57-68.

51. Grompe, M., 1999, Therapeutic liver repopulation for the treatment of metabolic liver diseases. *Hum Cell* **12**, 171-178.

52. Vessey, C.J., and Hall, P.D.L.M., 2001, Hepatic stem cells: a review. *Pathology* **33** (2), 130-148.

53. Susick, R., *et al.*, 2001, Hepatic progenitors and strategies for liver cell therapies. *Ann NY Acad Sci* **944**, 398-413.

54. Lazaro, A., *et al.*, 1998, Generation of hepatocytes from oval cell precursors in culture. *Cancer Res* **58**, 5514-5519.

55. Sherley, J.L., 2002, Adult stem cell differentiation: what does it mean? *Proc IInd J EMBS/BMES Conf*, Houston, TX, USA, pp.741-742.

56. Lee ,H.S., Crane, G.G., Merok, J.R., *et al.*, 2003, Clonal expansion of rat hepatic stem cell lines by supression of asymmetric cell kinetics (SACK). *Biotech Bioeng* **83** (7), 760-771.

57. Bretzel, R.G., 2003, Pancreatic islet and stem cell transplantation in diabetes mellitus: results and perspectives. In *Tissue Engineering, Stem Cells and Gene Therapies*, (Y.M. Elçin, Ed.), AEMB Series, Vol. 534, Kluwer Academic-Plenum Publishers, New York, Boston, Dordrecht, London, Moscow, pp. 69-96.

58. Lumelsky, N., Blondel, O., Laeng, P., Velasco, I., Ravin, R., and McKay, R., 2001, Differentiation of embryonic stem cells to insulin-secreting structures similar to pancreatic islets. *Science* **292**, 1389-1394.
59. Assady, S., Maor, G., Amit, M., Itskovitz-Eldor, J., Skorecki, K.L., and Tzuckerman, M., 2001, Insulin production by human embryonic stem cells. *Diabetes* **50**, 1691-1697.
60. Zulewski, H., Abraham, E.J., Gerlach, M.J., Daniel, P.B., Moritz, W., Muller, B., Vallejo, M., Thomas, M.K., and Habener, J.F., 2001, Multipotential nestin-positive stem cells isolated from adult pancreatic islets differentiate ex vivo into pancreatic endocrine, exocrine, and hepatic phenotypes. *Diabetes* **50**, 521-533.
61. Ramiya, V.K., Maraist, M., Arfors, K.E., Schatz, D.A., Peck, A.B., and Cornelius, J.G., 2000, Reversal of insulin-dependent diabetes using islets generated in vitro from pancreatic stem cells. *Nat Med* **6**, 278-282.
62. Bonner-Weir, S., Taneja, M., Weir, G.C., Tatarkiewicz, K., Song, K.H., Sharma, A., and O'Neill, J.J., 2000, In vitro cultivation of human islets from expanded ductal tissue. *Proc. Natl Acad Sci USA* **97**, 7999-8004.
63. Soria, B., Andreu, E., Berna, G., Fuentes, E., Gil, A., Leon-Quinto, T., Martin, F., Montanya, E., Nadal, A., Reig, J.A., Ripoll, C., Roche, E., Sanchez-Andres, and J.V., Segura, J., 2000, Engineering pancreatic islets. *Eur J Physiol* **440**, 1-18.
64. Cheung, A.T., Dayanandan, B., Lewis, J.T., Korbutt, G.S., Rajotte, R.V., Bryer-Ash, M., Boylan, M.O., Wolfe, M.M., and Kieffer, T.J., 2000, Glucose-dependent insulin release from genetically engineered K cells. *Science* **290**, 1959-1962.
65. Lee, H.C., Kim, S.J., Kim, K.S., Shin, H.C., and Yoon, J.W., 2000, Remission in models of type 1 diabetes by gene therapy using a single-chain insulin analogue. *Nature* **408**, 483-488.
66. Shaw, J.A.M., Delday, M.I., Hart, A.W., and Docherty, K., 2002, Secretion of bioactive human insulin following plasmid-mediated gene transfer to non-neuroendocrine cell lines, primary cultures and rat skeletal muscle *in vivo*. *J Endocrinol* **172**, 653-672.
67. Riu, E., Mas, A., Ferre, T., Pujol, A., Gros, L., Otaegui, P., Montoliu, L., and Bosch, F., 2002, Counteraction of type 1 diabetic alterations by engineering skeletal muscle to produce insulin: insights from transgenic mice. *Diabetes* **51**, 704-711.
68. Ferber, S., Halkin, A., Cohen, H., Ber, I., Einav, Y., Goldberg, I., Barshack, I., Seijfers, R., Kopolovic, J., Kaiser, N., and Karasik, A., 2000, Pancreatic and duodenal homeobox gene 1 induces expression of insulin genes in liver and ameliorates streptozotocin-induced hyperglycemia. *Nat Med* **6**, 568-572.
69. Kahn, A., 2000, Converting hepatocytes to β-cells - a new approach for diabetes? *Nat Med* **6**, 505-506.
70. Temple, S., 2001, The development of neural stem cells. *Nature* **414**, 112-117.
71. Temple, S., 1989, Division and differentiation of isolated CNS blast cells in microculture. *Nature* **340**, 471-473.
72. Cattaneo, E., and McKay, R., 1990, Proliferation and differentiation of neuronal stem cells regulated by nerve growth factor. *Nature* **347**, 762-765.
73. Stemple, D.L., and Anderson, D.J., 1992, Isolation of a stem cell for neurons and glia from the mammalian neural crest. *Cell* **71**, 973-985.
74. Reynolds, B.A., and Weiss, S., 1992, Generation of neurons and astrocytes from isolated cells of the adult mammalian central nervous system. *Science* **255**, 1707-1710.
75. McKay, R., 1997, Stem cells in the central nervous system. *Science* **276**, 1707-1710.
76. Rao, M.S., 1999, Multipotent and restricted precursors in the central nervous system. *Anat Rec* **257**, 137-148.
77. Gage, F.H., 2000, Mammalian neural stem cells. *Science* **287**, 1433-1438.

78. Weissman, I.L., 2000, Stem cells: units of development, units of regeneration, and units of evolution. *Cell* **100**, 157-168.
79. Bianco, P., and Robey, P.G., 2001, Stem cells in tissue engineering. *Nature* **414**, 118-121.
80. Wilkins, L.M., Watson, S.R., Prosky, S.J., Meunier, S.F., and Parenteau, N.L., 1994, Development of a bilayered living skin construct for clinical applications. *Biotechnol Bioeng* **43**, 747-756.
81. Rodriguez, H., Jaruga, P., Birincioglu, M., Barker, P.E., O'Connell, and Dizdaroglu, M., 2003, Oxidative DNA damage biomarkers used in tissue engineered skin. In *Tissue Engineering, Stem Cells and Gene Therapies*, (Y.M. Elçin, Ed.), AEMB Series, Vol. 534, Kluwer Academic-Plenum Publishers, New York, Boston, Dordrecht, London, Moscow, pp. 129-135.
82. Gentzkow, G.D., Iwasaki, S.D., Hershon, K.S., Mengel, M., Prendergast, J.J., Ricotta, J.J., Steed, D.P., and Lipkin, S., 1996, Use of dermagraft, a cultured human dermis, to treat diabetic foot ulcers. *Diabetes Care* **19**, 350-354.
83. Sabolinski, M.L., Alvarez, O., Auletta, M., Mulder, G., and Parenteau, N.L., 1996, Cultured skin as a 'smart material' for healing wounds: experience in venous ulcers. *Biomaterials* **17**, 311-320.
84. Pellegrini, G., Ranno, R., Stracuzzi, G., Bondanza, S., Guerra, L., Zambruno, G., Micali, G., and De Luca, M., 1999, The control of epidermal stem cells (holoclones) in the treatment of massive full-thickness burns with autologous keratinocytes cultured on fibrin. *Transplantation* **68**, 868-879.
85. Quarto, R., Mastrogiacomo, M., Cancedda, R., Kutepov, S.M., Mukhachev, V., Lavroukov, A., Kon, E., and Marcacci, M., 2001, Repair of large bone defects with the use of autologous bone marrow stromal cells. *N Eng J Med* **344** (5), 385-386.
86. Pellegrini, G., et al., 1997, Long-term restoration of damaged corneal surfaces with autologous cultivated corneal epithelium. *Lancet* **349**, 990-993.
87. Tsai, R.J., Li, L.M., and Chen, J.K., 2000, Reconstruction of damaged corneas by transplantation of autologous limbal epithelial cells. *N Engl J Med* **343**, 86-93.
88. Lysaght, M.L., Frydel, B., Gentile, F., Emerich, D., and Winn, S., 1994, Recent progress in immunoisolated cell therapy, *J Cell Biochem* **56**, 196-203.
89. Elçin, A.E., Elçin, Y.M., and Pappas, G.D., 1998, Neural tissue engineering: adrenal chromaffin cell attachment and viability on chitosan scaffolds. *Neurol Res* **20**, 648-654.
90. Elçin, Y.M., Elçin, A.E., and Pappas, G.D., 2003, Functional and morphological characteristics of bovine adrenal chromaffin cells on macroporous poly(DL-lactide-co-glycolide) scaffolds. *Tissue Eng* **9** (5), 1047-1056.
91. Björklund, A., 2000, Cell replacement strategies for neurodegenerative disorders. *Novartis Found Symp* **231**, 7-15.

Cartilage Tissue Engineering

GAMZE TORUN KÖSE[*] and VASIF HASIRCI[#]

[*]Department of Genetics and Bioengineering, Yeditepe University, Istanbul 34755, TURKEY;
[#]Department of Biological Sciences, Biotechnology Research Unit, Middle East Technical University, Ankara 06531, TURKEY

1.   INTRODUCTION

Cartilage damage is of great clinical consequence given the tissue's limited intrinsic potential for healing. Current treatments for cartilage repair are less than satisfactory, and rarely restore full function or return the tissue to its native normal state. Methods such as autografting, allografting and or use of man-made materials and devices have been developed for tissue repair but each has some major disadvantages[1]. Even then, a large number of surgical procedures that require tissue or organ substitutes to repair or replace damaged or diseased organs and tissues are performed every year[2].

Autografting involves harvesting a tissue from one location in the patient and transplanting it to another part of the same patient. Autologous grafts typically produce the best clinical results since rejection is not an issue. Examples of commonly used autografting procedures include coronary bypass (i.e., vein grafts are removed from the leg and then transplanted to the heart as a conduit for blood flow around blocked coronary arteries), and spinal fusion (i.e., a bone graft from the patient's hip is used to stabilize a segment of the patient' spine). However, it has several associated problems including the additional surgical costs for the harvesting procedure, and infection and pain at the harvesting site. For example, harvesting an ileac crest graft (i.e., the protruding bony section of the patient's hip) for bone grafting can cost between $1,000 to $9,000/procedure for the harvesting operation and the additional hospital stay. The morbidity at the harvest site can be tremendous with problems such as pain, infection, and blood loss requiring blood transfusion adding the associated risks of transfusion reaction and blood borne infection. As another example, there are approximately 300,000 coronary by-passes/year and harvesting saphenous veins results in wound infection in 20% of the cases in addition to a significant post operative pain in the leg.

Allografting involves harvesting tissue or organs from a donor and then transplanting it to the patient. The donor might have recently died and donated heart, kidney, liver, bone, pancreas, etc. Living donors might also be used to donate lungs and kidneys. While transplantation technology has dramatically improved over the past several decades with the advent of anti-rejection drugs, the immediate problem is shortages in donor availability. For example, in 1998, there were 22,170 organ transplants in the U.S. but 4,855 patients died while waiting for suitable donor organs.

Man-made materials and devices have been created by engineers and scientists to replicate, augment or extend functions performed by biological systems. Examples range from artificial hearts and valves to prosthetic hip and breast implants. Many of these systems have had an enormous positive impact. The materials used in these therapies, however, are subject to fatigue, fracture, toxicity, inflammation, wear, and do not remodel with time (i.e., a metal bone implant can not grow with the patient and cannot change shape in response to the loads placed upon the implant). Also, they do not behave physiologically like true organs or tissues. Thus, devices like an artificial heart (or left ventricular assist device) are best suited as temporary therapies until a donor organ becomes available.

2.   TISSUE ENGINEERING

The rapidly emerging field of tissue engineering holds great promise for the generation of functional tissue substitutes. Current tissue engineering approaches are mainly focused on the restoration of pathologically altered tissue structure based on the transplantation of cells in combination with supportive matrices and biomolecules. The ability to manipulate and reconstitute tissue structure and function *in vitro* has tremendous clinical implications and is likely to have a key role in cell and gene therapies in the coming years[1].

The general approach involves a biocompatible, structurally and mechanically sound scaffold, with an appropriate cell source, which is loaded with bioactive molecules that promote cellular differentiation and/or maturation. Cell actions and their responses to various environmental cues, including mechanical, electrical, structural and chemical, are mediated by protein based molecules loosely referred to as growth factors. This technique involves seeding highly porous biodegradable scaffolds with donor cells and/or growth factors, then culturing and implanting the scaffolds to induce and direct the growth of new tissue. The goal is for the cells to attach to the scaffold, replicate, differentiate (i.e., transform from a non-specific state into a cell exhibiting the functions of the target tissue), and then organize into normal healthy tissue as the scaffold degrades. This method has been used to create various tissue analogs including skin, cartilage, bone, liver, nerve, vessels, etc.

For example, hepatocyte transplantation on implantable devices is a tissue engineering approach to improve the treatment of liver diseases and the efficacy of *ex vivo* gene therapy. Strategies involve hepatocyte attachment to microcarriers, encapsulation, and transplantation on biodegradable polymer scaffolds[3,4].

The challenge of tissue engineering of blood vessels with the mechanical properties of native vessels and with nonthrombogenicity is demanding. One approach was to seed autologous endothelial cells on ePTFE, expanded poly(tetrafluoroethylene), for construction of a synthetic graft[5]. In order to address functional requirements and practical issues, a nondegradable polyurethane scaffold seeded with smooth cells were endothelialized using fluid shear stress and this tissue engineered vascular graft was found to remain patent with a neointima for up to four weeks[6]. Efforts are also underway to repopulate natural biomaterials with autologous cells as an ideal template for the design of vascular grafts[7]. These results show that tissue engineering of vascular grafts has a real potential for application in clinical situations.

Among the strategies for reconstruction of the cornea by tissue engineering is the production of a complete cornea for meeting the demands of corneal substitutes for those individuals affected with severe corneal wounds as well as establishment of *in vitro* model systems as tools for further physiological, toxicological, and pharmacological studies and gene expression studies[8-10].

Researchers at Children's Hospital (affiliated with Harvard Medical School) have been able to generate kidney-like constructs that possess some of the filtering capabilities of fully functional kidneys. Additionally, researchers at the University of Toronto are working on a long-term project to synthesize a human heart within 10 to 20 years. Considering that in 1999, cardiovascular diseases cost the U.S. around $326.6 billion dollars in direct and indirect costs, this is a project that could have huge impact on the health care industry in the US and around the world.

A nano-HAp/collagen (nHAC) composite that mimics the natural bone to some extent both in composition and microstructure was employed as a matrix or the tissue engineering of bone[11]. The porous nHAC scaffold provided a microenvironment resembling the natural tissue, and cells within the structure eventually acquired a 3D polygonal shape. In addition, new bone matrix was synthesized at the interface of bone fragments and the composite. Hsu et al[12] used microspheres of hydroxyapatite/reconstituted collagen as a support for osteoblast growth and concluded that these microspheres could be used as filling materials for bone defects.

Köse et al[13,14] observed the formation of bone in *in vitro* conditions by culturing rat marrow stromal osteoblasts in biodegradable, macroporous poly(3-hydroxybutyric acid-co-3-hydroxyvaleric acid) (PHBV) matrices over a period of 60 days. Both visual and biochemical assays showed that cultured osteoblasts retained their phenotype throughout the duration of the experiment.

3.   CARTILAGE TISSUE ENGINEERING

Joint pain due to cartilage degeneration is a serious problem affecting people of all ages. Cartilage is a relatively simple tissue and does not contain blood vessels or nerves. These properties makes it a suitable target for tissue engineering. Genzyme Tissue Repair in Cambridge, MA (USA)[15] has received FDA approval for their product Carticel - an autologous engineered tissue derived from a patient's own cells for the repair of traumatic knee-cartilage repair. Considerable efforts are still being expended to improve the biomaterials used for the regeneration of cartilage.

The goal of *in vitro* cartilage tissue engineering is to achieve mechanical properties comparable to those of native cartilage. Various scaffold materials have been tested for that purpose, including both the naturally derived (e.g. collagen, gelatin, hyaluronic acid, biodegradable polymers like poly(3-hdroxybutyric acid-co-3-hydroxyvaleric acid) (PHBV) etc.) and synthetic polymers (e.g. polylactic acid (PLA), polyglycolic acid (PGA), poly(lactic acid-co-glycolic acid) (PLGA), etc.) for cartilage tissue engineering.

3.1   Carriers in Tissue Engineering

Cell carriers or scaffolds considered for use in tissue engineering need to possess certain characterisctics such as biodegradability, and appropriate surface chemistry that allows attachment of cells or allows surface modification for improved cell attachment. In addition the carrier must provide suitable chemical and biological cues to help proliferation and differentiation of the seeded cells in order to create a viable construct.

3.1.1   Naturally Derived Polymers

Collagen matrices have been found to have the proper molecular cues to stimulate new collagen production by transplanted cells as compared with other scaffold types[16]. Caplan et al[17] reported that when mesenchymal stem cells were seeded into collagen gels implanted in osteochondral defects in rabbits, embryogenesis was recapitulated and both bone and hyaline cartilage were formed, although mechanical properties of the regenerated tissues were significantly less than normal and there was some evidence of degeneration after 24 weeks.

Yaylaoglu et al[18] produced a highly porous HAp (hydroxyapatite) /collagen construct before seeding with chondrocytes. The treatment increased the stability of the foam in the medium and cell proliferation was observed in the form of ECM deposition.

Marijnissen et al[19] showed promising possibilities to generate structurally regular neo-cartilage using multiplied chondrocytes in alginate in combination with a fleece of polylactic acid /polyglycolic acid.

A chondrocyte-collagen composite was prepared by Fujisato et al[20] and the effect of basic fibroblast growth factor (bFGF) on cartilage regeneration by its subcutaneous implantation in nude mouse was studied. It was observed that chondrocytes can proliferate and mature quickly, especially when the scaffold is impregnated with bFGF. It was also reported that bFGF promotes cartilage repair in vivo[21] and inhibits the terminal differentiation of chondrocytes and calcification[22].

Perka et al[23] seeded human articular chondrocytes either in alginate, in a mixture of alginate and fibrin, or in a fibrin gel after the extraction of the alginate component (porous fibrin gel) over a period of 30 days. It was observed that a mixture of 0.6% alginate with 4.5% fibrin promoted sufficient chondrocyte proliferation and differentiation, resulting in the formation of a specific cartilage matrix.

By combining chitosan with chondroitin sulfate A, the major GAG molecule present in glycosaminoglycan, Sechriest et al[24] have formed a hydrogel material that mimics the GAG-rich ECM of the articular chondrocyte. Bovine primary articular chondrocytes, when seeded onto a thin membrane of CSA-chitosan, maintained many characteristics of the differentiated chondrocytic phenotype.

Köse et al[25] observed cartilage formation using chondrocytes and biodegradable, macroporous matrix material, poly(3-hydroxybutyric acid-co-3-hydroxyvaleric acid) (PHBV) both in vitro and in vivo conditions. Repair of a full thickness cartilage defect in rabbit model was achieved using autologous chondrocytes on this material.

3.1.2 Synthetic Polymers

Firm, hyaline-like cartilage has been observed after six weeks when undifferentiated perichondrial cells were seeded onto a PLLA mesh and implanted in the femoral condyles of rabbits[26]. Similar cartilage morphology was observed when PGA porous non-woven scaffolds were seeded with bovine chondrocytes and cultured in vitro for 12 weeks. In this study, mechanical properties such as compressive modulus and aggregate modulus were similar to that of normal bovine cartilage[27]. Both types of degradable polyesters tended to increase proteoglycan synthesis compared to a collagen scaffold[16]. Initially (less than two months) when seeded with bovine chondrocytes cell growth on PGA matrices was approximately twice as high as that on PLLA matrices, but after six months, total cellularity was found to be similar. Initial differences were attributed to the fact that PLLA degrades much more slowly and allowed less space for cell proliferation[28].

PLLA has been found to be less toxic to human chondrocytes than PGA in studies maintaining a pH over 12 days[29].

It was demonstrated that perichondrocytes were capable of attaching to and surviving within porous D,D-L,L-PLA composite, and this composite graft affected a consistent, cartilaginous-appearing repair when implanted into drilled osteochondral defects in a rabbit model[26].

Polyethylene oxide (PEO) has been tested as an injectable matrix material for chondrocyte transplantation. Sims et al[30] reported that 12 weeks after subcutaneous delivery of bovine chondrocytes in a PEO gel into nude mice cartilage that was similar to natural bovine cartilage had been produced.

Caterson et al[31] have demonstrated that bone marrow derived cells seeded within a three dimensional polylactide/alginate amalgam polymers can be effectively induced to undergo chondrogenic differentiation upon treatment with TGF-β1.

A new artificial cartilage (a three dimensional fabric (3-DF) comprising ultrahigh molecular weight polyethylene fiber with a triaxial three dimensional structure and coated with hydroxyapatite) was used to repair large osteochondral defects in rabbit knees[32]. Hyaline-like cartilage formation was observed to a certain extent on the surface of the 3-DF, and the surface damage of the patella opposed to the 3-DF was minimal.

4.   CELL GUIDANCE

Microengineered scaffold based guided tissue generation is one enabling technology for the emerging field of tissue engineering. Advances in micropatterning methodologies have made it possible to create structures with precise architecture on the surface of cell culture substrata.

4.1   ECM, Cell Attachment and Cell Orientation

The microenvironment of an engineered tissue must be properly regulated during the process of tissue development to induce the appropriate pattern of gene expression in cells forming the new tissue. The expression of genes by cells in engineered tissues may be regulated by multiple interactions with the microenvironment, including interactions with the adhesion surface, with other cells and with soluble growth factors, and mechanical stimuli imposed on the cells[33]. Synthetic ECM should provide the appropriate combination of these signals.

The ECM to which cells adhere can regulate the cell phenotype. Cell adhesion to ECM is mediated by cell-surface receptors. The integrin transmembrane receptors found on the cell surface bind to relatively short amino acid sequences, Arg-Gly-Asp (RGD), of extracellular matrix (ECM) molecules. For many years, ECM was thought to serve only as a structural support for tissues. Bissel et al[34] proposed the model of "dynamic reciprocity" between the ECM on one hand and the cytoskeleton and nuclear matrix on the other. In this model, ECM interacts with the receptors on the cell surface, which transmit signals across the cell membrane to molecules in the cytoplasm. These signals initiate a cascade of events through the cytoskeleton into the nucleus, resulting in the expression of specific genes, whose products in turn affect the ECM in various ways.

It is the continuous interaction between the cells and surrounding matrix environment that leads to formation of patterns, the development of form and acquisition and maintenance of differentiated phenotype during embryogenesis. Similarly, during wound healing these interactions contribute to the process of clot formation, inflammation, granulation, tissue development and remodeling. ECM molecules interact with their receptors and transmit signals directly or indirectly to second messengers that in turn unravel a cascade of events leading to the coordinated expression of a variety of genes involved in cell adhesion, migration, proliferation, differentiation and death[35].

The challenge of tissue engineering for the future is to develop tissue substitutes that restore the normal physiological functions of living tissues, in addition to structural features. It has become clear that both chemical and mechanical determinants play important roles in tissue formation in which biochemical signals mediate the process and mechanical forces play a regulatory role. In other words, chemical regulators mediate tissue morphogenesis while the signals that are actually responsible for dictating tissue pattern are often mechanical in nature. In order to engineer functional structural tissues (e.g., cartilage), the correct mechanical stimuli may be provided through guided tissue engineering on patterned surfaces.

Biological cells in a tissue are arranged in distinct patterns; their orientation and alignment depends on the tissue[36]. If the tissue is to be repaired, the new cells must be aligned and positioned correctly. The reconstruction of organs is even more demanding; here cells of different types have to be aligned correctly with respect to each other, and the whole complex of tissue cells, blood vessel cells and nerve cells has to work correctly together. During development, cues are provided to the proliferating cells to dictate their final position and orientation. These cues can be chemical in nature as in presence of adhesion promoting proteins or purely physical, as the response of cells to the topography of their surroundings. An understanding and use of cues that influence cell positioning and alignment is crucial in tissue engineering. Cellular and subcellular function can be significantly improved through the incorporation of micro- and nanoscale features. A tissue engineered construct that has a well-controlled microstructure is expected to better maintain cell morphology, differentiation and functionality over long periods of time[37]. General approach to organizing cells on exogenous matrices is to use physical (topographical), chemical or biological cues.

4.2   Chondrocytes and Other Cells On Patterned Surfaces

There have been many studies showing topographical and chemical control of cells and tissues and the techniques of microfabrication have been extended to create more complex, controlled biomaterial surfaces. The arrangement of cells in controlled two- and three-dimensional patterned surfaces have been shown to have beneficial effects on cell differentiation, maintenance, and functional longevity. Micropatterning techniques have been employed to introduce topographic features to the surface of smooth cell culture substrata such as silicon, glass, and titanium. Such features are affective in controlling the number, spatial distribution, and function of cells[38].

One of the first studies using microfabricated substrates to study cellular behavior was performed by Brunette et al[39]. This group demonstrated the use of a silicon mask etching technique to prepare grooved surfaces to control the direction of outgrowth of human gingival explants. In another study, better osteointegration and induction of mineralized tissue using surface microtexture were observed. In that study, titanium coated epoxy replicas of different micromachined, grooved or pitted, surfaces that ranged between 30 and 120 μm deep, as well as smooth control surfaces were implanted percutaneously and fixed to the parietal bone of rats[40]. Mineralization was found to be more frequent with increase in depth of pit and decrease in depth of grooves, while was rare on smooth surfaces, indicating the importance of surface topography.

Many researchers have shown that surface topography can influence cell migration and orientation. Eisenbarth et al[41] demonstrated that osteoblasts and fibroblast-like cells in contact with ground cp-titanium aligned and spread in the direction of the surface structure and had higher density of focal contacts, better cytoskeleton organization and stronger actin filaments that resist detaching forces more effectively. It has been demonstrated by Deutsch et al[42] that 3D surface topography of membranes significantly affects in vitro cardiac myocyte orientation and attachment.

By creating culture surfaces with topographic features corresponding to cellular dimensions (10-50 μm), cells seem to exhibit a more in vivo-like cellular morphology [37]. The microtopography provides directional growth for cells and thus, can recreate tissue architecture at the cellular and subcellular level in a reproducible manner, facilitating the culture and maintenance of differentiated state.

Petersen et al[38] cultured chondrocytes on polysaccharide (agarose) gel with microstructures on its surface (15-65 μm wide; 40 μm deep), and this method proved to be effective in maintaining key aspects of the chondrogenic phenotype: rounded cell morphology and significant production of type II collagen. Differentiation in chondrocytes is maintained through restriction of cell spreading. The maintenance of dense cell populations on scaffolds following seeding was greatly improved in micropatterned as compared with featureless regions.

Invaginations found in the native basal lamina of the skin were mimicked by micropatterning collagen and gelatin membranes, which influenced the differentiation of keratinocytes seeded on them through enhanced stratification[43].

Microtechnology also enable us to apply defined chemistries, found in the extracellular environment, to the microenvironment of the cells. Chemical patterning of adhesive and nonadhesive regions onto a substrate in order to control spatial organization of cells on 2D constructs is possible in this way. Adhesion and orientation of the bovine aortic endothelial cells and PC12 nerve cells on the 70 and 50 μm wide lines of anchored RGD and IKVAV peptides, respectively, on the PLA-PEG surfaces was achieved with the use of microfluidic patterning to generate the adhesive strips on nonadhesive PLA-PEG surfaces[44]. Directionally controlled neurite outgrowth was stimulated by the IKVAV micropattern, with neurites extending between groups of cells often hundreds of microns apart. Another group investigated the effect of initial attachment and spreading of bone derived cells on the rate of matrix mineralization[45]. Cell adhesive N-(2-aminoethyl)-3-aminopropyl trimethoxysilane (EDS) strips (50 μm) were generated in alteration with nonadhesive dimethyldichlorosilane (DMS) strips (100 μm). Cells were organized on the EDS regions within 30 min, but after 15 days in culture, cells had grown over both EDS and DMS regions, while mineralization occurred only on the EDS strips, emphasizing the role of initial shape in determining differentiated function. Spatial control of cell-to-cell interactions in co-cultures is also possible with 2D micropatterning.

When hepatocytes were co-cultured with fibroblasts by, an increase in hepatocyte/fibroblast interaction was achieved through micropatterning, and improvement of cell function over randomly oriented cell cultures was found[46]. Variation of initial heterotypic cell-to-cell interactions was found to modulate long-term bulk tissue function.

Combinations of 2D and 3D micropatterning (application of both chemical and physical cues) are used for more effective guidance of cellular organization. Miller et al[47] investigated the effect of substrate-mediated chemical and physical guidance on the growth and alignment of Schwann cells in vitro. Laminin was selectively adsorbed onto the grooves of micropatterned biodegradable films (PLA), and was found to improve adhesion of the cells on the substrates. Microgrooves were found to cause alignment of Schwann cells along their direction, where groove width was the effective parameter. Groove widths closer to the cellular dimensions were found to be optimal for alignment of the Schwann cells.

In another study, rat osteoblasts were cultured on hydroxyapatite or titanium coated smooth and micromachined grooved substrates[48]. Osteoblasts aligned on both micropatterned substrates and produced significantly more bone-like nodules than on smooth surfaces. There was a statistically significant interaction that is synergism, between topography and chemistry in the formation of mineralized nodules.

5.   CONCLUSION

Biomaterials had a phenomenal growth in the 1970's and 1980's with the introduction of new polymers, new delivery systems involving macromolecules such as enzymes and nucleic acids, and miniaturization helped by the developments in the microelectronics. In the 1990's introduction of cells into the biomaterials opened up a new approach for the development of better and adaptable implantable biomaterials for tissue repair. In the 2000's the use of cells including chondrocytes has already taken root and the impetus is in the creation of better artificial ECM for better attachment, proliferation and differentiation of cells seeded on carriers in order to achieve construction of complex tissues involving complex organization and a multitude of cell types. Among the methods most employed are the application of chemical and physical patterns to guide cell attachment and differentiation. Also introduced are the use of factors that influence differentiation of the cells to help develop a better tissue in vitro. The methods of cell printing and complex scaffold creation through 3D manufacturing processes are all similar attempts for better in vitro tissue production. The next decade is promising to be very exciting as a result of these developments.

REFERENCES

1. Risbud, M.V., and Sittinger, M., 2002, Tissue engineering. *TRENDS in Biotechnology* 20: 351-356.
2. Temenoff, J.S., and Mikos, A.G., 2000, Review: tissue engineering for regeneration of articular cartilage. *Biomaterials* 21: 431-440.
3. Davis, M.W., and Vacanti, J.P., 1996, Toward development of an implantable tissue engineered liver. *Biomaterials* 17: 365–372.
4. Kim, S.S., Kaihara, S., Benvenuto, M.S., Kim, B.S., Mooney, D.J., Vacanti, J.P., Anderson K., and Atkinson, J.P., 2000, Small intestinal submucosa as a small-caliber venous graft: A novel model for hepatocyte transplantation on biodegradable scaffolds with direct access to the portal venous system. *MRS Bulletin* 25: 33–37.

5. Deutsch, M., Meinhart, J., Fischlein, T., Preiss, P., and Zilla, P., 1999, Clinical autologous *in vitro* endothelialization of infrainguinal ePTFE grafts in 100 patients: a 9 year experience. *Surgery* 126: 847-855.

6. Ratcliffe, A., 2000, Tissue engineering of vascular grafts. *Matrix Biology* 19: 353–357.

7. Schmidt, C.E., and Baier, J.M., 2000, Acellular vascular tissues: natural biomaterials for tissue repair and tissue engineering. *Biomaterials* 21: 2215–2231.

8. Minami, Y., Sugihara, H., and Oono, S., 1993, Reconstruction of cornea in three-dimensional collagen gel matrix culture. *Invest. Ophthamol and Vis. Sci,* 34: 2316–2324.

9. Germain, L., Auger, F.A., Grandbois, R., Giasson, M., Boisjoly, H., and Guerin, S.L., 1999, Reconstructed human cornea *in vitro* by tissue engineering. *Pathobiology* 67: 140–147.

10. Germain, L., Carrier, P., Auger, F.A., Salesse, C., and Guerin, S.L., 2000, Can we produce a human corneal equivalent by tissue engineering?. *Progress in Retinal and Eye Research* 19(5): 497–527.

11. Du, C., Cui, F.Z., Zhu, X.D., and de Groot, K., 1999, Three dimensional nano-HAp/collagen matrix loading with osteogenic cells in organ culture. *J.Biomed.Mater.Res.* 44: 407.

12. Hsu, F.Y., Chueh, S.C., and Wang, Y.J., 1999, Microspheres of hydroxyapatite/reconstituted collagen as supports for osteoblast cell growth. *Biomaterials* 20: 1931.

13. Torun Köse, G., Ber, S., Korkusuz, F., and Hasırcı, V., 2003, Poly(3-hydroxybutyric acid-co-3-hydroxyvaleric acid) Based Tissue Engineering Matrices. *J.Mat.Sci.:Mat. in Medicine* 14: 121-126.

14. Torun Köse, G., Korkusuz, F., Korkusuz, P., Purali, N., Özkul, A., and Hasırcı, V., 2003, Bone generation on PHBV matrices: an in vitro study. *Biomaterials*, 24: 4999-5007.

15. Ellis, D.L., and Yannas, I.V., 1996, Recent advances in tissue synthesis *in vivo* by use of collagen-glycosaminoglycan copolymers. *Biomaterials* 17: 291–299.

16. Grande, D.A., Halberstadt, C., Naughton, G., Schwartz, R., and Manji, R., 1997, Evaluation of matrix scaffolds for tissue engineering of articular cartilage grafts. *J.Biomed.Mat.Res.* 34: 211-220.

17. Caplan, A.I., Elyaderani, M., Mochizuki, Y., Wakitani, S., and Goldberg, V.M., 1997, Principles of cartilage repair and regeneration. *Clin.Orthop.Rel.Res.* 342: 254-269.

18. Yaylaoglu, M.B., Yıldız, C., Korkusuz, F., and Hasırcı, V., 1999, Novel osteochondral implant.*Biomaterials* 20: 1513-1520.

19. Marijnissen, W.J.C.M., van Osch, G.J.V.M., Aigner, J., Verwoerd-Verhoef, H.L., and Verhaar, J.A.N., 2000, Tissue engineered cartilage using serially passaged articular chondrocytes. Chondrocytes in alginate, combined in vivo with a synthetic (E210) or biologic biodegradable carrier (DBM). *Biomaterials* 21: 571-580.

20. Fujisato, T., Sajiki, T., Liu, Q., and Ikada, Y., 1996, Effect of basic fibroblast growth factor on cartilage regeneration in chondrocyte seeded collagen sponge scaffold. *Biomaterials* 17(2): 155-162.

21. Cuevas, P., Burgos, J., and Baird, A., 1988, Basic fibroblast growth factor (FGF) promotes cartilage repair *in vivo*. *Biochem.Biophys.Res.Commun.* 156: 611-618.

22. Kato, Y., and Iwamoto, M., 1990, Fibroblast growth factor is an inhibitor of chondrocyte terminal differentiation. *J.Biol.Chem.* 265: 5903-5909.

23. Perka, C., Spitzer, R.S., Lindenhayn, K., Sittinger, M., and Schultz, O., 2000, Matrix-mixed culture: New methodology for chondrocyte culture and preparation of cartilage transplants. *J.Biomed.Mater.Res.* 49: 305-311.

24. Sechriest, V.F., Miao, Y.J., Niyibizi, C., Westerhausen-Larson, A., Matthew, H.W., Evans, C.H., Fu, F.H., and Suh, J.K., 2000, GAG-augmented polysaccharide hydrogel: A novel biocompatible and biodegradable material to support chondrogenesis. *J.Biomed.Mater.Res.* 49: 534-541.

25. Torun Köse, G., Korkusuz, F., Özkul, A., Soysal, Y., Özdemir, T., Yıldız, C., and Hasırcı, V., 2002, Tissue engineered cartilage on PHBV matrices, Research project (MISAG-158, TUBITAK), Middle East Technical University, Turkey.

26. Chu, C.R., Coutts, R.D., Yoshioka, M., Harwood, F.L., Monosov, A.Z., and Amiel, D., 1995, Articular cartilage repair using allogenic perichondrocyte seeded biodegradable porous polylactic acid (PLA): a tissue engineering study. *J.Biomed.Mater.Res.* 29: 1147-1154.

27. Ma, P.X., Schloo, B., Mooney, D., and Langer, R., 1995, Development of biomechanical properties and morphogenesis of in vitro tissue engineered cartilage. *J.Biomed.Mater.Res.* 29: 1587-1595.

28. Freed, L.E., Marquis, J.C., Nohria, A., Emmanual, J., Mikos A.G., and Langer, R., 1993, Neocartilage formation in vitro and in vivo using cells cultured on synthetic biodegradable polymers. *J.Biomed.Mater.Res.* 27: 11-23.

29. Sittinger, M., Reitzel, D., Dauner, M., Hierlemann, H., Hammer, C., Kastenbauer, E., Planck, H., Burmester, G.R., and Bujia, J., 1996, Resorbable polyesters in cartilage engineering: affinity and biocompatibility of polymer fiber structures to chondrocytes. *J.Biomed.Mater.Res.* 33: 57-63.

30. Sims, C.D., Butler, P.E.M., Casanova, R., Lee, B.T., Randolph, M.A., Lee, W.P.A., Vacanti, C.A., and Yaremchuk, M.J., 1996, Injectable cartilage using polyethylene oxide polymer substrates. *Plast.Reconstr.Surg.* 98: 843-850.

31. Caterson, E.J., Nesti, L.J., Li, W.J., Danielson, K.G., Albert, T.J., Vaccaro, A.R., and Tuan, R.S., 2001, Three dimensional cartilage formation by bone marrow derived cells seeded in polylactide/alginate amalgam. *J.Biomed.Mater.Res.* 57: 394-403.

32. Nehrer, S., Breinan, H.A., Ramappa, A., Hsu, H.P., Minas, T., Shortkroff, S., Sledge, C.B., Yannas, I.V., and Spector, M., 1998, Chondrocyte seeded collagen matrices implanted in a chondral defect in a canine model. *Biomaterials* 19: 2313-2328.

33. Kim, B.S., and Mooney, D.J., 1998, Development of biocompatible synthetic extracellular matrices for tissue engineering. *TIBTECH* 16: 224-230.

34. Bissel, M.J., Hall, H.G., and Parry, G., 1982, How does the extracellular matrix direct gene expresssion?. *J.Theor.Bio.* 99: 31-68.

35. Green-Martins, M., 1997, The dynamics of cell-ECM interactions with implications for tissue engieering. In *Principles of Tissue Engineering* .(R.P.,Lanza, R., Langer, W.L.Chick, eds.), Texas USA, pp.23-37, (1997).

36. Curtis, A., and Rielhe, M., 2001, Tissue engieering: the biophysical background. *Phys.Med.Biol.* 46: R47-R65.

37. Desai, A.T., 2000, Micro- and nanoscale structures for tissue engineering constructs. *Medical Eng.and Phys.* 22: 595-606.

38. Petersen, E.F., Spencer, R.G.S., and McFarland, E.W., Microengineering neocartilage scaffolds. *Biotechnology and Bioengineering* 78(7): 802-805.
39. Brunette, D.M., Kenner, G.S., and Gould, T.R., 1983, Grooved titanium surfaces orient growth and migration of cells from human gingival explants. *J.Dent.Res.* 62(10): 1045.
40. Chehroudi, B., Brunette, D.M., and McDonell, D.M., 1997, The effects of micromachined surface on formation of bonelike tissue on subcutaneous implants as assessed by radiography and computer image processing. *J.Biomed.Mater.Res.* 34: 279-290.
41. Eisenbarth, E., Linez, P., Biehl, Velten, V.D., Breme, J., and Hildebrand, H.F., 2002, Cell orientation and cytoskeleton organisation on ground titanium surfaces. *Biomolecular Eng,* 19: 233-237.
42. Deutsch, J., Motlagh, D., Russell, B., and Desai, T.A., 2000, Fabrication of microtextured membranes for cardiac myocyte attachment and orientation. *J.Biomed.Mater.Res.Appl.Biomater,* 53(3): 267-275.
43. Pins, G.D., Toner, M., and Morgan, J.R., 2000, Microfabrication of an analog of the basal lamina: bicompatible membranes with complex topographies. *FASEB J.* 14: 593-602.
44. Patel, N., Padera, R., Sanders, G.H.W., Cannizzaro, S.M., Davies, M.C., Langer, R., Roberts, C.J., Tendler, S.J.B., Williams, P.M., and Shakesheff, K.M., 1998, Spacially controlled cell engineering on biodegradable polymer surfaces. *FASEB J.* 12: 1447-1454.
45. Healy, K.E., Thomas, C.H., Rezania, A., Kim J.E., McKeown, P.J., Lom, B., and Hockberger, E.P., 1996, Kinetics of bone cell organization and mineralization on materials with patterned surface chemistry. *Biomaterials* 17: 195-208.
46. Bhatia, S.N., Balis, U.J., Yarmush, M.L., and Toner, M., 1999, Effect of cell-cell interactions in preservation of cellular phenotype: cocultivation of hepatocytes and nonparenchymal cells. *FASEB J.* 13: 1883-1900.
47. Miller, C., Shanks, H., Witt, A., Rutkowski, G., and Mallapragada, S., 2001, Oriented Schwann cell growth on micropatterned biodegradable polymer substrates. *Biomaterials* 22: 1263-1269.
48. Perizzolo, D., Lacefield W.R, .and Brunette, D.M., 2001, Interaction between topography and coating in the formation of bone nodules in culture for hydroxyapatite- and titanium-coated micromachined surfaces. *J.Biomed.Mater.Res.* 56: 494-503.

Towards Dissecting The Complexity of The AP-1 (Activator Protein 1) Family

RESAT UNAL*,#, FALK WEIH*, HUBERT SCHORLE¶ and
PETER HERRLICH*,¥
*Research Center Karlsruhe, Institute of Toxicology and Genetics, D-76021 Karlsruhe,
GERMANY; #Middle East Technical University, Department of Biological Sciences,
Biotechnology Research Unit, 06531 Ankara, TURKEY; ¶Bonn University, Department of
Developmental Pathology, D-53105 Bonn, Germany, ¥Institute of Molecular Biotechnology,
D-7708 Jena, GERMANY

1. INTRODUCTION

The transcription factors called Activator Protein-1 (AP-1) collectively are dimers of two subunits derived from the Fos, Jun and ATF families[1]. Dimerization is mediated by the bzip domain of the subunits. For instance, c-Jun can dimerize with c-Fos or ATF-2. Each AP-1 dimer selects a closely related but distinct element in target promoters[2, 3]. Nevertheless the function of these numerous AP-1 dimers is yet ill understood. One approach to dissecting the different roles has been chosen in overexpression studies: transfection or infection of c-Jun mutants whose dimerization interphase was altered such that their dimerization abilities are restricted. To explore the function of such restricted dimers in vivo, we prepared targeting vectors that could be used for insertional mutagenesis, that is replacement of the endogeneous gene by the mutant gene. In paralel we used c-Jun null murine fibroblasts to introduce the Jun expressing genes as expression constructs and tested their target promoter specificity.

AP-1 family of dimeric factors plays a central role in the regulatory network of eucaryotic gene expression. It has been reported that AP-1 plays

Biomaterials: From Molecules to Engineered Tissues, edited by
N. Hasırcı and V. Hasırci, Kluwer Academic/Plenum Publishers, 2004

important roles in cellular growth, differentiation, tumor formation and development[4].

The heterogeneity in dimer composition of AP-1 complex is high which is caused by the fact that multiple AP-1 subunits can be expressed at the same time. The members of AP-1 are characterized by the presence of a leucine zipper required for dimerization and the basic region involved in the recognition of DNA motifs, the heptameric AP-1 binding site TRE (TPA responsive elemet)[2] where Jun/Fos (like) dimers bind to or the octameric CRE (cAMP responsive element) where Jun/ATF (like) dimers bind to[3]. The octameric elemet differs only by one nucleotide from the heptameric AP-1 binding site.

2. RESULTS AND DISCUSSION

The use of Jun homodimers do not help to understand the role of the Jun target genes involved in transformation since Jun homodimers are able to bind both the Jun/Fos recognition sites which are the classical TRE site sor the Jun/ATF recognition sites which are the CRE sites. To understand the role of Jun target genes involved in transformation, mutants are made where c-Jun specifically bound to c-Fos or ATF-2.

Targeting vectors carrying c-Jun mutant sequences in the bzip region for specific dimerization with either members of the Fos family or ATF-2 are constructed by site directed PCR mutagenesis. In one of the construct, c-Jun preferentially dimerizes with c-Fos whereas in the other one, c-Jun preferentially dimerizes with ATF-2. The detection of the presence of the diagnostic restriction enzyme sites provided by site directed PCR mutagenesis is shown in Figure 1.

Upon the insertion of LTNL sequence (thymidine kinase and neomycine selection marker casettes between two lox sites) into a unique sequence in the promoter of c-Jun, targeting vectors are completed and electroporated into embryonic stem cells. Positive clones for homologous recombination are picked under selection conditions in cell culture media. Genotyping for positive clones is performed by southern blot analysis and PCR based strategies. The detection of the homologous recombinationbetween the wild type and targeting vector is shown schematically in Figure 2.

The cells which survived in the presence of neomycine containing medium were the cells which are identified as positives. These clones are electroporated with Cre expressing vector, in order to remove LTNL. The genomic alteration after removal of the LTNL is shown in Figure 3.

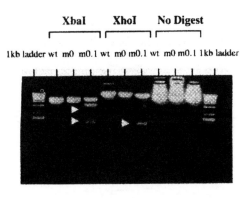

Figure 1. Ethidium Bromide stained 0.8% agarose gel. Lanes 1 and 11 show the DNA size marker (1 kb ladder).Lanes 2,3 and 4 show the digestion of the wild type, "m0 (fos favouring)" and "m0.1 (ATF-2 favouring)" mutants with Xba I respectively. Lanes 5, 6 and 7 show the digestion of the wild type, "m0 (fos favouring)" and "m0.1 (ATF-2 favouring)" mutants with Xho I respectively. The arrow indicates the band which are observable in lanes 4 and 7 but not observable in lanes 3 and 6. This shows that the Xba I and Xho I diagnostic sites are the restriction sites which are present in the "m0.1" mutant but absent in the "m0" mutant as it is expected. Lanes 8,9 and 10 show the undigested wild type c-Jun "m0" and "m0.1".

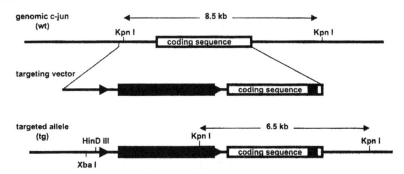

Figure 2. Gene targeting of c-Fos and ATF-2. Targeting of the "m0.1" and "m0" targeting vectors to their cognate locus. Diagram representing the targeting vector (upper lane)and its method of integration into the genomic c-Jun (middle lane) locus. The targeting vector is represented schematically. The bold line in the 5' of the targeting vector corresponds for the promoter. The black arrowheads correspond to the 34 bp lox sites and the black box between these two sites corresponds for the LTNL selection marker. The white box in the downstream of the promoter and the LTNL correspond for the coding sequence and the red colored square in this box corresponds for the sequence encoding the basic leucine zipper region where the point mutations are introduced. The digested DNAs of targeted alleles are hybridized via an external probe. The digestion og genomic c-Jun with Kpn I and hybridization with the external probe results in the generation of a large wild type fragment (middle lane). The homologous recombination between the enomic c-Jun and the targeting vector reduces the size of the mutated fragment obtained after the Kpn I digestion and hybridization by southern analysis (lower lane).

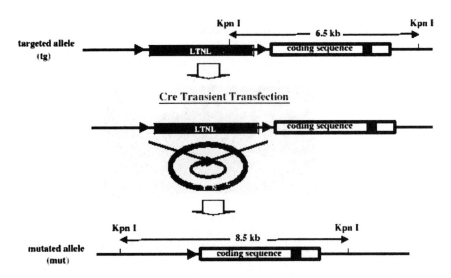

Figure 3. Removal of the LTNL marker cassette. Schematic representation of Cre transient transfection and the products of Cre transient transfection. Through the Cre transient transfection, the LTNL is recognized from the both lox sites and removed. The presence of the LTNL in the targeted allele between two lox sites is shown in the upper part of the figure. Cre transient transfection results in the removal of the LTNL from the promoter and the remaining of a single lox site as shown in the lower part of the figure. The removal of the LTNL leads to the increase of 6.5 kb fragment that can be obtained via Kpn I digestion by southern analysis to 8.5 kb fragment after LTNL removal.

Gancyclovir presence results in dead of cells containing the thymidine kinase casette. The clones which survived due to the removal of the thymidine kinase cassette are identified as positives to be used for genotyping. These clones are picked, their DNA is isolated and the occurrence of specific genomic integration is confirmed by southern blot analysis. Positive clones are injected into blastocytes to obtain transgenic mice. Figure 4 shows the chimeric animals generated in this study and derived from KPA Es cells.

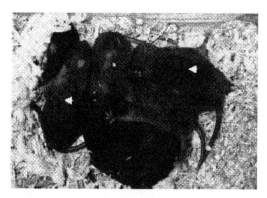

Figure 4. Chimaeric animals derived from ''m0.1'' KPA derived cells. The mice indicated with white arrows have lighter coat coloures, so are higher chimeras.

Stably transfected fibroblast cell cultures were obtained by retroviral transfer. The fibroblasts were derived from c-Jun null mice and transfected with expression constructs encoding the mutant Jun variants specific for dimerization with Fos or ATF-2. Transfection methods, RNA extraction, cDNA synthesis and reverse transcriptase PCR (RT-PCR) techniques are used to analyse the transcription of different target genes, e.g. Fra-1 and c-Jun.

The aim of the study was to generate transgenic mice predominantly carrying c-Jun with altered dimerization specificity. The targeting vectors were sequenced and tested in cell culture. The succesful insertinal recombination into one allele of different ES cells was proven by Southern blotting. The cells were injected into blastocytes and chimeric mice produced. These will be crossed to ensure germline transmission and to obtain homozygote animals carrying in both alleles the mutant sequence.

In the cell culture model the traget promoter specificity was proven by RT-PCR. The Fos-seaking mutant of Jun transcribed efficiently the Fra gene, but not the Jun-Neo gene. The ATF-2 preferring Jun mutantselected the promoter of c-Jun (which in the knock-out cells encoded a Jun-Neomycine-resistance fusion RNA). Transcription of the Fra gene by the Fos favouring Jun mutant could be enhanced by the addition of phorbol ester-which enhances the transcription of c-Fos and causes increased levels of Fos-Jun heterodimers. Transcription from the c-Jun promoter could be stimulated by irradiation with UVC.

3.   CONCLUSION

Refined Mouse Technologies, e.g the Cre mediated insertion and replacement of genes, are the method of choice for the dissection of complex phneotypes. In case of the Jun mutants, homozygote animals will clarify which function of Jun (and its dimer partner) is important at which stege of embryonic development or in the adult animal. The cell culture studies proved the target promoter specificities of the Jun mutants. These cell clones are now ready for use e.g. in an analysis of the interaction between AP-1 and the glucocorticoid receptor.

REFERENCES

1. Angel, P., and Karin, M., 1991, The role of Jun, Fos and the AP-1 complex in cell proliferation and transformation. *Biochim Biophys Acta* 1072 (2-3), 129-157 Review.
2. De Cesare, D., Vallone, D., Caracciolo, A., Sassone-Corse, P., Nerlov, C., and Verde, P., 1995, Heterodimerization of c-Jun with c-Jun with ATF-2 and c-fos is required for positive and negative regulation of the human urokinase enhancer. *Oncogene* 11(2), 365-376.
3. Hai, T.W., Liu, F., Coukos, W.J., and Green, M.R., 1989, Transcription factor ATF c-DNA clones: an extensive family of leucine zipper proteins able to selectively form DNA-binding heterodimers. *Genes Dev.* 3(12B), 2083-2090.
4. Curran, T. and Franza, B.R., Jr., 1988, Fos and Jun: the AP-1 connection. *Cell* 55(3), 395-397.

Health Informatics

NAZIFE BAYKAL
Middle East Technical University Informatics Institute, 06531 Ankara, TURKEY

1.   INTRODUCTION

In this chapter important aspects of a rapidly emerging discipline, Health Informatics, is presented. Medical Informatics and Bioinformatics have been the two main subdivisions of Health Informatics. In the first section, we will introduce "Medical Informatics", in the second "Bioinformatics", and in the third and final section, we will discuss a new discipline combining the both Medical and Bioinformatics, the Bio-Medical Informatics.

Today, we are witnesses of the rise of the "Information age" [1]. Advances in the computer and communications technologies facilitated the integration of different disciplines. Until recently, there was not a smooth transition between different disciplines and therefore the researchers performed their studies within a single discipline. Now the transitions have become much smoother between different disciplines.

Since health care environment has applications in a huge number of areas, there are several different disciplines involved in this domain. One of the most important disciplines involved is the Information Technologies. This need could be clearly understood considering the complexity level of the problems related to this domain and the increasing cost of health related issues. Each year, the amount of data collected and stored is doubled while the need for reusing, retrieving, acquiring, and processing the information to reach a critical decision escalates. Informatics is the application of information science, computer science, mathematics, and their associated technologies to a discipline or domain. Consequently, there is an immense

Biomaterials: *From Molecules to Engineered Tissues,* edited by
N. Hasırcı and V. Hasırcı, Kluwer Academic/Plenum Publishers, 2004   337

demand for a Medical Informatics discipline that offers a systematic application of methods of information processing for the use of computers in the medical domain.

Another important discipline is the Bioinformatics. It covers the fields of biology and informatics and is concerned with the development of techniques for the collection and manipulation of biological data using computational methods, theories, and computer-based techniques. It plays a key role in the acquisition, processing and analysis of individual genetic information as there are huge amounts of data to be handled.

Sometimes, medical informatics and bioinformatics are united under biomedical informatics roof. It is a field that studies research, development, or application of computational tools and approaches for expanding the use of biological and health data.

2.   HEALTH INFORMATICS

2.1   Overview

There are explosive advances in the areas of both the communications and medical technologies. Furthermore, there is a huge increase in the new knowledge created by the medical science. This causes the process of the data a very complex task[2]. However, information technologies enable us to process the data in a more efficient manner. All of the above reasons have led to the combination of the medical and informatics disciplines: Medical Informatics (MI).

Medical Informatics is a science of techniques, theories, and methods that form a basic research discipline. This discipline is a heterogeneous field, composed mainly of the information technology, medicine and, health care[3]. But there are many other disciplines that are also closely related to medical informatics such as biomedical engineering, electronics engineering, computer sciences, and social sciences. The detailed representation of the interaction of these disciplines can be seen on Figure 1.

Ultimately, medical informatics is the study of how we organize ourselves to create and run healthcare organizations. In other words, it is the study of how medical knowledge is created, shaped, shared, and applied.
There are several definitions of Medical Informatics in the literature. Three different definitions have been given in Figure 2.

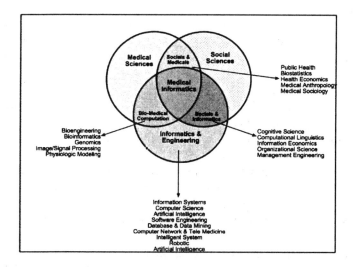

Figure 1. Interaction of scientific fields with medical informatics.

"Medical Informatics is a developing body of knowledge and set of techniques concerning the organization *and management of information in support of medical research, education, and patient care. Medical informatics combines medical science with several technologies and disciplines in the information and computer sciences and provides methodologies by which these can contribute to better use of medical knowledge and ultimately to better medical care."*
Proceedings of the Symposium on Medical Informatics Association of American Medical Colleges

"Medical Informatics comprises the theoretical and practical aspects of information processing and communication, based on knowledge and experience derived from processes in medicine and health care."
Van Bemmel J.H., The structure of medical informatics. Med Inform 1984; 9:175-80

"Medical Informatics is the rapidly developing scientific field that deals with the storage, retrieval, and optimal use of biomedical information, data, and knowledge for problem solving and decision making."
Shortliffe et al., Medical Informatics: Computer Applications in Healthcare, Addison Wesley, 1990.

Figure 2. Definitions of medical informatics.

There are explosive advances in the areas of both the communications and medical.

Development of improved methods of information navigation has been the driving force for the emergence Medical Informatics as a discipline. The first physician who implemented specific methods for information storage was Dr. John Shaw Billings, the Surgeon General of the Army and founding editor of the Index Medicus. In 1890, Dr. Billings described an electromechanical device that would tabulate the census automatically through the use of punch cards. Herman Hollerith, a government statistician, materialised this concept. Fifty-six Hollerith machines were used to process census information for 62 million people in 1890.

The first use of computers in the medical environment was achieved by Robert Ledly in the mid 1950s for dental projects. Another important milestone is the development of the Health Evaluation Through Logical Processing (HELP) system, in the early 1960s, a clinical decision support process.

In the late 1960s, the first hospital information system was implemented at London Hospital. Medical Informatics discipline was formally defined in 1974 in "Education in Informatics of Health Personnel". In 1975, Shortliffe and Buchanan developed an expert system called MYCIN[4].

In the early 1980s one of the most important expert systems, Internist-1 was developed. In 1994, the Good Electronic Health Record (GEHR) project, which was a European Union project, was completed. In 2001 the first transatlantic surgical operation was performed between France and New York.

The 1990s and 2000s can be described as a period where, artificial neural networks, fuzzy logic, genetic algorithms, data mining techniques and various telemedicine applications have been implemented in Medical Informatics.

2.2   Medical Informatics Applications Areas

Medical Informatics has an extensive application area, influencing many medical activities. These application areas are summarized in Table 1.

2.3   Core Topics in Medical Informatics

Today, medical informatics is in a rapidly growing and extending phase. In the near future, the applications will be implemented in almost all of the health environment. The most important topics of research are the electronic health records, standardization in medical informatics, security considerations, virtual reality, and telemedicine

Table 1. Application areas of medical informatics

CLINICAL DATABASES AND STORAGE	• Data mining • Development of data warehouses
ELECTRONIC PATIENT RECORDS	• The storage of patient records in electronic environment in order to track the patient information. • Development of large databases in order to keep track of diseases and research.
HEALTH INFORMATICS	• Hospital Information Systems (HIS), Clinical Information systems, PACS systems. • Intelligent electronic health cards.
Clinical Decision Support Systems (DSS)	• Expert systems • Artificial Intelligence applications
MEDICAL INFORMATICS STANDARDS	• HL7, SNOMED, ICD, DICOM
VIRTUAL REALITY APPLICATIONS	• Virtual reality applications for surgical operations. • Planning before operation
EDUCATION	• Distance education • Virtual reality applications

2.3.1   Electronic Health Records

The continuous increasing of the health care cost has forced the authorities to seek solutions for improvement of the healthcare quality and productivity. These efforts have led to the development of new health care models, restructuring of organizations and redesign of healthcare business. Electronic health record (EHR) systems (or electronic patient records, EPR) stand on the basis of this extensive reorganizing procedure. Electronic health records enable the continuity of the health care process by sharing the patient information via information networks[5].

Although EHRs are widely used today, there is still considerable use of the traditional paper based health records. The traditional paper-based patient records used in the clinical setting generally contain the notes from clinicians and other care providers. A comparative representation of a paper based medical record and an electronic patient report is depicted in Figure 3.1 and 3.2. Figure 3.1 represents a paper based medical report while the Figure 3.2 shows an electronic patient record.

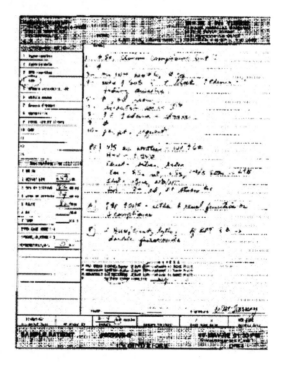

Figure 3.1. An example of a paper based medical record.

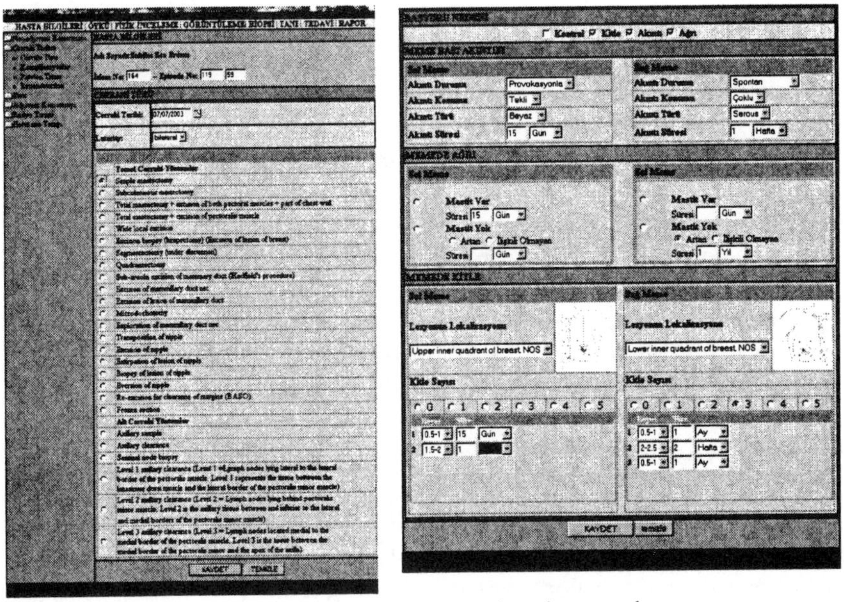

Figure 3.2. Examples of electronic patient records.

Many problems are encountered during the use of paper based medical records[6]. Some of them are:

- Paper files can only be in one location at a time and sometimes they cannot be found at all.
- Handwriting may be poor and illegible, data may be missing, or notes may be too ambiguous to allow proper interpretation.
- The record cannot actively draw the care provider's attention to abnormal laboratory values, contraindications for drugs, or allergies of patient.

The EHR is an electronic database of health care data about an individual. This database is a record of all health-related activities in which the individual has been engaged and the health phenomena experienced by the individual. Electronic Health Record systems form a basis for all other applications of Medical Informatics such as decision support systems, expert systems, and intelligent systems.

The information in the EHRs is collected from clinicians (recorded observations, summaries, health care plans), staff (nurses, social workers), raw data (test results) and patients. The required information fields of an EHR are demographic data, medical history, physical condition, laboratory results, treatment processes, prescriptions, progress information and medical images of the patient.

Recording the health data in an electronic environment has many advantages. Firstly, since the records are stored electronically, handwriting misinterpretations are avoided. Secondly, the data are stored more formally, are accessible in any time from anywhere, and can be used by more than one user. The shorter process time means saving time and money.

The main challenge of the individual use of EMR has been integration into the busy clinical workflow. A related challenge is determination of the optimal computing device for the clinical setting. At the organizational level, main challenges have been managing complex informatics applications and the computer networks upon which they run. Maintaining large networks and training a myriad of health care providers who use them are expensive. Another challenge is the protection of patient privacy and confidentiality.

2.3.2   Hospital Information Systems (HIS)

Hospitals have become overloaded with information and this fact increases the operating costs considerably. In order to improve health care management, information systems have become essential. Hospital Information Systems (HIS) have been introduced to solve these problems and they have a widespread usage all around the world.

The main function of a HIS is to support hospital activities on operational, tactical, and strategic levels. In other words, a HIS uses computers to collect, store, retrieve, and process data for all hospital affiliated activities. Here the term data refers both to the patient medical data such as diagnostic and treatment activities and the hospital administrative data such as admission, discharge, and information transfer.

In the 1990s, Clinical Information Systems (CIS) embedded in HIS began to emerge as an alternative to the HIS. They are patient focused systems and are generally used for administrative purposes.

Furthermore, Primary Care Information Systems constitute an important need for a better health care system. Primary care is the area of general health care where patients are treated before they are referred to a specialist or a hospital. In other words, the patient enters the health care system from the primary care unit. In order to be efficient, primary care information must be integrated with other information systems in health care area.

2.3.3   Standardization in Medical Informatics

It is impossible to process, interpret, and apply the health informatics data unless the information is adapted to Health Informatics Standards that are accepted by the majority of the community[7]. Health care units have a complex nature as they require high data processing and the same data may be used in different applications necessitating the exchange of data between different systems. Also using different programming languages and data formats by different systems in health care units make data exchange more expensive and difficult. Consequently, standards must be adapted and applied for efficient use of informatics in health care.

Health informatics standards can be studied under four major categories:

- Vocabulary standards
- Content and structure standards
- Messaging standards
- Security standards

Vocabulary standards define the medical terms and show how these terms will be used in health records. The aim of creating the vocabulary standards is to develop a common terminology for practitioners to represent medical events. Thus a medical situation will be represented by only one term. Using the same terminology raises the level of communication between the people working in the health sector and increases the efficiency of the usage of the health records.

The main objective of the content and structure standards have is to explicitly define the data elements in Electronic Health Records. This purpose includes determining important data elements (e.g. temperature, blood pressure), data type, field length, and content of the each data element[8]. The data provided must be optimised: not too high to complicate the practitioners' job, while it should be enough to meet the users' needs.

Information security has become a very important issue in today's electronic environment. It turns out to be a more sensitive subject when health information security is considered. Although there is an extensive consensus on developing the security standards we are still at an early stage of the whole process.

Health information has heterogeneous and complex data format characteristics. Moreover all the information is used either at the electronic environment or paper-based systems. We have to ensure that such information can be used, shared, stored and exchanged across all the systems that are operating in the health care sector[9]. Messaging standards enable the information sharing between two or more computers.

CEN, HL7, IEEE, DICOM and ISO are the leading organizations for the standardization of health care informatics. CEN (Comite Europeen de Normalization) is a federation of official national standards bodies of the twenty European countries. In 1990, the technical committee of CEN/TC 251 was set up for the standardization of health informatics as a part of CEN[10]. HL7[11] is an USA origin organization established in 1987. Its main focus is on messaging standards. Although HL7 has its headquarters in USA, it has affiliations in various countries. DICOM[12] is an institute actively working on the transmission of medical images since 1983. ISO is a worldwide organization, with participants from 147 countries that are the official national standards bodies of those countries. ISO/TC 215, established in 1998, is the technical committee of ISO working on the standardization of health informatics. Since 1998 an active collaboration between ISO/TC 215 and CEN/TC 251 has been established. The 'Institute of Electrical and Electronics Engineers' (IEEE)[13] has been developing standards in its area for more than 100 years. IEEE had worked coherently with CEN/TC 251 until the establishment of ISO/TC 215. Today ISO/TC 215 has undertaken the role of IEEE.

The main aim of the standardization of health informatics is to provide the interoperability and harmony between the individual systems that are operating in the health sector. Today, there are both global and individual efforts in this issue. Although a considerable progress has been achieved in the field of standardization of health informatics, the standards are not extensively used.

2.3.4   Security in Medical Informatics

"Security" is to protect certain assets against certain threats and attacks. It is deceptive to think that information systems in healthcare do not need sophisticated security countermeasures simply because there is no profit unlike banking or military. On the contrary, some real-world examples demonstrate just the opposite. For instance, in both United States and Britain there are people who earn their living by only violating medical record privacy of others[14].

Security protocols used in Medical Informatics are usually built on lower level cryptographic tools. There are mainly six tools are used for this purpose[14]:

- Symmetric encryption
- Public-key encryption
- Message authentication codes
- Random number generators
- One-way hash functions
- Digital signature schemes

Considerable effort has been put in for security issues in medical informatics, but there is still no satisfactory achievement yet.

2.3.5   Decision Support Systems

It is a must to improve the efficiency of the decision making process for a better working health care system. This can be realized by systems that use information systems technologies. It is "Decision Support Systems" (DSS) that imitate human decision-making process to facilitate the diagnostic process.

The clinical use of DSSs for the clinical domain, "Clinical Decision Support Systems" (CDSS) began in the 1970s. Shortliffe has defined the clinical decision support system as: "A clinical decision support system is any computer program designed to help health care professionals make clinical decisions"[15]. Clinical decision support systems were first popularized in 1970s. Shortliffe developed "MYCIN"[4] which is a consultation system that helps the physician to select the appropriate antibiotic. The researchers were attracted to the field of artificial intelligence for the development of MYCIN. Another important CDSS is the HELP system. It was an integrated hospital information system developed at LDS Hospital in Salt Lake City. HELP demonstrated the integrative function of a decision support system with different systems.

Major aims of decision support systems are to assist the health care professionals, improve patient care, and reduce costs. Furthermore they manage clinical complexity and administrative complexity and also support clinical research.

Decision support systems merge the data with the scientific information in order to achieve this task. The fundamental components of a decision support system are:

- A comprehensive knowledge base containing scientific information.
- An electronic patient record system to reach the patient data effectively.
- An inference engine that implements decision rules to make a decision.
- User interfaces that enable the easy use of the whole system.

Today the decision support systems are used for a wide variety of applications.

2.3.6   Virtual Reality and Telemedicine

Virtual reality refers to a collection of methods that represent a computer-generated world in three-dimensional perspective, where users can interact within this world[3].

One of the most important differences between the virtual reality applications in medical environment from other application domains is that the virtual world often functions as a real world in health care. Some medical areas where virtual reality is used to improve health care are:

- Surgery
- Medical education, modeling, and training
- Psychiatry
- Architectural planning of medical facilities
- Telemedicine and telesurgery
- Anatomical imaging

The virtual reality technology is still in a growing phase. There are obstacles such as the insufficient speed offered by the current hardware technology. However, there is no doubt that they will be widely implemented in the near future.

Another core topic in medical informatics is the "Telemedicine" applications. Telemedicine can be defined as the use of electronic information and telecommunication technologies to enable the exchange of medical information and services across distances.

Telemedicine is a valuable tool that improves the effective use of resources and allows experts to reach patients in rural areas. It also provides cost-effectiveness since it eliminates the travel and accommodation expenditures and prevents the replication of resources.

By telemedicine applications, a patient can be treated by a physician at a distance. Similarly physicians can make consultations although they are separated by thousands of kilometers. Patient records, audios, videos, and images can be transferred in telemedicine applications. Also patient interviews, consultations, surgeries across distances, educational and administrative services are now available using videoconferencing equipment.

Today telemedicine is in a state of research and development. Although there are many implementations on several different systems, it is premature to claim that telemedicine technologies are extensively used in the world. On the other hand, providing health care by telemedicine is getting popular. There are applications in almost all fields of medicine such as mental health, home care, cardiology, radiology, surgery, medical education, patient education, neurology and dermatology. First transatlantic telesurgery that has been performed in September 2001 is one of the milestones of telemedicine. Surgeons located in US performed a successful laparoscopic operation on a patient in France with the aid of the robotic technology.

Despite the considerable efforts and progress, there are obstacles to achieve the global implementation of the telemedicine. Most countries do not allow physicians to practice unless they are licensed in their country. Absence of laws and standards create problems. Also technological barriers and the lack of telecommunication infrastructure in rural areas slow down the improvements in telemedicine. Internet, as a worldwide network, is an everyday target for attacks. It is evident that security and reliability are important issues that need to be resolved.

3.   BIOINFORMATICS

3.1   Introduction

In a cell, 3 billion ladder steps of supercoiled deoxyribonucleic acid (DNA) are packed and located at the core or the nucleus. These packages are known as chromosomes. Chromosomes contain genes which encode certain proteins.

Protein synthesis starts with the synthesis of ribonucleic acid (RNA) from DNA. This process is called the transcription. In this process, the code of the proteins in DNA is transferred to the cytoplasm of the cells from the nucleus. In the cytoplasm, amino acids are recruited one by one according to the code in the RNA. This process is called the translation, meaning the synthesis of proteins via RNA. Amino acids are the building blocks of proteins and proteins are the blueprints of individuals as they are responsible for catalysation of chemical reactions, formation of structures, and signal transmission.

DNA *is the genetic material responsible for inheritance as the the information encoded in DNA is passed on to other cells during division.*

The flow if information: "DNA $\rightarrow$ RNA $\rightarrow$ protein"— is called the Central Dogma (Figure 4). Because of Central Dogma principle errors in DNA are transferred to proteins leading to death (mortality) or disease (morbidity). These type of diseases are called the genetic diseases.

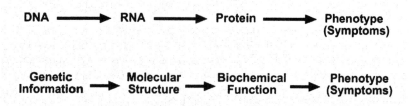

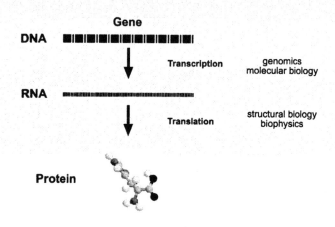

Figure 4. The flow of genetic information: Central dogma

3.2   General Concepts

Since 1960s computers have been used to handle huge amounts of data in biological research. The analysis of protein and nucleic acid sequences depends heavily on the computational power. The need for the evaluation and processing of vast amounts of data led to the development of a new discipline called the bioinformatics, which has been established as an academic discipline for the last fifteen years. Its roots lie in the research areas such as biological sequence analysis, structural biology, and molecular information repositories.

Below two definitions of bioinformatics are given.

Bioinformatics is a field of biology concerned with the development of techniques for the collection and manipulation of biological data, and the use of such data to make biological discoveries or predictions. This field encompasses all computational methods and theories applicable to molecular biology and areas of computer-based techniques for solving biological problems including manipulation of models and datasets (MeSH definition).

Research, development, or application of computational tools and approaches for expanding the use of biological, medical, behavioral or health data, including those who acquired, store, organize, archive, analyze, or visualize such data (National Institutes of Health, 2000/7/17 Working Definition for Bioinformatics).

Genomics, proteomics, and pharmocogenomics are very important study areas in bioinformatics.

Genomics is the systematic study of the complete DNA sequences (GENOME) of organisms (MeSH definition). Structural and functional genomics are two main subdivisions of genomics.

Functional genomics is the development and application of global (genome-wide or system-wide) experimental approaches to assess gene function by making use of the information and reagents provided by structural genomics. Gene expression analysis, gene annotation, genetic and biochemical pathway construction and analysis are some of the tools for the studies.

Structural genomics groups proteins into different clusters of similar structures based on their sequences. Then, based on the known structure of at least one protein in a cluster and using a computational technique called homology modeling, an educated guess can be made about the shapes of other proteins in the family. Protein structure determination, protein structure modeling and prediction, computational proteomics (protein sequencing via mass spectrometry) are the basic methodologies of structural genomics.

Currently genomics research favors functional genomics which has great potential to make immediate impact on clinical medicine based on new molecules and diagnostic procedures. Based on the results of research on genomics, new molecules and diagnostic procedures will reach the market in the near future. In this promising field, both industry and academia have to compete in a market characterised by a shortage of bioinformatics professionals.

Proteomics can be defined as the qualitative and quantitative comparison of proteomes under different conditions to further unravel biological processes. The dream of having human genomes completely sequenced has been realized. However, the understanding of probably half a million human proteins encoded by some 35000 genes is still a long way away and the hard work to unravel the complexity of biological systems is yet to come. A new fundamental concept called proteome (PROTEin complement to a genOME) has recently emerged that should drastically help phenomics to unravel biochemical and physiological mechanisms of complex multivariate diseases at the functional molecular level. This new discipline, proteomics, has been initiated that complements physical genomic research.

Pharmacogenetics is the study of genetic causes of individual variations in drug response.

Pharmacogenomics broadly involves genome-wide analysis of the genetic determinants of drug efficacy and toxicity. Primary candidate genes of interest include those encoding for drug receptors, metabolizing enzymes, and transporters. However, selection of optimal drug therapy may also involve disease susceptibility genes indirectly affecting drug response. Moreover, pharmacogenomics includes the identification of suitable targets for drug discovery and development.

Bioinformatics is a rapidly developing field that brings the world of genomics and proteomics closer to the clinical desktop. The work of bioinformatics has contributed substantially to the rapid mapping and current understanding of the human genome[16]. The Human Genome Project was an ambitious study aimed to identify the sequence analysis of tens of thousands of genes in humans. Advances in computing power and bioinformatics resulted in the completion of Human Genome Project (HGP) several years ahead of the schedule. Human Genome Project opens up possibilities for the collaboration of Bioinformatics with Medical Informatics that provide new insights and create a synergy for challenges needed to create new genomic applications in medicine.

Development of web-based databases have greatly facilitated the exchange of data among remote researchers fostering collaboration and information integration.

The technology of DNA sequencing has become less of an immediate concern for the typical practising biologist. Currently, simulations and modeling of cellular and subcellular messaging, which has the promise of improving our understanding of intracellular signaling, has become a popular research area in bioinformatics. We are most probably at the beginning of the "post-genomic era" in which the questions about how we acquire sequence data will be more strategic and political than scientific.

3.3   Research Issues in Bioinformatics

Bioinformatics builds mathematical models of the evolution, protein folding, protein function processes to infer relationships between components of biological systems. Some important research areas are given below.

- Structural and functional genomics
  - Gene prediction
  - Sequence analysis

- Protein structure
- Gene/Protein function prediction
- Microarray and Proteomics
    - Image analysis
    - Data mining of gene expression profile
- Biomedical databases
    - Efficient uses of these databases (fast/accurate search, comprehensive annotation,…etc)
    - Integration across different databases of genes, proteins, pathways, etc.
- Literature data mining
    - Mainly against MEDLINE for gene-protein-disease relationship

3.3.1 Clinical Applications

- Diagnostic tools
    - Gene chips, protein chips
    - Early detection of infectious agents, congenital disorders, enzyme activity, susceptibility to disease and eventually idiosyncracy.
- Drug Discovery/ Development
    - Gene expression profile to predict drug functions and side effects
    - Protein structure and candidate molecules for potential drugs

4.   BIOMEDICAL INFORMATICS

Bio-Medical Informatics is a new discipline which combines the medical informatics and bioinformatics in order to combine and process the complex data obtained from the clinical setting and the molecular biology research.

The mission of biomedical informatics is to provide the technical and scientific infrastructure and knowledge to allow evidence-based, individualized healthcare using all relevant sources of information. These sources include the classical information as currently maintained in the health record, as well as new tissue and molecule-based information. Interactions between medical informatics and bioinformatics could have a great influence on future health research and as well as the continuity and individualization of health care. This section addresses current applications and core themes that motivate this new synergy: Biomedical Informatics.

4.1   General Concepts

Basically medical informatics is the application of informatics to clinical sciences, whereas the bioinformatics is the application of informatics to basic medical or biological research. Classical epidemiology and clinical research on the one hand, and genomic research on the other, alone are no longer enough to achieve significant advances in genomic medicine, and a new integrative approach is required. The integration of all the data and information generated at all levels requires synergy of bioinformatics and medical informatics[17]. Currently, there are a number of initiatives that combine elements of the two areas up to the point of an integrated approach on databases, standards, analysis, applications and education (Figure 5).

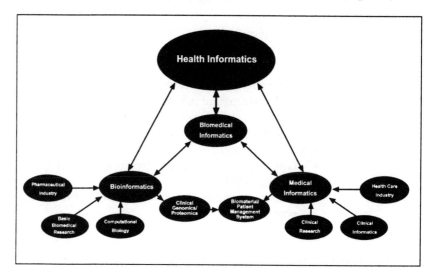

Figure 5. Medical Informatics, Bioinformatics and Biomedical Informatics

Biomedical informatics can be defined as the application of the science of informatics to the domains of biology and health and provides the framework for developing and sharing new biomedical knowledge. This new studies biomedical information and knowledge: their structure, acquisition, integration, management, and optimal use. It requires a multidisciplinary research in, application development for, and administrative approaches to all aspects of health care delivery, biomedicine, and public health. Biomedical informatics adopts, applies, evaluates, modifies, and expands results from a variety of disciplines including information science, computer science, library science, cognitive science, business management and organization, statistics and biometrics, mathematics, artificial intelligence, operations research, economics, and of course, basic and clinical health sciences.

The opportunities for rapid technology transfer between the medical informatics and bioinformatics is one of the most compelling reasons for housing these two subdisciplines under the same roof (Figure 6). There are still barriers between the two fields, so there are also opportunities for accelerating progress in both fields by knowing the problems and literature in each. For example, probabilistic methods in clinical informatics applied to biological structure, knowledge representation methodologies from clinical informatics have inspired efforts and contributions in bioinformatics and the boundary between medical informatics and bioinformatics is blurring[18].

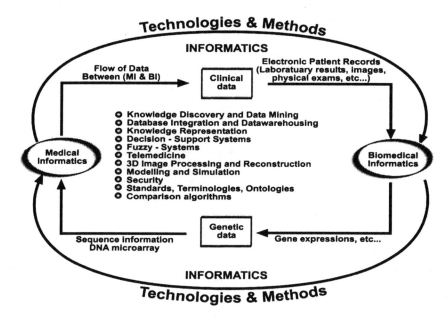

Figure 6. Technology transfer between medical informatics and bioinformatics

One of the ultimate goals of both bioinformatics and medical informatics is to have robust computational models of physiology that will enable us to model, store, retrieve, and analyze the effects of disease, medications, and the environment[18].

Models of cellular behavior and patterns of information exchange between molecules and within functional biological systems will be essential for predicting function and dysfunction. Molecular modeling of disease must include a detailed understanding of the systems theory and knowledge management in the subcellular environment[16].

As the biological world enters the post-genome age, the interplay between basic biological data (sequences, structures, pathways, and genetic networks) and clinical information systems is of utmost importance. Genes and structures are useful only in the context of the functions and phenotypes that they produce, and so we look toward continuing interaction and, perhaps, unification of the two fields[18].

The complete sequencing of the human genome has been influential in beginning of the post-genomic era. There are new perspectives for the study of complex multigenic diseases, which are more common than monogenic diseases. Our belief that to fully grasp the mechanisms of disease we need to understand its proteomics base and the genetic failure behind it, and to integrate the knowledge generated in the laboratory to clinical settings intensifies as our knowledge about the human genome increases. The new genetic and proteomic data has brought forth the possibility of developing new targets and therapies based on new findings, of implementing newly developed preventive measures, and also of discovering new research approaches to old problems[17]. To accomplish the work it is important that we are able to deal with the large amount of data generated in the laboratory by functional genomics and proteomics and that we integrate this data into electronic health records. We need to merge bioinformatics with the tools and techniques of clinical informatics such as knowledge representation, data mining, automated diagnosis, and information retrieval, with clinical information (e.g. electronic health records, clinical decision systems, image- and signal-processing). Only then we can correlate essentially genotypic information with expressed phenotypic information. Biomedical informatics is the emerging technology that aims to put these two worlds together in order to participate in the discovery and creation of novel diagnostic and therapeutic methods.

New discoveries can increase the success rate against some diseases. Most diseases are multifactorial, a genetic problem being only partially responsible for the resulting disease. Patients' lifestyle usually have a significant influence on the development of certain diseases. We need to know not only the genotype, which polymorphisms and haplotype apply to which disease but also the phenotype. It is here where both classical clinical and epidemiological research come into place[17].

To fully enhance our understanding of disease processes, to develop better therapies to fight and cure diseases, and to develop strategies to prevent them, there is a need for the synergy of the disciplines involved, which are medicine, biology, informatics, medical informatics, bioinformatics, biochemistry, pharmacology, epidemiology[17]. This synergy gives rise to biomedical informatics that has an integrated approach and its subdivisions such as genomic medicine.

A combination of the expertise of medical informatics in developing clinical applications and the focused principles that have guided bioinformatics could create a synergy between the two areas of application. Such interaction could have a great influence on future health research and the ultimate goal, namely continuity and individualization of health care[19].

The term biomedical informatics is increasingly used— as in the title of the present and many new academic programs[19]. Both European and United States organizations have tried to analyze separately the synergy between Two principal factors will drive the collaboration. First, the results of research in molecular biology will increasingly move toward clinical research and clinical practice providing a natural, shared issue: the application in daily practice. Second, the methodologies used by bioinformatics and medical informatics will prove to have many similarities allowing exchange of experience between the fields of medical informatics and bioinformatics (see Figure 6).

4. 2   Physiome Project

Physiome Project, conducted jointly by Physiome Sciences and the University of Auckland in New Zealand aims to understand and describe the physiology and pathophysiology of the human organism [20]. Insights and data obtained from this project are intended to be used for improvement of the human health. Simulation of complex models of living structures from cells to organs will provide quantitative functional descriptions of human and other biological systems. Scientists, medical professionals, teachers and industry throughout the world will have the opportunity to implement obtained information in practice. In order to overcome the high and complex computing demand IBM's new eServer* POWER4-based supercomputers will be used to simulate complex living models as well as commonly encountered diseases such as diabetes, hypertension and asthma.

REFERENCES

1. Satava R. M., 2002, The Bio-Intelligence Age: Surgery After the Information Age, Ann. Meeting of *The Society for Surgery of the Alimentary Tract*, San Francisco, California.
2. Hovenga E., Kidd M., and Cesnik B. 1998, *Health Informatics: An Overview*, Churchill Livingstone, Australia.
3. Bemmel J. H., and Musen M. A., 1997, Handbook of Medical Informatics, Springer, NY.
4. Shortliffe E., 1976, Computer–Based Medical Consultations: MYCIN. New York Elsevier/North Holland.

5. Lukoff Y. J., 2002, Standards for Computerized Clinical Data: Current Efforts and Future Promise, *The Permanente Journal*, 6: 71-75.
6. Shortliffe E. H., and Perreault L. E. (Eds) 2001, *Medical Informatics: Computer Applications in Health Care and Biomedicine*, 2nd Ed. Springer, New-York.
7. De Moor J. E. 1993, Standardization in Health Care Informatics and Telematics in Europe: CENT TC 251 Activities, *Progr. in Standardization in Health Care Informatics*, 6, 1-14.
8. National Electronic Health Records Taskforce, 2000, Health Information Standards. Report to Health Ministers by National Electronic Health Records Taskforce, App D page 5.
9. Takeda H., Matsumura Y., Kuwata S., Nakano H., Sakamoto N., and Yamamoto R., 2000, *Architecture for Networked Electronic Patient record Systems*, Inter. Journal of Medical Informatics, 60, 161-167.
10. CEN/TC 251, 2003, http://www.centc251.org/
11. HL7, 2003, http://www.hl7.org/
12. DICOM, 2003, http://www.medical.nema.org/dicom.html
13. IEEE, 2003, http://www.ieee.org/.
14. Schneier B., 2000, *Secrets and Lies,* Wiley Computer Publishing,
15. Shortliffe E., Buchanan B., Knowledge Engineering for Medical Decision Making: A review of computer-based decision aids, *Proc.of the Institute of Electrical and Electronic Engineers*, 67, 1207-1224.
16. Elkin P.L, 2003, Primer on Medical Genomics, Part V: Bioinformatics, *Mayo Clin Proc.*; 78:57-64.
17. *"Synergy between Medical Informatics and Bioinformatics: Facilitating Genomic Medicine for Future Healthcare"* April 29th, 2003, EC-IST 2001-35024, BIOINFOMED Study *"Prospective Analysis of the Relationships and Synergy Between Medical Informatics and Bioinformatics"*.
18. Altman R.B., 2000, The Interactions Between Clinical Informatics and Bioinformatics: A Case Study, *Journal of the American Medical Informatics Association*, 7, 5
19. Maojo V., Iakovidis I., Martin-Sanchez F, Crespo J, and Kulikowski C, 2001, Brief Report: Medical Informatics and Bioinformatics: European Efforts to Facilitate Synergy, *Journal of Biomedical Informatics*, 34, 423–427 .
20. http://www.physiome.org/

Ethics And Animal Use In Biomedical Research

İSMAİL HALUK GÖKÇORA
Department of Pediatric Surgery, Ankara University, School of Medicine, Cebeci Hastanesi, Dikimevi, 06100 Ankara, Turkey

1.   INTRODUCTION

The recognition for humane and responsible biomedical research by scientists, professional societies, scientific organizations and governmental agencies have enabled a better understanding of the resources, support, ethics and aims in this area. Animal experiments remain as the foundation for biomedical research and are the best hope in reaching a solution to the problems that face both the animals themselves, and human beings. Finding the cause of a certain disease, its treatment and prevention require animal subjects either as a unique cell or the organism in its complete form. The aim is to find a cure for pain, disability, and functional disorder of a system or perhaps death. Denial of better quality and longer health for mankind and animals in the future would add up to the present problems, should we abandon research making use of laboratory animals. Public and scientific awareness must be encouraged by practical and effective information dissemination. The ethics, rules and regulations that involve animal research to obtain medical benefits and improvement to our lives should be available and enforced thoroughly both in research and in education. Every effort and precaution taken to avoid experimental animal suffering and inhumane practice or mistreatment is a must [1-16].

For animals, as well as for people, preventive medicine is the least expensive, yet the most effective way in reaching a huge population. It is through the techniques and products of biomedical research where preventive medicine has greatly benefited both animals and humans with

Biomaterials: *From Molecules to Engineered Tissues,* edited by
N. Hasırcı and V. Hasırci, Kluwer Academic/Plenum Publishers, 2004   359

proper nourishment, sanitation, vaccination, and breeding methods. Medications and measures to prevent and treat infestations or infections such as parasites (hookworms, tapeworms, etc.) and microbes (cholera, rabies, hepatitis, smallpox, diphtheria, tetanus, whooping cough, etc.) developed through research have saved and are still saving millions. Striking similarities between human and animal physiologic systems place animals as very valuable and irreplaceable insights into the human system. Thus, animal models have been critical in understanding the basic biology of micro-organisms, and recently in determining how the immune system and genomics and proteomics work. Animals have played a vital role in detecting desirable and unwanted features of newly developed vaccines. Strict regulations prevent vaccine from being used directly in humans until it passes tests for safety evaluations in research animals. Furthermore immunomodulation (control of the immune system) in humans is possible through determination of antigeneity, route of infection, the dose of vaccine required for optimal response, all of which are derived from animal models [15-21].

Animal experiments constitute the foundation of biomedical research. Finding a cause of a certain disease, its prevention and treatment require animal subjects either as a unique cell or as the organism in its complete form. The most prestigious scientific award since 1901, the Nobel Prize, recognizes the outstanding achievements such as chemistry, medicine; all of which have used animals in research. The Guinea pig, hamster, rat, mouse, snake, reptiles, frog, toad, amphibians, chick, chicken, sea slug, crab, squid, fish, fruit-fly, bee, pigeon, birds, worms, rabbit, dog, cat, pig, sheep, cow, horse, chimpanzee, monkey, various animal cells are to name a few. Hence a layout of some biomedical research winning Nobel prizes awarded since 1901 and examples of animals used in the corresponding research is presented in the following table (Table 1) by the Foundation for Biomedical Research, Washington, D.C., USA.

Analgesia for laboratory research animals is the usual practice in surgical or traumatic experiments, whereas euthanasia is mostly used when a thorough examination of the animal's tissues are required. The purpose is to reduce or abolish pain or anxiety related suffering. Of utmost importance is not to affect the results of the experiment, by using such analgesic agents. The animal in pain will not feed or drink, but will lose weight. Furthermore, it will position itself in the cage in a strange manner and be unresponsive to external stimuli. Euthanasia is administered either by chemical means as overdose (lethal) analgesics or by physical means such as decapitation, cervical dislocation, liquid nitrogen freezing, microwave heat or excessive bleeding. Researchers, too, must take into consideration the dangers to themselves of various chemicals or physical agents used in such experiments. Discarding the corpses of experimental animals must also follow well-established rules; for example: either being cremated or buried in limed ditches[4].

Table 1. Examples of animals used in biomedical research winning Nobel prizes since 1901[18]

Year	Scientist/s	Animal/s	Contributions made to biomedical science
1901	von Behring	Guinea pig	development of diphtheria antiserum
1902	Ross	Pigeon	understanding of malaria cycle
1903	Pavlov	Dog	animal responses to various stimuli
1905	Koch	Cow, Sheep	studies of pathogenesis of tuberculosis
1906	Golgi, Cajal	Dog, Horse	characterization of central nervous system
1907	Laveran	Bird	protozoa as a cause of disease
...			
...			
1996	Doherty, Zinkernagel	Mouse	immune system detection of virus-infected cells
1997	Prusiner	Hamster, Mouse	discovery and characterization of prions
1998	Furchgott, Ignarro, Murad	Rabbit	NO, signalling in cardiovascular system
1999	Blobel	Various animal cells	proteins, intrinsic signalling for transport
2000	Carlsson, Greengard, Kandel	Mice, Guinea pigs, Sea slugs	signal transduction in the nervous system
2002	Brenner, Horvitz, Sulston	Roundworm	genetic regulation of organ development and programmed death.
2003	MacKinnon, Agre	Bacteria	discovery of potassium and water channels in the cell membrane

Rats and mice that are specially bred for laboratory account for 95% of the animals used in research. Most of the remaining animals are rabbits, dogs, cats, Guinea pigs, hamsters, piglets, sheep, horse, cattle, fish and insects. Non-human primates, dogs and cats only account for less than half a per cent (0.5%) of the animals used in biomedical research[8, 9,20]. Since the subject is very important, many publications concerning the ethical use of laboratory animals in biomedical practice have been propagated during the last few decades. Some examples of textbooks published for animal experimentations are given in Table 2.

Table 2. Examples of textbooks concerning use and ethics on animal use in biomedicine

The Biology of The Laboratory Rabbit.	Eds: Weisbroth S., Flatt R., Kraus A., Academic Press, 1974, ISBN: 0-12-742150-5
The Laboratory Rat: Biology and Disease: Research Application, Vol 1-2	Eds: Baker H., Lindsey R., Weisbroth S., Academic Pres, 1979, ISBN: 0-12-074901-7 and 0-12-074902-5
Formulary For Laboratory Animals.	Eds: Hawk C.T., Leary S., Iowa State University Press, 1999, ISBN: 0-8138-2469-9

Humans depend on animals for energy, food, clothing and research, yet there should be more interest for the responsible use of animals in biomedical research. Hence, C. Everet Koop (Paediatric Surgeon, former US Surgeon General) has said: "Virtually every major medical advance for both humans and animals have been achieved through biomedical research by using animal models to study and find a cure for disease, and through animal testing to prove the safety and efficacy of a new treatment. Without the use of animals in research, continuation of medical milestones will be stiffled"[12]. The rationale underlying the use of animals in biomedical research is that a living organism provides an interactive, dynamic system that can be observed and manipulated experimentally in order to investigate mechanisms of normal function and of disease[14]. While the use of animal models is a common component, its role and its scientific and ethical justification are frequently not addressed in the education and training of junior researchers.

2. HOW ANIMAL RESEARCH BENEFIT ANIMALS

Vaccines for feline leukaemia, rabies, distemper and parvovirus are available only because of animal research. Treatments for heartworm, certain cancers, and hip dysplasia have also been developed using animals. Endangered species of animals have been preserved and increased in number with the knowledge and technique developed through animal research. Although benefits of biomedical research are usually described in terms of life and death for humans and animals, there are economic benefits as well: the government and citizens save in medical costs and lost wages for money spent on biomedical research. Every American dollar spent in applied research and clinical trials saves 8 dollars in the treatment costs (National Health Institutes, USA) [12].

Should difficult ethical judgments arise in using a certain animal for a specific research, these are to be faced with a proper logical and scientific approach if other animals and humans are to benefit from healthcare advances resulting from that experiment[5, 17].

Having served as the secretary for the Research Granting Committee in Health Sciences at The Scientific and Technical Research Council of Turkey (TUBITAK) for over six years (1994-2001), the author had some experience in many confrontations and disagreements among the scientists, institutions, hospitals, the universities and the international organizations involved in such biomedical research. World Health Organization (WHO) and United Nations Educational, Scientific and Cultural Organization (UNESCO) and other organizational bodies, such as research bodies and institutions, universities and medical research foundations including the European Medical Research Council (EMRC, a standing committee under the European Scientific Foundation (ESF) in Europe), National Institutes of Health (NIH, USA), and the Royal Society (UK), have all agreed to the importance of the results and the benefits we have all obtained from antibiotics and insulin to blood transfusions, solid organ transplantation and treatments of cancer or HIV, oral contraceptives and in-vitro fertilization to hormone replacement therapy. In short, practically every medical achievement that has been gained depends directly or indirectly on animal use in biomedical research[15].

Biological weapons and security in laboratories need to be also addressed, since bacteria, viruses, prions, or microbial toxins can be misused or perhaps intentionally used against humans and animals. Hence, bio-terrorist activities are under close scrutiny, especially in the past few years. Guarding against such activity has intensified interest into and protection from smallpox, anthrax, plague, botulism, tularaemia, hemorrhagic fever viruses, and most recently SARS (Serious Adult Respiratory Syndrome, originating in China and South-East Asia). A threat to human and animal life, these stockpiles of lethal disease of mass destruction require a thorough understanding and control of such sites. Identification, preventive measures and developing vaccines against such attacks can only be achieved through scientific knowledge obtained with animal research[16, 21].

Women and nature having been exploited by the patriarchal system, have been taken into consideration by most philosophers following the recognition of environmental problems during the 1970s. A strong relationship between the oppression of females and degradation and destruction of nature exist, where in turn a feministic perspective can be helpful in fighting against or solving such aggression. Cultural and social *ecofeminism* has gained much stride towards a world that we all accept as our home, peacefully living with the nature surrounding us (*Francaise d'Eaubonne)*[22]. One of the major revolutions of the history of civilization was realized by women, with the domestication of plants during the Neolithic age enabling mankind to switch from a parasitic to a productive economy (*Mother nature*). Yet, with the formation of towns and political organizations leading to a "man cult", which in turn both animals and women became slaves of this patriarchal system[23].

Integration to the world involves commitment and adaptation to developments in knowledge, productivity, equality, freedom of belief and speech, just and judicial government, democracy, clarity, ability to justify and ethical values[17]. A brief history of rules and tendencies for animal experimentation is given in Table 3.

Table 3. History of rules and tendencies for animal experimentation

Laws of Khammurabi (Sumerians, 1750 BC)	point out the rights for individuals. But is well kept in the area of "eye for an eye" understanding
The Oath of Hippocrates (470 BC)	has led from euthanasia patients' rights. Thus, the saying, "Trust in the doctor to honouring of the patient and his/her rights."
St. Francesco d'Assisi	has further pointed out that, animals are considered as sibling creatures and that their welfare are left to humans
James Watt	as the father of the steam-engine has indirectly saved animals from labour
P.J.M. Flourens (Physiologist, France,1840)	has used anaesthesia for animal experiments
John Hunter	is considered the father of experimental medicine and biology, transplantation
Alexis Carrel	has further added to animal experimentation by enabling vascular anastomosis
Arthur Schopenhauer (1977)	has summarized the humanistic approach to animals

3.   QUERIES CONCERNING ANIMAL RESEARCH

Queries to be answered before embarking on animal experimentation are as follows:

- Are animals a source for spare parts for human beings?
- Is human health and long-life worth the effort spent for biomedical research applied on animals in spite of the suffering?

While one needs to be concerned about the above questions there are facts to be considered for ethical evaluation in animal experimentation. These are presented in Table 6.

Table 6. Facts to be considered for ethical evaluation in animal experimentation [1-3,5,8-14,18]

Computer technology has enabled chemical, biological, plant and animal models for experimentation.
As biomedical experiments benefited both man and the beast there has been more trust on animal research.
Benefit for the human society has become more and more apparent.
Hence, benefit for the animals / plants / other living creatures
Frankness and open-minded research developed into knowledge based research
Awareness of pain and death has become a worldwide trend.
And what we as humans owe to animals has become a worthy ethical topic.

What do we expect of a world where humans have no or little respect for others? Humans affect the balance of nature, cause global inequality and global warming. Unethical biomedical research during the second world war such as experimentation on Jews by the Nazi rule and the 1970 Tuskegee syphylis research in the USA led to the 1947 *Nuremberg bylaws* and 1964 *Helsinki Declaration*[1,6,21-23].

Views Concerning Animal Use in Biomedical Research[24]

"The advancement of biological knowledge and the development of improved means for the protection of the health and the well-being both of man and of animals require recourse to experimentation on intact live animals on a wide variety of species"

1982 CIOMS (Council for International Organizations of Medical Sciences)

"Animals are essential for biomedical research. Animals offer the best hope of finding the cause, treatment, and prevention for many diseases that inflict pain, disability and death. The study of animals remains a necessary prelude to using human subjects. Our choice is clear: animals must continue to play a vital role in medical research. Future generations of humans and animals cannot be denied better health."
American Academy of Paediatrics

"Animal use in biomedical research is essential for continued medical progress."
41st World Medical Assembly, 1989, World Medical Association

---

The use of animals in biomedicine is a moral issue as well. People from religious centres and scientists as well have not yet resolved the unease of taking the life of animals in pursuit of scientific knowledge and better medical care. Broadly, all major religions believe that human life is more valuable than animal life, that humans have a God-given authority over animals, that humans may eat or use animals for work, that humans should not be cruel to animals and cause their pain or suffering, that humans should be kind to animals. And they all actively support the use of animals in biomedical research or are tolerant towards those who conduct such research with a high degree of concern for the welfare of the animals.

4.   THEOLOGICAL ASPECTS OF ANIMAL RESEARCH

Religious centres and scientists as well have not yet resolved the unease of taking the life of animals in pursuit of scientific knowledge and better medical care.

Broadly, all major religions believe that human life is more valuable than animal life, that humans have a God-given authority over animals, that humans may eat or use animals for work, that humans should not be cruel to animals and cause their pain or suffering, that humans should be kind to animals. And they all actively support the use of animals in biomedical research or are tolerant towards those who conduct such research with a high degree of concern for the welfare of the animals.

Animal Research From the Point of View of Religions

"A man is worth many sparrows, but not one sparrow can die unnoticed in God's World."

Matthew 10 vv.29-31, New Testament Church of England

"Provided they remain within reasonable limits medical and scientific experiments on animals are morally acceptable since they may help to save human lives or advance therapy."

Catholic Church

"A very good case can be made out for vivisection of animals provided safeguards are taken to reduce the pain to a minimum. Here benefits to medical progress are considerable and the price worth paying."

Judaism

"Some research on animals may yet be justified, given the traditions of Islam. Basic and applied research in the biological and social sciences, for example, will be allowed, if the laboratory animals are not caused pain or disfigured, and if human beings and other animals would benefit because of the research. "

Islam

" Actions shall be judged according to intention. Any kind of medical treatment of animals and experiments on them becomes ethical and legal or unethical and illegal according to the intention of the person who does it."

Islam

"A human life is distinguished from animal life due to its heavy responsibilities... "

Hinduism and Sikhism

"I undertake the rule of training not to do any harm to any living (breathing) thing."

Buddhism

---

5. ANIMAL RIGHTS AND BIOMEDICAL RESEARCH

To date an extra twenty years and nine months have been added to human life by conquests established through the use of animals in biomedical research. Yet a tragedy and a waste of resources have come into being by the combat of anti-vivisectionists who claim to protect animal rights. Since

Darwin's establishment of "the origin of species" (1859) two basic ideas confront each other for use of animals in biomedical research[7,12-14,17]:.

1. *anti vivisectionists:* animals have the same rights and privileges as humans and should never be used for research purposes.

2. *animal welfare advocates:* animal experimentation can be allowed under ethical principles.

Opponents point of view

"Even painless research is fascism, supremacist, because the act of confinement is traumatising in itself." "Six million Jews died in concentration camps, but six billion broiler chickens will die this year in slaughterhouses"

   Ingrid Newkirk (President, PeTA, People for Ethical Treatment of Animals)

"The life of an ant and that of my child should be granted equal consideration"
Michael W. Fax (Advisor to the Humane Society of United States)

"Arson, property destruction, burglary and theft are acceptable crimes when used for the animal cause"
Alex Pacheco (People for Ethical Treatment of Animals, PeTA)

"It would be great if all the fast-food outlets, slaughterhouses, these laboratories and the banks who fund them exploded tomorrow"
Bruce Friedrich

"In a war you have to take up arms and people will get killed, and I can support that kind of action by petrol bombing and bombs under cars, and probably at a later stage, the shooting of vivisectors on their doorsteps"
Tim Daley (Animal Liberation Front)

---

6.   PHILOSOPHICAL UNDERSTANDING OF ANIMAL RESEARCH

Cartesian dualism and modern concepts in philosophy and social sciences about man, medicine and ethics[27].
Rene Descartes> Cartesian dualism:
   Man is a natural being / Man is a rational and free-will being
   Thus the tension between medicine/ ethics
       In other words; fact/ value
   Tension cannot be decreased without the support/coordination of medicine with philosophy and social sciences.

Bio security identifies the preventive measures to control harmful consequences of biotechnology and its products on human life[2]. Basic and applied sciences, which concern the security and ethics, are given in Table 4.

Table 4. Basic and Applied Sciences which Concern the Security and Ethics

Basic sciences	Applied sciences
Chemistry	Medicine
Genetics	Agriculture
Biochemistry	Ecology
Microbiology	Food
Physiology	Bio quality
Computer sciences	Fermentation

7.   THE PRINCIPLES OF HUMANE EXPERIMENTAL TECHNIQUE

Russell and Burch have formatted the principles of ethical use of animals in biomedical research shown in the following lines as the "three R's"; replacement, reduction and refinement, which have been well accepted by the scientific community.

Replacement; use of lower animals with one of higher development
use of dummies, in-vitro or computer models

Reduction;   use of least possible number of animals leading to
understandable results and knowledge

Refinement; minimization of suffering during experimentation

These enhance the use of individual molecules, mathematical and computerized models, cells, tissue cultures, invertebrates, indirect laboratory tests, in vitro tests, transgenic or chimeral animals, maximal use of researchers knowledge and skills, best possible Project suggestions and management... more respect for other living creatures. Hence, through these principles the number of animals used for biomedical research in the British Isles have dropped from 5.5 millions to 2.3 millions during 1973 - 1997.

Objectives of Institutional Animal Care and Use Committee are: (IACUC)[8,9]:
understanding the ethical pros and cons for use of animals in research
information of resources and regulations regarding the care and use of animals
experience in relating to cases drawn from biomedical engineering
understanding and validating of animal models

Basic criteria for IACUC approval:[8,9]
Research has the potential for new information
    teach skills or concepts that cannot be obtained using an alternative
    will generate knowledge that is scientifically and socially important
    is designed such that animals are treated humanely.

Any project using animals in research should[7] have a rationale for using animals (benefits to be gained); a description of procedures involving animals, a research design; with observational checklists for determining endpoints, and reduce (minimize) the number of animals needed in research. Maximizing the data gained from an individual animal; avoiding the duplication of prior research unnecessarily, using sound statistical methods in design and analysis are solid principles to be followed in the ethical use of animals in biomedical research. Refined experimental procedures to minimize pain and distress, confining with the standards for acceptable means of euthanasia are also most important.

In conclusion, continued biological, biomedical, veterinary, medical or even forestry, marine, agricultural and horticultural progress in the world, in other words, life in a universal scale depends upon animal research, properly and ethically performed.

REFERENCES

1. Özgür, A., Animal use in scientific research and alternatives. *Bioethics Congress*, May 2003, Bursa, Turkey
2. Kaplan S., Çobanoğlu N., Bioethical approach to biotechnology: Bio security. *Bioethics Congress*, May 2003, Bursa, Turkey
3. Müftüoğlu, A, An enquiry on the philosophical foundations of the relation between medicine and ethics. *Bioethics Congress*, May 2003, Bursa, Turkey
4. Altuğ, T., Karaca, Ç., Bayrak, İ., Analgesia and euthanasia of animals in research. *Bioethics Congress*, May 2003, Bursa, Turkey
5. Yılmaz, A., Social, economical, environmental, ethical and political outcomes of biotechnology and genetic engineering. *Bioethics Congress*, May 2003, Bursa, Turkey
6. Aksoy, N., Edisan, Z., Çevik, E., Aksoy, Ş., Ethics in international research. *Bioethics Congress*, May 2003, Bursa, Turkey
7. Bird, S.J., *The ethics of using animals in research.* http://www.onlineetihcs.org
8. Americans for Medical Progress, *Animal research.* http://www.amprogress.org
9. OLAW (Office of Laboratory Animal Welfare): *Public Health Service Policy on Humane Care and Use of Laboratory Animals. Health Research Extension Act 01 1985*, NIH 2002
10. FBR (Foundation for Biomedical Research), http://www.fbresearch.org
11. ACLAM (American College of Laboratory Animal Medicine), http://www.aclam.org
12. The Foundation for Animal Use Education, *Animals in Biomedical Research.* http://www.animaluse.org
13. *Your child may have been saved through animal research,* American Academy of Pediatrics, 1987, http://www.acadped.org

14. *Do we care about research animals?* 2000.National Institutes of Health. Public Health Service: NIH Pub. No. 79-355.
15. Gökçora, İ.H., 2003, in *Periodical Publishing in Medical Sciences* (O. Yılmaz, ed) TÜBİTAK Pub. House, Ankara, pp 1-3
16. *Regulations concerning ethical committee on animal experiments,* 2003, Baskent University Press, Ankara, Turkey
17. Tansal, S., Türkiye Etik Değerler Merkezi (TEDMER) Vakfı: *For a better Turkey,* Yılmaz Basımevi, İstanbul, 2002
18. Nobel prizes: The payoff from animal research, Foundation for Medical Research:. http://www.fbresearch.org
19. *Draft report on The Animal Protection Act and views of the Commission for Ecology of the Turkish Republic, 1999,* Turkish Grand National Parliament, Official Gazette, Ankara.
20. Sümbüloğlu, V., Geyik, P., Sümbüloğlu, K., Ethics for the use of statistics in research, 1999, in *Medical Ethics Research* (B. Arda, ed.), Öncü Pub., Ankara, p: 19-22
21. von Engelhardt D, 1997, *Ethik im Alltag der Medizin Spektrum der Disziplinen zwischen Forschung und Therapie,* Birkhauser Pub, Berlin (Turkish: Namal A: Tıbbın Gündelik Yaşaminda Etik: Araştırmadan Terapiye Disiplinler Yelpazesi. Nobel Tıp Kitabevleri Ltd Sti, Ankara, 2000)
22. Egemen, M.N., Çobanoğlu, N., Ecofeminist approaches to bioethics, *Bioethics Congress,* May 2003, Bursa, Turkey
23. Başağaç, R.T., Özkul, T., The slaves of patriarchy – women and animals, *Bioethics Congress,* May 2003, Bursa, Turkey
24. What organizations, scientists, major religions, opponents say about animal research, Foundation for Medical Research, http://www.fbresearch.org
25. Rosner, F., 1986, *Modern medicine and Jewish ethics.* Ktav Pub House Inc., New York, pp 317-333
26. Our responsibility for the living environment, Church House Press, London, 1986, pp 29-32
27. Masri, B.A., Al-Hafiz, A., 1989, *Animals in Islam.* Athene Trust, London, p 17

Index